AF356599

Advanced Greenhouse Horticulture

Advanced Greenhouse Horticulture: New Technologies and Cultivation Practices

Editor

Athanasios Koukounaras

MDPI • Basel • Beijing • Wuhan • Barcelona • Belgrade • Manchester • Tokyo • Cluj • Tianjin

Editor
Athanasios Koukounaras
Aristotle University of Thessaloniki
Greece

Editorial Office
MDPI
St. Alban-Anlage 66
4052 Basel, Switzerland

This is a reprint of articles from the Special Issue published online in the open access journal *Horticulturae* (ISSN 2311-7524) (available at: https://www.mdpi.com/journal/horticulturae/special_issues/greenhouse_hort).

For citation purposes, cite each article independently as indicated on the article page online and as indicated below:

LastName, A.A.; LastName, B.B.; LastName, C.C. Article Title. *Journal Name* **Year**, *Volume Number*, Page Range.

ISBN 978-3-0365-0350-9 (Hbk)
ISBN 978-3-0365-0351-6 (PDF)

Cover image courtesy of Athanasios Koukounaras.

Contents

About the Editor

Athanasios Koukounaras is an Assistant Professor in the Lab of Vegetable Crops, Department of Horticulture and Viticulture, Aristotle University of Thessaloniki, Greece. He is specialized in issues related to plant physiology, vegetables production, nutritional quality and postharvest handling. He has published several articles in peer-reviewed journals in the fields of agriculture and environment, horticulture and biotechnology, to date, and he has participated in several national and EU research projects.

 *horticulturae*

Editorial

Advanced Greenhouse Horticulture: New Technologies and Cultivation Practices

Athanasios Koukounaras

School of Agriculture, Aristotle University of Thessaloniki, 54124 Thessaloniki, Greece; thankou@agro.auth.gr

Received: 19 December 2020; Accepted: 22 December 2020; Published: 24 December 2020

Abstract: Greenhouse horticulture is one of the most intensive agricultural systems, with the advantages of environmental parameter control (temperature, light, etc.), higher efficiency of resource utilization (water, fertilizers, etc.) and the use of advanced technologies (hydroponics, automation, etc.) for higher productivity, earliness, stability of production and better quality. On the other hand, climate change and the application of high inputs without suitable management could have negative impacts on the expansion of the greenhouse horticulture sector. This special issue gathers twelve papers: three reviews and nine of original research. There is one review that focuses on irrigation of greenhouse crops, while a second surveys the effects of biochar on container substrate properties and plant growth. A third review examines the impact of light quality on plant–microbe interactions, especially non-phototrophic organisms. The research papers report both the use of new technologies as well as advanced cultivation practices. In particular, new technologies are presented such as dye-sensitized solar cells for the glass cover of a greenhouse, automation for water and nitrogen deficit stress detection in soilless tomato crops based on spectral indices, light-emitting diode (LED) lighting and gibberellic acid supplementation on potted ornamentals, the integration of brewery wastewater treatment through anaerobic digestion with substrate-based soilless agriculture, and application of diatomaceous earth as a silica supplement on potted ornamentals. Research studies about cultivation practices are presented comparing different systems (organic-conventional, aeroponic-nutrient film technique (NFT)-substrate culture), quantitative criteria for determining the quality of grafted seedlings, and of wild species as alternative crops for cultivation.

Keywords: soilless culture; light; protected cultivation; vegetables; ornamental

---

1. Introduction

Horticulture is characterized by a wide range of cultivation systems (i.e., open field, soilless, protected, greenhouse, organic, indoor) and a plethora of fruit, vegetable and ornamental species. Among these, greenhouse horticulture is one of the most intensive agricultural systems, focusing on the production of high-value products. Control of environmental parameters (temperature, light, etc.), higher efficiency of resource utilization (water, fertilizers, etc.) and the use of high-tech systems (hydroponic, automation, etc.) provide opportunities for higher yields, earliness, stability of production and better quality. The value of the global greenhouse horticulture market for 2019 was 30 billion US dollars and it has been projected to increase annually around 9% for the next five years [1]. Increasing demand for greenhouse horticultural crops could be explained by the growth of the world population as well as adaptation to negative environmental impacts of future water availability and climate change scenarios.

Climate change especially could have a significant effect on greenhouse horticulture since the environmental parameters inside greenhouses are dependent on outside conditions such as temperature and solar radiation. On the other hand, the use of high inputs without suitable management could have negative environmental impacts. Thus, suitable cultivation practices and new technologies will be needed to achieve the maximum benefits from greenhouse horticultural production.

2. Special Issue Overview

This Special Issue collects current research findings dealing with a wide range topics related to greenhouse horticulture. The papers can be broadly organized into two main subjects: (i) five original research articles about new technologies, and (ii) three reviews and four original research articles about cultivation practices.

2.1. New Technologies on Advanced Greenhouse Horticulture

Greenhouse horticulture has been one of the pioneering sectors of agriculture in the use of new technologies. This has increased due to increasing globalization as well as the requirement for more efficient use of resources and more sustainable farming practices. Castro et al. [2] review new technologies such as improved cover materials, light-emitting diode (LED) lighting, alternative nutrient resources, and sensors that are expected to contribute to more digital, automatic, and advanced greenhouse horticultural production.

In greenhouse horticulture, there is a significant need for energy cost reduction, since energy constitutes a substantial fraction of the total production costs. Motivated by the above, Ntinas et al. [3] evaluated and quantified a higher crop yield and improved quality of hydroponic tomato cultivated in a greenhouse, using the innovative technology of dye-sensitized solar cells (DSSC) for the glass cover of the greenhouse. All energy harvested from the sun with the DSSC was used to cover part of the electrical consumption of the DSSC greenhouse as a renewable energy source, in comparison to a reference greenhouse that had a glass cover and a conventional electrical source that was provided by a grid (CONV greenhouse). The results from the DSSC greenhouse during the summer season were satisfactory, since shading had a positive effect on the qualitative characteristics of the tomato fruit of two commercial table tomato hybrids. Furthermore, surplus light, especially during summer when it is not needed for plant growth, was used for electrical generation by the photoselective DSSCs installed in the greenhouse roof covering. In future work, the DSSC greenhouse will be evaluated year round, focusing on wintertime.

Water and nitrogen deficit stresses are among the most critical growth limiting factors in crop production. Usually quite complex and problematic methods have been used to quantify the impact of water and nitrogen deficit stresses on plants. Elvanidi et al. [4] developed a model based on the classification tree (CT) method to analyze complex reflectance index datasets in order to provide visual assessments of plant water and nitrogen deficit stress. The results showed that the combination of MSAVI (Modified soil-adjusted vegetation index), mrNDVI (Modified red edge normalized difference vegetation index), and PRI (Photochemical reflectance index) had the potential to determine water and nitrogen deficit stress with 89.6% and 91.4% classification accuracy values for the training and testing samples. These results are promising for further design of a smart decision support system for better climate and irrigation management.

Light is one of the most important factors related to plant growth and development and therefore, using artificial lighting is common practice in commercial greenhouses with the use of light emitting diode (LED) technology rapidly replacing traditional lighting sources. Gibberellic acid (GA_3) is a hormone found in plants, which is produced in low amounts, and therefore synthetic GA_3 is commonly used in commercial agriculture to manipulate growth and development. Mills-Ibibofori et al. [5] evaluated the above parameters and combinations on potted ornamentals. The results showed that light and GA_3 have a synergistic relationship with each other regarding plant and floral development. More research needs to be conducted with diverse species using an array of LED lights with different spectra in combination with the plant hormone GA_3 to control plant growth and flowering, as effects may be species dependent.

Urban agriculture is experiencing a resurgence in popularity in many parts of the world and can play a key role in recycling urban waste streams, promoting nutrient recycling, and increasing sustainability of food systems. Riera-Vila et al. [6] investigated the integration of brewery wastewater treatment through anaerobic digestion with substrate-based soilless agriculture. The yield of three crops in the digested wastewater treatments was higher than with raw wastewater or a no fertilizer control, indicating that nutrients in the brewery wastewater can be recovered for food production and diverted from typical urban waste treatment facilities. The work showed how to solve some of the agronomic and technological challenges of this integration, but additional complementary research is needed pertaining to the economic, legal, and social challenges of this decentralized urban system.

In horticultural plants, silica (Si) is a nonessential element; however, its role as a needed supplement in soilless media is gaining interest. Therefore, Mills-Ibibofori et al. [7] conducted a study to determine the effects of diatomaceous earth (DE) as a Si supplement on three potted ornamentals under well-watered and water-stressed conditions. Several growth and flowering characteristics were improved, depending on the rate and application method, by application of DE. Benefits of DE included increased height, width, shoot dry weight, stem, and flower diameter. Future studies should further evaluate application of DE on a range of crops and stress conditions.

2.2. Cultivation Practices on Advanced Greenhouse Horticulture

Vox et al. [8] reviewed how greenhouse horticultural cultivation is a very intensive form of agriculture and could be environmentally unfriendly; therefore, application of novel, suitable cultivation techniques is essential to prevent problems developing from this type of production.

Probably the most sensitive daily work of the growers is irrigation management of the crops, balancing water availability with crop need. Decisions concerning irrigation schedules may be based on a grower's estimation of need on a daily basis for soil-based crops, while estimates for soilless systems require shorter time intervals. Nikolaou et al. [9] presented a comprehensive review of irrigation management in soil and soilless crop production in greenhouses, as well as the need for the development of a commercial irrigation controller unit in order to model and monitor the soil-plant-atmosphere utilizing artificial intelligence.

Recently, there has been an increasing demand for biochar application, carbon-rich material made from biomass, in horticulture crop production systems. Hunag and Gu [10] reviewed biochar production systems, biochar effects on container substrates characteristics, and biochar effects on plant growth when used in container substrates. Further studies will be needed for assessment of biochars as a result of promising results to date to fine-tune the pyrolysis process and incorporate formulae for diverse container substrates.

As mentioned previously, light is extremely important for the plants, however, covering materials alter available light for the plants in greenhouses, which may became a light intensity problem in greenhouses at circumpolar regions. Alsanius et al. [11] reviewed the impact of light quality on plant–microbe interactions, such as bacterial and fungal pathogens, biocontrol agents, and the phyllobiome. Relevant molecular mechanisms regulating light-quality-related processes in bacteria are described and knowledge gaps are discussed with reference to ecological theories.

Control of the protected environment of a greenhouse makes them suitable for organic production. Golubkina et al. [12] compared nine new leek cultivars under organic and conventional systems in a greenhouse. The results showed that organic cultivation resulted in higher dry matter, sugar, ascorbic acid and potassium content but lower nitrates in the pseudo-stems than conventional cultivation, but with the same cultivar ranking as for conventional management. Observation of the strong relationships between quality, antioxidant and mineral components in leek plants could provide wide possibilities in breeding programs for both conventional and organic management systems in greenhouses.

To maximize the benefits of the greenhouse environment, the choice of a soilless cultural system (SCS) is a common practice. Common forms of SCS are hydroponic but also aeroponic and aquaponics. Li et al. [13] compared an aeroponic to a hydroponic nutrient film technique (NFT) system for their effects on growth and root characteristics of lettuce. The results showed that aeroponic cultivation significantly improved root growth parameters (root biomass, root/shoot ratio, total root length, root area, and root volume). However, greater root growth in the aeroponic system did not lead to greater shoot growth compared with hydroponic culture due to the limited availability of nutrients and water. Further research is necessary to determine suitable pressure, droplet size, and misting intervals in the aeroponic system to improve the continuous availability of nutrients and water for lettuce cultivation.

The use of grafting is an environmentally friendly technology, safe for consumers and users, and may prevent biotic and abiotic disorders, and it is extensively used for watermelon. However, to enjoy the advantages of grafting, use of high quality grafted seedlings is a prerequisite. Based on the above, Bantis et al. [14] performed a study to set critical limits for objective measurements of grafted watermelon seedling quality categories as well as to suggest the most accurate and convenient among them for application by the end users (industry and growers). They concluded that crucial parameters were leaf and cotyledon area of scions, stem diameter, shoot and root dry weights as well as shoot DW/L, and Dickson's quality index as good indicators for categorizing grafted watermelon seedlings.

Recently, there has been an increased interest by consumers in alternative crop species with unique visual appearances and tastes as well as a source of bioactive compounds. Motivated by the above, Guarise et al. [15] evaluated two wild populations of hedge mustard as a potential leafy vegetable. A wide range of physiological parameters and chemical substances (leaf pigments, chlorophyll a fluorescence, sugars, ascorbic acid, total phenols, anthocyanins, nitrate) were determined. The results demonstrated that the two wild populations of *Brassicaceae* can be successfully grown in a greenhouse with nutritional value and quality characteristics similar to the most common commercial leafy vegetables. Further investigation will be required for evaluating postharvest quality and suitable storage conditions.

3. Conclusions

Greenhouse horticultural production is a very intensive method, and therefore the need for adopting new technologies and advanced cultivation practices are preconditions for successful production in a very competitive global environment. Moreover, the environmental impact of the above is more critical than ever to avoid negative impacts while achieving the goal of satisfactory food production for the increasing world population while improving agricultural sustainability.

Funding: This research received no external funding.

Acknowledgments: We gratefully acknowledge all the authors that participated in this Special Issues.

Conflicts of Interest: The authors declare no conflict of interest.

References

1. Available online: https://www.marketdataforecast.com/market-reports/greenhouse-horticulture-market (accessed on 14 December 2020).
2. Castro, A.J.; López-Rodríguez, M.D.; Giagnocavo, C.; Gimenez, M.; Céspedes, L.; La Calle, A.; Gallardo, M.; Pumares, P.; Cabello, J.; Rodríguez, E.; et al. Six Collective Challenges for Sustainability of Almería Greenhouse Horticulture. *Int. J. Environ. Res. Public Health* **2019**, *16*, 4097. [CrossRef] [PubMed]

3. Ntinas, G.K.; Kadoglidou, K.; Tsivelika, N.; Krommydas, K.; Kalivas, A.; Ralli, P.; Irakli, M. Performance and Hydroponic Tomato Crop Quality Characteristics in a Novel Greenhouse Using Dye-Sensitized Solar Cell Technology for Covering Material. *Horticulturae* **2019**, *5*, 42. [CrossRef]

4. Elvanidi, A.; Katsoulas, N.; Kittas, C. Automation for Water and Nitrogen Deficit Stress Detection in Soilless Tomato Crops Based on Spectral Indices. *Horticulturae* **2018**, *4*, 47. [CrossRef]

5. Mills-Ibibofori, T.; Dunn, B.L.; Maness, N.; Payton, M. Effect of LED Lighting and Gibberellic Acid Supplementation on Potted Ornamentals. *Horticulturae* **2019**, *5*, 51. [CrossRef]

6. Riera-Vila, I.; Anderson, N.O.; Hodge, C.F.; Rogers, M. Anaerobically-Digested Brewery Wastewater as a Nutrient Solution for Substrate-Based Food Production. *Horticulturae* **2019**, *5*, 43. [CrossRef]

7. Mills-Ibibofori, T.; Dunn, B.L.; Maness, N.; Payton, M. Use of Diatomaceous Earth as a Silica Supplement on Potted Ornamentals. *Horticulturae* **2019**, *5*, 21. [CrossRef]

8. Vox, G.; Teitel, M.; Pardossi, A.; Minuto, A.; Tinivella, F.; Schettini, E. Sustainable Greenhouse Systems. In *Sustainable Agriculture: Technology, Planning and Management*; Salazar, A., Rios, I., Eds.; Nova Science Publishers Inc.: New York, NY, USA, 2010; pp. 1–79.

9. Nikolaou, G.; Neocleous, D.; Katsoulas, N.; Kittas, C. Irrigation of greenhouse crops. *Horticulturae* **2019**, *5*, 7. [CrossRef]

10. Huang, L.; Gu, M. Effects of Biochar on Container Substrate Properties and Growth of Plants—A Review. *Horticulturae* **2019**, *5*, 14. [CrossRef]

11. Alsanius, B.W.; Karlsson, M.; Rosberg, A.K.; Dorais, M.; Naznin, M.T.; Khalil, S.; Bergstrand, K.-J. Light and Microbial Lifestyle: The Impact of Light Quality on Plant–Microbe Interactions in Horticultural Production Systems—A Review. *Horticulturae* **2019**, *5*, 41. [CrossRef]

12. Golubkina, N.A.; Seredin, T.M.; Antoshkina, M.S.; Kosheleva, O.V.; Teliban, G.C.; Caruso, G. Yield, Quality, Antioxidants and Elemental Composition of New Leek Cultivars under Organic or Conventional Systems in a Greenhouse. *Horticulturae* **2019**, *5*, 39. [CrossRef]

13. Li, X.; Li, Q.; Tang, B.; Gu, M. Growth Responses and Root Characteristics of Lettuce Grown in Aeroponics, Hydroponics, and Substrate Culture. *Horticulturae* **2018**, *4*, 35. [CrossRef]

14. Bantis, F.; Koukounaras, A.; Siomos, A.; Menexes, G.; Dangitsis, C.; Kintzonidis, D. Assessing Quantitative Criteria for Characterization of Quality Categories for Grafted Watermelon Seedlings. *Horticulturae* **2019**, *5*, 16. [CrossRef]

15. Guarise, M.; Borgonovo, G.; Bassoli, A.; Ferrante, A. Evaluation of Two Wild Populations of Hedge Mustard (*Sisymbrium officinale* (L.) Scop.) as a Potential Leafy Vegetable. *Horticulturae* **2019**, *5*, 13. [CrossRef]

 *horticulturae* 

Article

Performance and Hydroponic Tomato Crop Quality Characteristics in a Novel Greenhouse Using Dye-Sensitized Solar Cell Technology for Covering Material

Georgios K. Ntinas, Kalliopi Kadoglidou *, Nektaria Tsivelika, Konstantinos Krommydas, Apostolos Kalivas, Parthenopi Ralli and Maria Irakli

Hellenic Agricultural Organization-Demeter, Institute of Plant Breeding and Genetic Resources, Thermi, 57001 Thessaloniki, Greece; georgiosntinas@gmail.com (G.K.N.); riatsivel@gmail.com (N.T.); kos_krom@hotmail.com (K.K.); kalivasapostolis@yahoo.gr (A.K.); pralli@ipgrb.gr (P.R.); irakli@cerealinstitute.gr (M.I.)

* Correspondence: kkadogli@agro.auth.gr; Tel.: +30-23-1047-1110

Received: 21 March 2019; Accepted: 15 May 2019; Published: 1 June 2019

Abstract: In this study, we evaluated crop productivity and physiology during the hydroponic cultivation of medium-sized and cherry tomato crops, using two experimental greenhouses. Of the greenhouses, one used dye-sensitized solar cell (DSSC) technology for covering material, whilst the other, a conventional one (CONV), was covered using diffusion glass as a control. The effect of the colored lighting that resulted from the DSSC glass filtering on the physiological response of the crops was examined by measuring the plant transpiration rate and leaf chlorophyll content. Furthermore, we evaluated potential differences in the concentration of phytochemical compounds, such as ascorbic acid, lycopene, and quality characteristics. Tomato plants in the DSSC greenhouse presented lower early and total yields, as well as lower chlorophyll content, stomatal conductance, photosynthetic rate, and transpiration rate values, especially in the medium-sized fruits, as compared to the CONV greenhouse. The DSSC greenhouse showed significantly higher values of bioactive compounds for both the cherry and medium-sized tomato, with increases in the ascorbic acid, lycopene, β-carotene, and total carotenoids concentration, which ranged from 6% to 26%. Finally, for both the hybrids, the 2,2′-azino-bis-3-ethylbenzthiazoline-6-sulphonic acid (ABTS) and 1,1-diphenyl-2-picrylhydrazyl (DPPH) tests showed circa 10% and 5% increase, respectively, in the DSSC greenhouse.

Keywords: semi-transparent photovoltaic modules; hydroponics; tomato; bioactive ingredients; lycopene; antioxidant capacity

1. Introduction

Across the world, controlled environment agricultural systems function as a dynamic production process. Such systems provide shelter for the crops against the direct influence of external weather conditions, and they offer the opportunity to modify the indoor climate to create an environment that is optimal for crop growth and production, both in terms of quality and quantity [1,2]. Following the first energy crisis in the seventies, during which limited energy supplies led to an increase in energy prices, greenhouse energy consumption has again become a major issue. There is a significant need for energy cost reduction, since energy constitutes a substantial fraction of the total production costs.

In Mediterranean countries, energy consumption for control of environmental conditions constitutes approximately 20–30% of the total production costs, with a higher percentage in the northern countries. Moreover, due to the rising interest in climate change and greenhouse gas emissions (GHG), the use of fossil fuels is part of the political agenda. Therefore, the greenhouse industry is facing economic, political, and social pressure to use renewable energy sources, reduce energy usage and CO_2 emissions, and improve greenhouse energy efficiency via technological innovations [3–5]. Different types of photovoltaic (PV) modules for greenhouse roofs to harvest solar radiation are available [6–8]. In this case, there is competition for light between PV modules and plants, which can lead to decreased production [9]. However, excess light, especially during the summer, is unnecessary for plant growth, where channeling of the excessive light toward electricity generation is an option. Furthermore, high air temperatures inside the greenhouse during summer can decrease the photosynthetic rate, and, therefore, plant growth and crop yield [10,11].

Roslan et al. [12] presented new developments in dye-sensitized solar cells (DSSC) for greenhouse shading and electricity production. The authors' reported that DSSC had not been applied in a greenhouse setting for plant growth and energy saving. The application of photoselective DSSC in greenhouse covering can regulate environmental conditions and enhance the quantitative and qualitative characteristics of greenhouse products [13].

Two important parameters of tomato (*Solanum lycopersicum* L.) quality are the pH and titratable acidity. Tomatoes are high-acid foods, such that active thermal treatments are not required for the destruction of microorganisms that contribute to food spoilage. The pH of tomato fruits ranges from 4.0 to 4.5, such that the lower the pH, the greater the so-called "tartness", a factor by which some consumers judge the quality of the tomato fruit [14]. Citric acid is the basic acid found in tomatoes and it contributes to the titratable acidity. Total soluble solids (°Brix) and °Brix / titratable acidity are the common indicators that express the taste of a tomato. °Brix values are measured using a refractometer and they indicate the percentage (%) of the dissolved solids in a solution, which in tomato paste is mainly the total sugars (glucose, fructose), acids (citric and malic), and other components in a lower proportion (phenols, amino acids, ascorbic acid, and inorganic salts) [15].

Tomatoes are a good source of nutrients and bioactive compounds, such as carotenoids (lycopene, β-carotene, and lutein), vitamin C, and polyphenolic compounds, which are thought to be health-promoting factors with antioxidant properties [16,17]. The nature and concentration of these compounds depends on the cultivation practices, environmental factors, crop variety, and maturity [18]. There is a great deal of variation between crop varieties in terms of fruit size and color, which can affect the nutritional properties of tomato. Lenucci et al. [19] found significant differences among 14 varieties of cherry tomato (ChT) and four medium-sized tomato (MST) hybrids in terms of the lycopene and β-carotene content, where varieties with more pigments had a higher concentration of lycopene. Consumption of tomato and tomato products has been associated with a lower risk of developing digestive problems and prostate cancer.

The lycopene content depends on the redness of the tomato fruit, and it is the main carotenoid and one of the key antioxidants found in fresh tomatoes and processed tomato products. Vitamin C, including ascorbic and dehydroascorbic acid, is important for the protection of tomatoes from the auto-oxidizing factors that may increase when ripening. The role of ascorbic acid in the prevention of diseases associated with oxidative damage occurs because of its ability to neutralize the action of free radicals in biological systems [20]. In addition to its antioxidant action, ascorbic acid is essential to life because of its many physiological effects. Plants and a vast majority of mammals, but not humans, are able to synthesize it. Meanwhile, the main sources of vitamin C are citrus fruits, tomatoes, and potatoes [21]. Ascorbic acid is relatively stable in tomatoes, due to the acidic conditions prevailing in the tissue. However, it easily decomposes due to oxidation, exposure to light, or high temperatures. Significant losses of ascorbic acid occur during the postharvest storage period. Therefore, lowering the temperature from ambient (20 °C) to cool (4 °C) or freezing (−18 °C) reduces the rate of loss of ascorbic acid.

Tomato fruit maturation is a complex process involving various morphological, physiological, biochemical, and molecular processes. These processes include the reduction of chlorophyll, the synthesis and storage of carotenoids (mainly lycopene and β-carotene) and aromatic compounds, changes in the metabolism of organic acids, and the softening of the fruit tissue that occurs in combination with increased CO_2 and ethylene production by the fruit. The quantity of other important antioxidants, such as ascorbic acid and phenolic compounds, changes during the maturation of tomato [22,23]. Specifically, Ilahy et al. [22] described that the increase in antioxidants did not follow the fruit ripening pattern, while Helyes and Lugasi [23] found that lycopene and total antioxidant capacity increased during maturation, while polyphenol content remained almost the same. All of these processes affect texture, color, flavor, aroma, but also the content of antioxidant compounds and the antioxidant effect of tomato fruit [24].

Motivated by the above concept, the purpose of this study was to evaluate and quantify the gain of crop production in terms of higher crop yield and improved quality of hydroponic tomato cultivation in a greenhouse, using the innovative technology of DSSC for the glass cover. All energy harvested from the sun with the DSSC was used to cover part of the electricity consumption of the DSSC greenhouse with a renewable energy source, in comparison to a reference greenhouse that had a glass cover and conventional electricity was provided by the grid (CONV greenhouse). Specific parameters related to plant physiology, such as transpiration, stomatal conductance, photosynthetic rate, and chlorophyll content were determined. Additionally, chemical analyses of harvested fruits were conducted to determine potential differences in the concentration of phytochemical compounds, such as ascorbic acid and lycopene, and quality characteristics between the raw products harvested from the DSSC greenhouse and the CONV one.

2. Materials and Methods

2.1. Hydroponic Tomato Cultivation Set Up in the CONV and DSSC Greenhouses

The experiment was conducted in a CONV and in a DSSC greenhouse, at the Institute of Plant Breeding and Genetic Resources in Thermi Thessaloniki (22°59.956′ E/40°32.281′ N, −1 m.a.s.l.) in the summer of 2015. After preliminary assessments—resistance studies of various hybrids—two commercial table tomato hybrids, "Oasis" (MST) and "Genio" (ChT), were selected for this experiment (Figure 1).

(a)   (b)

Figure 1. Table tomato hybrids selected for cultivation: (**a**) Hybrid "Oasis" (medium-sized tomato, MST); (**b**) hybrid "Genio" (cherry tomato, ChT).

The tomato plants were grown in double lines of 4 m length and distances of 33 cm (plant to plant) and 100 cm (double-row corridor). In total, there were nine double lines in each greenhouse, four of which were cultivated with medium-sized tomato (MST) and the other five with cherry tomato

(ChT) (Figure 2). The tomato seedlings, planted in rockwool cubes, were transplanted in their final position on rockwool substrates (Grotop Master Grodan, Denmark) when they reached the stage of 5–6 true leaves. In both greenhouses, tomato plants were fertirigated with Hoagland's solution (Table 1). A total of 2.8 L of HNO_3 was added in a separate tank containing 100 L of water. The pH and the EC of the nutrient solution were 5.5–6.0 and 3.0–3.5 dS m^{-1}, respectively. The EC and pH were measured using portable EC and pH meters (HI 8733 and HI 8424, Hanna Instruments, Inc., Woonsocket, RI, USA).

(a)   (b)

Figure 2. Plant arrangement in the CONV (Conventional) (**a**) and DSSC (Dye-Sensitized Solar Cell) (**b**) greenhouses. The difference in the shading within the greenhouses is visible.

Table 1. Chemical characteristics of the nutrient solutions used in the experiments.

Stock Solution (Tank A)		Stock Solution (Tank B)		Micronutrients (Tank C)	
Macronutrients	**(g/100 L)**	**Macronutrients**	**(g/100 L)**	**Micronutrients**	**(g/20 L)**
$Ca(NO_3)_2$	6800	K_2SO_4	2780	$MnSO_4$	585
KNO_3	1000	KNO_3	1320	$Na_2B_4O_7$	530
EDTA-Fe (13%)	200	$MgSO_4$	1300	$CuSO_4$	40
		KH_2PO_4	2340	$ZnSO_4$	270
				$(NH_4)_6Mo_7O_{24}$	25

Preventive application of abamectin against *Tetranychus urticae* control and *Bacillus thuringiensis* subsp. kurstaki strain SA-11 for *Heliothis armigera* control were carried out once in the middle of the experiment.

Microclimatic parameters inside both greenhouses (solar radiation, air temperature, and humidity) were recorded with a HOBO data logger (HOBO micro-station, Onset Inc., USA).

2.2. Evaluation of Cultivation in the Two Greenhouses

Evaluation of the cultivation of tomatoes in the two greenhouse compartments was performed by measuring yield and morphological, physiological, and qualitative characteristics, related to chemical, nutritional, and bioactive ingredients as well as antioxidant activity.

2.2.1. Fruit Yield and Morphological Characteristics

Fruit were harvested weekly by collecting only ripe fruits from each greenhouse and each hybrid. The sum of the first and the second harvest represented the early yield. In total, seven harvests took place, and fruits were weighed and ranked in quality classes based on size, weight, and shape directly after harvest. The International Union for the Protection of New Varieties of Plants (UPOV) descriptor was used to measure and evaluate morphological characteristics.

2.2.2. Plant Physiological Parameters

Transpiration, stomatal conductance, and photosynthetic rate of plants was measured 46 days after transplanting (DAT), by using an infrared gas analyzer (LCi-SD portable photosynthesis system, ADC BioScientific Ltd., Hertfordshire, UK). For this purpose, in each twin cultivation row (24 plants), five measurements were taken from the upper third of a fully developed leaf, between the fourth and fifth inflorescence. The photosynthetically active radiation (PAR) was 576 ± 100 and 387 ± 100 µmol (photon) m^{-2} s^{-1} in the CONV and DSSC greenhouses, respectively. The CO_2 concentration in the chamber was 390 ± 10 µmol (CO_2) mol^{-1}, the temperature in the chamber was set at 30 ± 2 °C, and the water reference as partial pressure was 35 ± 3 mbar. Additionally, the chlorophyll content of the leaves was measured on the fortieth day using a portable Chlorophyll Content Meter (CCM-200, Opti-Sciences, Inc., Hudson, NH, USA). For this purpose, ten measurements were taken from the terminal leaflet of a fully developed leaf, between the fourth and fifth inflorescences.

2.2.3. Fruit Physicochemical Parameters

From each harvest, representative fruit samples from each hybrid and each greenhouse were collected at each inflorescence stage, to determine the quality parameters described below. Prior to measurements, samples were cut into pieces and homogenized in a conventional blender in order to obtain the tomato pulp. For each measurement, three representative samples of tomato fruit, from each hybrid and truss, were taken in both greenhouses and each measurement was repeated three times (three replicates). Tomato pulp pH was measured with a portable pH meter (MW802, Milwaukee Instruments Inc., Rocky Mount, NC, USA) and dry weight was determined after oven drying (48 h, 72 °C). Free sugars and citric acid in tomato pulp were determined enzymatically using the Megazyme kit of glucose/fructose/sucrose and the citric acid, respectively, according to the corresponding protocol (Megazyme International, Wicklow, Ireland).

2.2.4. Fruit Antioxidant Compounds

Ascorbic acid was determined by extraction of tomato pulp with 20% metaphosphoric acid and titration with 2,6-dichloroindophenol. The ascorbic acid content was calculated according to a standard and expressed as mg ascorbic acid/100 g of tomato pulp [25]. Carotenoids were extracted with a mixture of hexane/ethyl acetate (50:50, *v/v*) from tomato pulp in a ratio of 1:10 with 0.1% butylated hydroxytoluene (BHT) in an ultrasonic bath for 5 min according to Irakli et al. [26]. The procedure was repeated until discoloration of the extract was reached. The total extracts were concentrated to dry weight on a rotary evaporator and then re-dissolved in 2 mL of dichloromethane/methanol/acetonitrile solution (30/30/40, *v/v/v*). Consequently, the extracts were filtered through 0.45 µm polytetrafluoroethylene (PTFE) membrane filters and an aliquot of 20 µL of this was introduced into a high-performance liquid chromatography (HPLC) system (Agilent Technologies, series 1200, Urdorf, Switzerland,) connected to a photodiode array detector (DAD). Separation was accomplished on an YMC C$_{30}$ column (250×4.6 mm; 5 µm) at a temperature of 15 °C. Elution was performed following a gradient elution program starting from a mixture of acetonitrile/methanol (85/15, *v/v*) at a speed of 1.5 mL/min and resulting in a mixture of acetonitrile/methanol/dichloromethane (30/20/60, *v/v/v*). The DAD was set at 450 nm for β-carotene and 475 nm for lycopene. The components were identified based on elution times and absorption spectra of reference standards.

2.2.5. Antioxidant Capacity

An aliquot of 2 g of tomato pulp was extracted two times with 20 mL of a 70% aqueous methanol mixture followed by ultrasonication for 15 min and centrifugation at 3000 rpm for 10 min. The supernatants were collected, homogenized, and methanolic tomato extracts were used for the evaluation of antioxidant capacity using the 2,2'-azino-bis-3-ethylbenzthiazoline-6-sulphonic acid (ABTS) and 1,1-diphenyl-2-picrylhydrazyl (DPPH) tests.

The ABTS radical scavenging activity assay was evaluated according to the protocol of Re et al. [27]. The ABTS$^{\bullet+}$ radical was produced by the oxidation of 7.4 mM ABTS (2,2′-azino-bis-3-ethylbenzothiazoline-6-sulfonic acid diammonium salt) with 2.45 mM potassium persulfate in water. The mixture was placed in the dark at room temperature for 16 h before use, and then the ABTS$^{\bullet+}$ solution was diluted with water to an absorbance of 734 nm of 0.70 ± 0.02. After the addition of 100 μL of extract to 3.9 mL of ABTS solution, the absorbance was measured against a blank at 734 nm after 4 min. The results were obtained by interpolating the absorbances on a calibration curve obtained with Trolox. The results were expressed as mg Trolox equivalents per 100 g of dry mass (mg TE/100 g).

The DPPH$^+$ radical scavenging ability assay was performed using tomato methanolic extracts mixed with DPPH$^+$ (2,2-diphenyl-1-picrylhydrazyl) reagent and absorbance was read at 517 nm according to Yen and Chen [28]. A volume of 0.15 mL extract was reacted with 2.85 mL 0.1 mM methanolic solution of DPPH. After 5 min, the absorbance at 516 nm was recorded. DPPH+ reagent was used as blank, and reduction percentage of free radical scavenging activity was monitored. Trolox was used as the standard. The results were expressed as mg Trolox equivalents (TE) per 100 g RB (mg TE/g).

2.3. Statistical Analysis

The statistical analysis of all experimental data (ANOVA) was carried out with the MSTAT-C version 1.41 statistical program (Michigan State University, East Lansing, MI). The experimental design was a completely randomized block (CRBD) split plot with greenhouse type (glass or DSSC) as main plots and hybrid species as subplots with five replications for the production and yield data and with six replications for the physiological parameters. For the comparison of the averages, the least significant difference criterion (Least Significant Difference Test, LSD test) was used at $P \leq 0.05$. The figures presented in the manuscript were created using an Excel spreadsheet. In each figure, the values are the average of the evaluated parameter, in both tomato hybrids (ChT and MST) and both greenhouses (CONV, DSSC). The averages of a parameter followed by different letters in the same column were statistically significantly different ($P < 0.05$).

3. Results and Discussion

Results from microclimate parameters along with quantitative and qualitative characteristics of the fruits in the two greenhouses are presented below. Specifically, these are air temperature, illuminance, relative humidity, early and total yield, chlorophyll content and transpiration rate, pH, citric acid %, dry matter %, total sugar concentration, and content of fruit bioactive components (ascorbic acid, lycopene, β-carotene, total carotenoids) and antioxidant capacity.

3.1. Microclimatic Conditions in the Examined Greenhouses

Microclimatic parameters inside a greenhouse can significantly affect physiological processes (such as photosynthesis) of plant tissues. These parameters are solar radiation, air temperature/humidity, and concentration of carbon dioxide in the air. Recorded data from these parameters were used as measurable variables by instruments quantifying photosynthesis and chlorophyll content. Therefore, they are essential in order to determine photosynthesis, stomatal conductance, transpiration rate, and leaf chlorophyll content. In Figure 3, the air temperature, illuminance, and relative humidity inside the greenhouse are presented for the total duration of the experiment, since they are the parameters that most influence physiological and chemical processes of tomato. Concerning air temperature, there was no difference between the two greenhouses (Figure 3a). However, the illuminance in the DSSC greenhouse was ca. 20% lower compared to the CONV greenhouse (Figure 3b). The shading effect of the DSSC was expected and, moreover, it is beneficial during the summer months. Higher values of light in the CONV greenhouse enhanced yield and physiological characteristics of tomato plants, while lower values in the DSSC greenhouse had a positive effect on fruit qualitative characteristics.

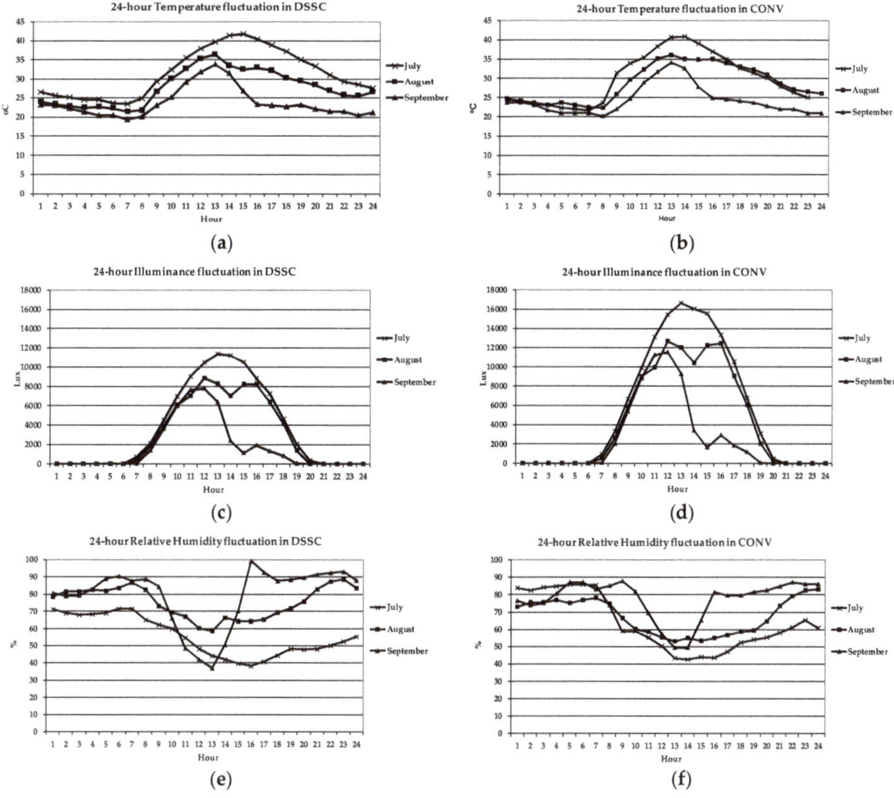

Figure 3. Air temperature, illuminance, and relative humidity inside the two greenhouses [CONV (Conventional), DSSC (Dye-Sensitized Solar Cell)]: (**a**) Air temperature (DSSC); (**b**) air temperature (CONV), (**c**) illuminance (DSSC); (**d**) illuminance (CONV); (**e**) relative humidity (DSSC); and (**f**) relative humidity (CONV).

3.2. Yield and Mean Weight of Early and Total Fruit Production

The results of early and total production parameters (yield and mean fruit weight) are presented in Figure 4, while a combined analysis of their variance is given in Table 2. Early yield was influenced by the greenhouse type (G), the hybrid species (H), and the G × H interaction, which was also statistically significant (Table 2). In particular, the MST performed differently in the two greenhouses, resulting in a significant yield decrease (by 50.8%) in the DSSC greenhouse (Figure 4a). However, no significant difference was observed in ChT yield in the two greenhouses. The total yield was also influenced by the type of greenhouse and hybrid (Table 2). In particular, for both hybrids a significant yield reduction (by 31% and 40% for ChT and MST, respectively) was observed in the DSSC greenhouse (Figure 4b), possibly due to lower light levels (Figure 3b).

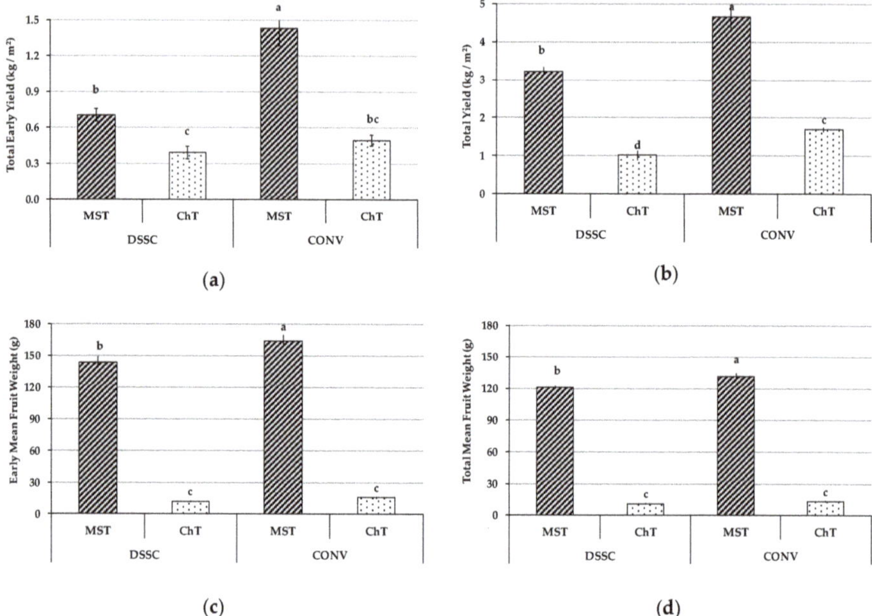

Figure 4. Production and yield data of the two tomato hybrids [MST (medium-sized) and ChT (cherry)] in the two greenhouses [DSSC (Dye-Sensitized Solar Cell) and CONV (Conventional)]: (**a**) Early yield; (**b**) total yield; (**c**) early mean fruit weight; (**d**) total mean fruit weight. Bars with different letters were significantly different by LSD at $P \leq 0.05$.

Table 2. Combined variance analysis of the production and yield data, as influenced by the greenhouse types [DSSC (Dye-Sensitized Solar Cell) and CONV (Conventional)] and the species of tomato hybrid [MST (medium-sized) and ChT (cherry)].

Source	df [z]	Significance of F Ratio			
		Early Yield		Total Yield	
		Early Yield	Fruit Weight	Yield	Fruit Weight
Replications	4	NS [y]	NS	NS	NS
Greenhouse (G)	1	**	NS	**	**
Error (a)	4				
Hybrid (H)	1	**	**	**	**
G × H	1	**	*	NS	NS
Error (b)	8				
CV %		23.87	9.30	17.59	5.64

[z] df, degree of freedom; CV, coefficient of variation; [y] NS, nonsignificant; * = $P < 0.05$ level of significance; ** = $P < 0.01$ level of significance.

Average fruit weight in the early production stage was affected by the type of hybrid, but not by the type of greenhouse (Table 2). More specifically, the average fruit weight of MST and ChT was 154 and 14 g, respectively, regardless of the greenhouse (Figure 4c). The MST had a 12.4% higher mean fruit weight in the CONV greenhouse than in the DSSC one, while the ChT fruit weighed the same in both greenhouses. The average fruit weight for total production was also influenced by the type of hybrid and greenhouse (Table 2). The average weight of MST and ChT tomato was 126 and 12 g, respectively, irrespective of the greenhouse (Figure 4d). The MST had a 7.5% higher average fruit weight in the CONV compared to the DSSC greenhouse, whereas ChT had the same average fruit weight in both greenhouses.

3.3. Physiological Parameters of Plants

Results of leaf chlorophyll content, transpiration rate, stomatal conductance, and photosynthetic rate are presented in Figure 5, while a combined analysis of their variance is given in Table 3. The chlorophyll content of leaves (chlorophyll content index, CCI), estimated 40 days after transplanting (DAT), differed between the plants of the two greenhouses and between the hybrids. However, the G × H interaction was not statistically significant. More specifically, in the CONV greenhouse, MST and ChT had a higher CCI than those of the DSSC greenhouse by 37% and 38%, respectively (Figure 5a). The CCI values did not differ between the hybrids grown in the DSSC greenhouse, while they differed significantly in the CONV one. Irrespective of the hybrid, the DSSC greenhouse plants presented lower CCI values compared to plants of the CONV greenhouse.

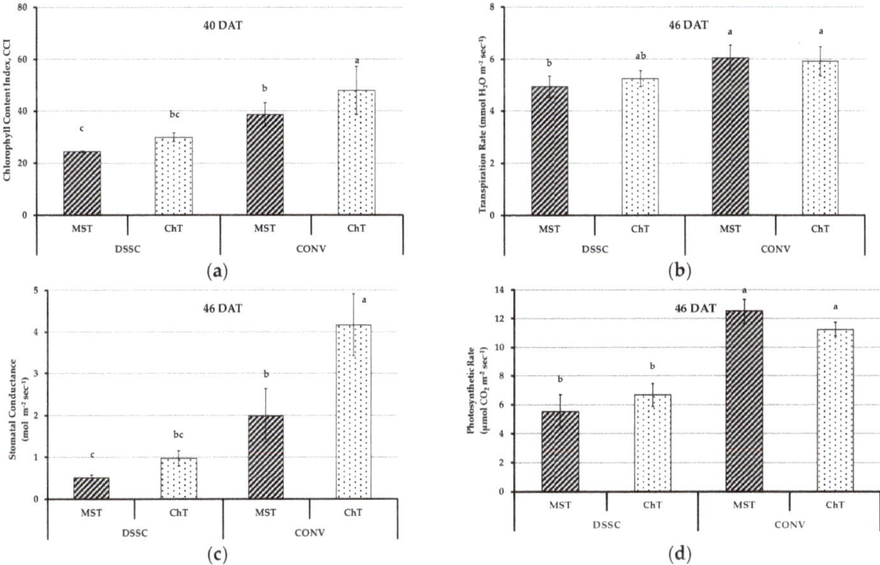

Figure 5. Physiological characteristics of the two tomato hybrids [MST (medium-sized) and ChT (cherry)] in the two greenhouses [DSSC (Dye-Sensitized Solar Cell) and CONV (Conventional)] 40 and 46 days after transplanting (DAT): (**a**) Leaf chlorophyll content; (**b**) transpiration rate; (**c**) stomatal conductance; (**d**) photosynthetic rate. Bars with different letters were significantly different by LSD at $P \leq 0.05$.

Table 3. Combined variance analysis of the physiological parameters, as influenced by the greenhouse types [DSSC (Dye-Sensitized Solar Cell) and CONV (Conventional)] and the species of tomato hybrid [MST (medium-sized) and ChT (cherry)].

		Significance of F ratio			
Source	**df [z]**	**Chlorophyll Content Index**	**Transpiration Rate**	**Stomatal Conductance**	**Photosynthetic Rate**
		40 DAT [y]	**46 DAT**	**46 DAT**	**46 DAT**
Replications	5	NS [x]	NS	NS	NS
Greenhouse (G)	1	*	*	**	**
Error (a)	5				
Hybrid (H)	1	**	NS	**	NS
G × H	1	NS	NS	NS	*
Error (b)	10				
CV %		19.38	13.31	12.71	12.74

[z] df, degree of freedom; CV, coefficient of variation; [y] DAT, days after transplanting; [x] NS, nonsignificant; * = $P < 0.05$ level of significance; ** = $P < 0.01$ level of significance.

The transpiration rate (measured on day 46) was only influenced by the type of greenhouse (G). The same applied to the G × H interaction (Table 3). The MST in the DSSC greenhouse had an 18% lower transpiration rate compared to the CONV greenhouse (Figure 5b); ChT did not present significant differences between the greenhouses.

Similarly, stomatal conductance differed between the plants of the two greenhouses and between the hybrids, but the G × H interaction was not significant (Table 3). The MST and ChT plants in the CONV greenhouse had higher stomatal conductance (74% and 76%, respectively) compared to the DSSC (Figure 5c).

The photosynthetic rate was mainly influenced by the type of greenhouse (Table 3). The MST and ChT in the DSSC greenhouse had a lower photosynthetic rate (55% and 40%, respectively) compared to the CONV greenhouse (Figure 5d).

Lower values of leaf chlorophyll content, transpiration rate, stomatal conductance, and photosynthetic rate were likely due to the shading effect in the DSSC greenhouse but also to stress from the high air temperature. Yamori et al. [10] and Tewolde et al. [11] reported that supplemental lighting did not lead to higher photosynthesis and yield of tomatoes in summer, when the irradiation from the sun was enough for plant cultivation, whereas it led to higher photosynthesis, growth, and yield in winter when the irradiation is limiting for plant growth. This means that even though light is the basic factor influencing photosynthesis and plant growth, surplus light, especially during summer when it is not needed for plant growth, can be used for electricity generation by photoselective DSSC installed in the greenhouse roof covering [12,13].

3.4. Fruit Quality Parameters and Bioactive Compounds

3.4.1. Physicochemical Parameters of Tomato Fruits

In Figure 6a,b, the pH and citric acid percentage of the two tomato hybrids are shown. MST presented significantly higher pH values than ChT, regardless of the greenhouse type (Figure 6a). No significant difference was observed in citric acid content between the two greenhouses for the hybrids. The greenhouse type did not significantly affect dry matter % of either ChT or MST (Figure 6c). Additionally, for total sugar content, no differences were observed between hybrids or greenhouse types (Figure 6d).

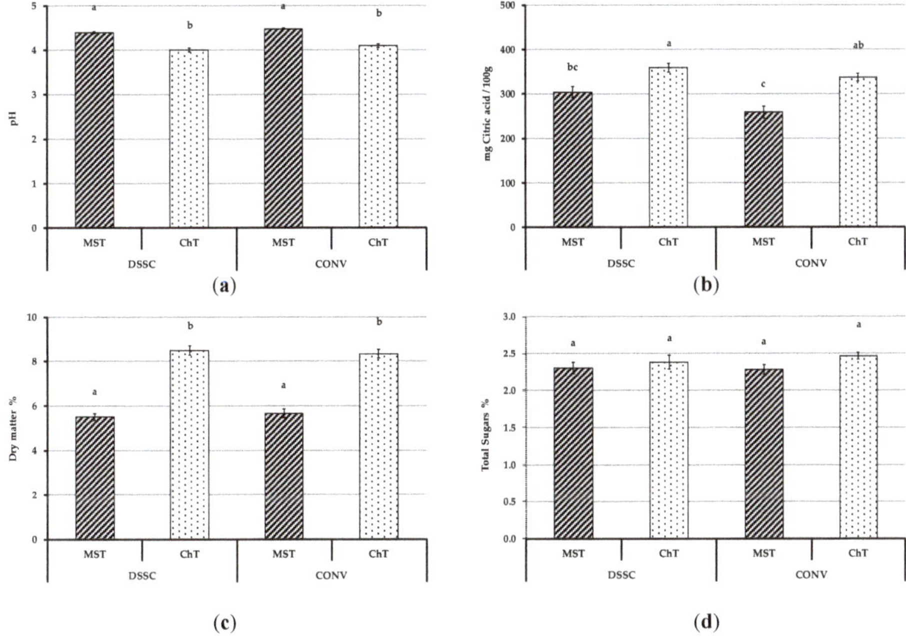

Figure 6. Physical parameters of fruits, independent of inflorescence, of the two tomato hybrids [MST (medium-sized) and ChT (cherry)] in the two greenhouses [DSSC (Dye-Sensitized Solar Cell) and CONV (Conventional)]: (**a**) pH; (**b**) citric acid %; (**c**) dry matter %; and (**d**) total sugars %. Bars with different letters were significantly different by LSD at $P \leq 0.05$.

According to the results of the variance analysis (Table 4), the effects of the greenhouse, hybrid, and inflorescence were significant for pH, while the interaction of all three factors was not significant.

Table 4. Variance analysis of physical parameters, bioactive ingredients, and antioxidant characteristics, as influenced by the greenhouse type [DSSC (Dye-Sensitized Solar Cell) and CONV (Conventional)] and the species of tomato hybrid [MST (medium-sized) and ChT (cherry)].

Source	Significance of F ratio									
	df [z]	pH	CA [y]	SUG	ASC	LUC	β-CAR	CARs	DPPH	ABTS
Replications	2	NS [x]	NS	NS	NS	NS	NS	NS	**	**
Greenhouse (G)	1	**	NS	NS	*	**	*	*	**	**
Error (a)	2									
Hybrid (H)	1	**	NS	NS	**	**	**	**	**	**
G × H	1	NS	**	NS	NS	NS	**	NS	**	**
Error (b)	4									
Inflorescence (I)	4	**	**	**	**	**	**	**	**	**
G × I	4	NS	**	**	NS	NS	**	NS	**	**
H × I	4	**	**	**	**	NS	**	*	**	**
G × H × I	4	NS	NS	**	NS	*	*	*	*	**
Error (c)	32									
CV %		1.5	6.2	3.7	11.1	18.8	11.3	16.7	5.9	3.2

[z] df, degree of freedom; CV, coefficient of variation; [y] CA, citric acid; SUG; total sugars; ASC, ascorbic acid; LUC, lycopene; β-CAR, β-carotene; CARs, total carotenoids; DPPH, 1,1-diphenyl-2-picrylhydrazyl assay, ABTS, 2,2′-azino-bis-3-ethylbenzthiazoline-6-sulphonic acid assay, [x] NS, nonsignificant; * = $P < 0.05$ level of significance; ** = $P < 0.01$ level of significance.

3.4.2. Bioactive Compounds of Tomato Fruits

No significant differences were found between the two greenhouses in ascorbic acid concentration of ChT and MST, but a slight decrease of 6% was observed in the DSSC greenhouse for all tomato samples analyzed (Figure 7a). It is noteworthy that ChT presented a higher ascorbic acid concentration (64%) compared to MST. Several factors contribute to high concentration of ascorbic acid in tomato: variety, high salinity of irrigation water, and solar radiation. Many studies have shown that high light intensity is associated with high vitamin C content in tomatoes [17]. Generally, high temperatures are required for the synthesis of ascorbic acid, and sunlight enhances the accumulation of additional ascorbic acid [24]. It was generally observed that the degree of significance was most affected by the hybrids and less by the greenhouse type, while most of their interactions were not significant (Table 4).

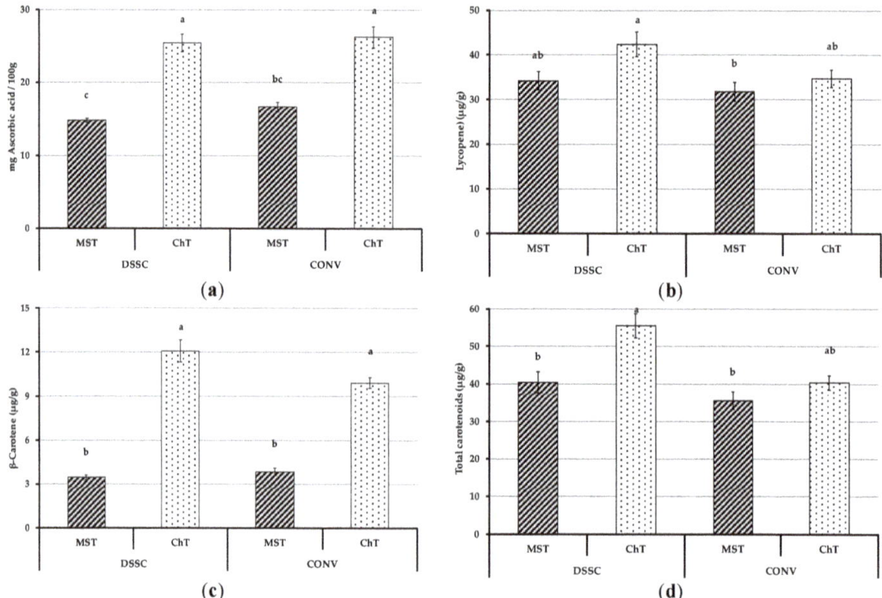

Figure 7. Content of bioactive ingredients of the fruits of the two tomato hybrids [MST (medium-sized) and ChT (cherry)] in the two greenhouses [DSSC (Dye-Sensitized Solar Cell) and CONV (Conventional)]: (**a**) ascorbic acid; (**b**) lycopene; (**c**) β-carotene; and (**d**) total carotenoids. Bars with different letters were significantly different by LSD at $P \leq 0.05$.

Concerning the concentrations of lycopene, β-carotene, and total carotenoids, significant differences were observed both between tomato hybrids and between inflorescences and greenhouses, as well as their interactions (Table 4). ChT presented higher values of β-carotene (11.0 µg/g) compared to MST (3.6 µg/g) in both greenhouses, whereas ChT had the higher value of lycopene (42.4 µg/g) in the DSSC greenhouse. In the DSSC greenhouse there was a 15% and 13% increase in the concentration of lycopene and β-carotene, respectively, compared to the CONV greenhouse (Figure 7b,c). Similar results were reported by Pek et al. [24], who found that tomatoes that were shaded had a higher concentration of lycopene. Similarly to β-carotene, ChT showed higher total carotenoid concentration (48 µg/g) compared to MST (38 µg/g) in both greenhouses (Figure 7d). In general, an increase of about 26% in concentration of total carotenoids was observed in the DSSC greenhouse for both hybrids.

3.4.3. Antioxidant Capacity

Concerning the antioxidant capacity, significant differences were observed between greenhouses, tomato hybrids, inflorescences, as well as their interactions (Table 4). For both hybrids, the ABTS and DPPH tests showed increases of about 10% and 5% in the DSSC greenhouse (Figure 8a,b). It is worth noting that the antioxidant capacity of the DPPH test was lower than ABTS, because the DPPH test was sensitive to obstructions due to the high concentration of carotenoids.

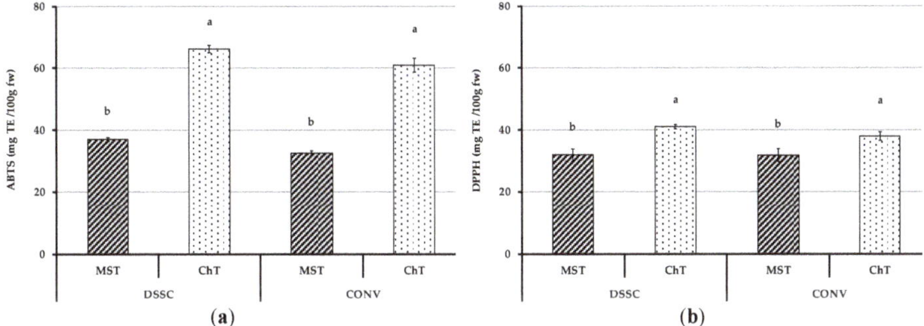

Figure 8. Antioxidant capacity of the fruits of the two tomato hybrids [MST (medium-sized) and ChT (cherry)] in the two greenhouses [DSSC (Dye-Sensitized Solar Cell) and CONV (Conventional)]: (**a**) ABTS (2,2′-azino-bis-3-ethylbenzthiazoline-6-sulphonic acid) assay; and (**b**) DPPH (1,1-diphenyl-2-picrylhydrazyl) assay. Bars with different letters were significantly different by LSD at $P \leq 0.05$.

4. Conclusions

Data concerning crop productivity and physiology were obtained through the hydroponic cultivation of tomato crops in two experimental greenhouses, one using the DSSC technology as covering material and a conventional one covered by glass as a control. Particular attention was given to the physiological response of the crops under the novel covering material, which resulted in a shading effect. For this reason, specific physiological parameters of the crop, such as plant transpiration rate and leaf chlorophyll content, were measured and monitored along with potential photomorphogenetic responses of the plants. Furthermore, the comparisons included chemical analyses of harvested tissues and fruits in order to determine potential differences in the concentration of phytochemical compounds, such as ascorbic acid, lycopene, and quality characteristics. The early yield and average fruit weight of the MST were affected by the DSSC greenhouse, but not the ChT. The total yield and average fruit weight of the MST were also affected by the DSSC greenhouse. As for the ChT, its total yield was affected by the DSSC greenhouse, but not average fruit weight. The DSSC greenhouse plants presented lower CCI and transpiration rate values, especially for the MST, compared to plants in the CONV greenhouse. However, the transpiration rate of the ChT did not present statistically significant differences between the two greenhouses. Additionally, the DSSC greenhouse showed significantly higher values of dry matter % than the CONV for both the ChT and MST, while no differences in the total sugar content were observed for the greenhouses. An increase of about 20% in concentrations of lycopene and β-carotene was observed in the DSSC greenhouse for the ChT and 10% for the MST. Finally, results showed that the ABTS test is the more appropriate method for measuring antioxidant capacity in tomato extracts. The results from the DSSC greenhouse during the summer season were satisfactory, since shading had a positive effect on the qualitative characteristics of the tomato fruits; however, lower values of physiological characteristics and yield were related to stress from high air temperatures. Furthermore, surplus light, especially during summer when it is not needed for plant growth, can be used for electricity generation by photoselective DSSCs installed in the greenhouse roof covering. In future work, the DSSC greenhouse will be evaluated during the entire year, especially during wintertime.

Author Contributions: Methodology, G.K.N., K.K. (Kalliopi Kadoglidou), N.T. and K.K. (Konstantinos Krommydas); investigation, K.K. (Kalliopi Kadoglidou), N.T. and K.Kr. (Konstantinos Krommydas); data curation, K.K. (Kalliopi Kadoglidou), K.K. (Konstantinos Krommydas) and M.I.; formal analysis, K.K. (Kalliopi Kadoglidou); writing—original draft preparation, G.K.N. and K.K. (Kalliopi Kadoglidou); writing—review and editing, G.K.N., K.K. (Kalliopi Kadoglidou), A.K., P.R. and M.I.

Funding: This research was funded by the Operational Programme "Competitiveness and Entrepreneurship", National Action "Cooperation 2011–Research Project 11 SYN _7 _298 "Energy autonomous smart greenhouse".

Conflicts of Interest: The authors declare no conflicts of interest.

References

1. Ntinas, G.K.; Morichovitis, Z.; Nikita-Martzopoulou, C. The influence of a hybrid solar energy saving system on the growth and the yield of tomato crop in greenhouses. *Acta Hortic.* **2012**, *952*, 723–730. [CrossRef]

2. Fernández, J.A.; Orsini, F.; Baeza, E.; Oztekin, G.B.; Muñoz, P.; Contreras, J.; Montero, J.I. Current trends in protected cultivation in Mediterranean climates. *Eur. J. Hortic. Sci.* **2018**, *83*, 294–305. [CrossRef]

3. Ntinas, G.K.; Neumair, M.; Tsadilas, C.D.; Meyer, J. Carbon footprint and cumulative energy demand of greenhouse and open-field tomato cultivation systems under Southern and Central European climatic conditions. *J. Clean. Prod.* **2017**, *142*, 3617–3626. [CrossRef]

4. Gruda, N.; Bisbis, M.; Tanny, J. Impacts of protected vegetable cultivation on climate change and adaptation strategies for cleaner production—A review. *J. Clean. Prod.* **2019**, *225*, 324–339. [CrossRef]

5. Gruda, N.; Bisbis, M.; Tanny, J. Influence of climate change on protected cultivation: Impacts and sustainable adaptation strategies-A review. *J. Clean. Prod.* **2019**, *225*, 481–495. [CrossRef]

6. Hassanien, R.H.E.; Li, M.; Dong Lin, W. Advanced applications of solar energy in agricultural greenhouses. *Renew. Sustain. Energy Rev.* **2016**, *54*, 989–1001. [CrossRef]

7. Yano, A.; Onoe, M.; Nakata, J. Prototype semi-transparent photovoltaic modules for greenhouse roof applications. *Biosyst. Eng.* **2014**, *122*, 62–73. [CrossRef]

8. Trypanagnostopoulos, G.; Kavga, A.; Souliotis, M.; Tripanagnostopoulos, Y. Greenhouse performance results for roof installed photovoltaics. *Renew. Energy* **2017**, *111*, 724–731. [CrossRef]

9. Marcelis, L.F.M.; Broekhuijsen, A.G.M.; Meinen, E.; Nijs, E.M.F.M.; Raaphorst, M.G.M. Quantification of the growth response to light quantity of greenhouse grown crops. *Acta Hortic.* **2006**, *711*, 97–114. [CrossRef]

10. Yamori, W.; Noguchi, K.; Hikosaka, K.; Terashima, I. Phenotypic Plasticity in Photosynthetic Temperature Acclimation among Crop Species with Different Cold Tolerances. *Plant Physiol.* **2010**, *152*, 388–399. [CrossRef] [PubMed]

11. Tewolde, F.T.; Lu, N.; Shiina, K.; Maruo, T.; Takagaki, M.; Kozai, T.; Yamori, W. Nighttime Supplemental LED Inter-lighting Improves Growth and Yield of Single-Truss Tomatoes by Enhancing Photosynthesis in Both Winter and Summer. *Front. Plant Sci.* **2016**, *7*, 448. [CrossRef]

12. Roslan, N.; Radzi, M.A.M.; Chen, G.; Jamaludin, D.; Hashimoto, Y.; Ya'acob, M.E. Dye Sensitized Solar Cell (DSSC) greenhouse shading: New insights for solar radiation manipulation. *Renew. Sustain. Energy Rev.* **2018**, *92*, 171–186. [CrossRef]

13. Kim, J.-J.; Kang, M.; Kwak, O.K.; Yoon, Y.-J.; Min, K.S.; Chu, M.-J. Fabrication and Characterization of Dye-Sensitized Solar Cells for Greenhouse Application. *Int. J. Photoenergy* **2014**, *2014*, 1–7. [CrossRef]

14. Jones, B.J. *Tomato Plant Culture: In the Field, Greenhouse, and Home Garden*, 2nd ed.; CRC Press: Boca Raton, FL, USA, 2007; ISBN 9781420007398.

15. Beckles, D.M. Factors affecting the postharvest soluble solids and sugar content of tomato (*Solanum lycopersicum* L.) fruit. *Postharvest Biol. Technol.* **2012**, *63*, 129–140. [CrossRef]

16. Friedman, M. Anticarcinogenic, cardioprotective, and other health benefits of tomato compounds lycopene, α-tomatine, and tomatidine in pure form and in fresh and processed tomatoes. *J. Agric. Food Chem.* **2013**, *61*, 9534–9550. [CrossRef]

17. Dumas, Y.; Dadomo, M.; Di Lucca, G.; Grolier, P. Effects of environmental factors and agricultural techniques on antioxidant content of tomatoes. *J. Sci. Food Agric.* **2003**, *382*, 369–382. [CrossRef]

18. Sánchez-Rodríguez, E.; Ruiz, J.M.; Ferreres, F.; Moreno, D.A. Phenolic profiles of cherry tomatoes as influenced by hydric stress and rootstock technique. *Food Chem.* **2012**, *134*, 775–782. [CrossRef] [PubMed]
19. Lenucci, M.S.; Cadinu, D.; Taurino, M.; Piro, G.; Dalessandro, G. Antioxidant Composition in Cherry and High-Pigment Tomato Cultivars. *J. Agric. Food Chem.* **2006**, *54*, 2606–2613. [CrossRef]
20. Borguini, R.G.; Ferraz, E.A.; Silva, D.; Ferraz, E.A.; Silva, D.A. Tomatoes and Tomato Products as Dietary Sources of Antioxidants Tomatoes and Tomato Products as Dietary Sources of Antioxidants. *Food Rev. Int.* **2009**, *25*, 313–325. [CrossRef]
21. Gómez-Romero, M.; Arráez-Román, D.; Segura-Carretero, A.; Fernández-Gutiérrez, A. Analytical determination of antioxidants in tomato: Typical components of the Mediterranean diet. *J. Sep. Sci.* **2007**, *30*, 452–461. [CrossRef]
22. Ilahy, R.; Hdider, C.; Lenucci, M.S.; Tlili, I.; Dalessandro, G. Antioxidant activity and bioactive compound changes during fruit ripening of high-lycopene tomato cultivars. *J. Food Compos. Anal.* **2011**, *24*, 588–595. [CrossRef]
23. Helyes, L.D.; Lugasi, A. Formation of certain compounds having technological and nutritional importance in tomato fruits during maturation. *Acta Aliment.* **2006**, *35*, 183–193. [CrossRef]
24. Pék, Z.; Helyes, L.; Lugasi, A. Color Changes and Antioxidant Content of Vine and Postharvest- ripened Tomato Fruits. *HortScience* **2010**, *45*, 466–468. [CrossRef]
25. AOAC. *Official Methods of Analysis of AOAC International*; AOAC: Rockville, MD, USA, 2000; ISBN 0935584544.
26. Irakli, M.; Chatzopoulou, P.; Kadoglidou, K.; Tsivelika, N. Optimization and development of a high-performance liquid chromatography method for the simultaneous determination of vitamin E and carotenoids in tomato fruits. *J. Sep. Sci.* **2016**, *39*, 3348–3356. [CrossRef]
27. Re, R.; Pellegrini, N.; Proteggenete, A.; Pannala, A.; Yang, M.; Rice-Evans, C. Antioxidant activity applying an improved ABTS radical cation decolorization assay. *Free Radic. Biol. Med.* **1999**, *26*, 1231–1237. [CrossRef]
28. Yen, G.; Chen, H. Antioxidant Activity of Various Tea Extracts in Relation to Their Antimutagenicity. *J. Agric. Food Chem.* **1995**, *43*, 27–32. [CrossRef]

 *horticulturae*

Article

Automation for Water and Nitrogen Deficit Stress Detection in Soilless Tomato Crops Based on Spectral Indi

Angeliki Elvanidi, Nikolaos Katsoulas * and Constantinos Kittas

Department of Agriculture Crop Production and Rural Environment, University of Thessaly, Fytokou Str., 38446 Volos, Greece; elaggeliki@gmail.com (A.E.); ckittas@uth.gr (C.K.)
* Correspondence: nkatsoulas@uth.gr; Tel.: +30-24-2109-3249

Received: 8 October 2018; Accepted: 16 November 2018; Published: 23 November 2018

Abstract: Water and nitrogen deficit stress are some of the most important growth limiting factors in crop production. Several methods have been used to quantify the impact of water and nitrogen deficit stress on plant physiology. However, by performing machine learning with hyperspectral sensor data, crop physiology management systems are integrated into real artificial intelligence systems, providing richer recommendations and insights into implementing appropriate irrigation and environment control management strategies. In this study, the Classification Tree model was used to group complex hyperspectral datasets in order to provide remote visual results about plant water and nitrogen deficit stress. Soilless tomato crops are grown under varying water and nitrogen regimes. The model that we developed was trained using 75% of the total sample dataset, while the rest (25%) of the data were used to validate the model. The results showed that the combination of *MSAVI*, *mrNDVI*, and *PRI* had the potential to determine water and nitrogen deficit stress with 89.6% and 91.4% classification accuracy values for the training and testing samples, respectively. The results of the current study are promising for developing control strategies for sustainable greenhouse production.

Keywords: remote sensing; hyperspectral; reflectance index; classification tree; machine vision

---

1. Introduction

Nowadays, the need to produce more food with fewer inputs (water, fertilizer, and land, among others) and zero effect on the environment has led to an increase in greenhouse production. However, to cultivate under greenhouse conditions is certainly not an easy task since performance of several farming practices are needed. In this manner, crop yield and quality optimisation will be further improved only if farmers integrate greenhouse control strategies, that is, measurements concerning the dynamic response of plants according to the greenhouse's spatial environment changes.

Within a greenhouse, however, it is a challenge to quantify the spatial impact of biotic or abiotic factors in plant growth. Usually, up to now, environmental patterns under greenhouses have been monitored and managed by sampling at a single position and by considering the indoor microclimate completely homogeneous [1]. This assumption, however, is not valid since an intense heterogeneity that must be taken into account actually occurs, especially in intensive production systems. A sensing system equipped with a multi-sensor platform moving over the canopy is the key to communicate the plant's real state and needs. Obviously, in such moving platforms, the continuous monitoring of the interactions between the microclimate and the physical conditions of the plants is performed using mostly non-contact and non-destructive sensing techniques [2].

Up to now, the methods used for monitoring plant physiology have been quite complex and problematic. Most of these measure plant physiology in a limited spatial scale and, in most cases, they require physical contact with the plants/soil or follow destructive sensing procedures, making their application as a commercial multi-sensor scale rather infeasible [2–4].

The current development of computational hyperspectral machine vision systems allows us to build a real-time plant canopy health, growth, and quality monitoring multisensory platform equipped with remote technologies [5–7]. Hyperspectral machine vision systems allow recording in large scale, the spatial interaction of sunlight with crop canopies and plant leaves providing valuable information about plant growth and health status [6,8]. In this way, changes observed in plant irradiance of visible (*VIS*) and near infrared (*NIR*) spectrum, indicate different types of plant stress. The reflectance variation in *VIS* spectrum, for instance, is recorded to assess a series of several pigments located in the mesophyll area, such as chlorophyll, carotenoids and xanthophyll. The reflectance variation, on the other hand, observed in the *NIR* spectrum, is recorded to obtain information about leaf water content stored in cavities of spongy parenchyma. Additionally, changes in *NIR* spectrum are used to assess the carbon content in different forms (sugar, starch, cellulose and lignin) in mesophyll cells and nutrient compounds (N, P, K) in mesophyll cells and palisade parenchyma [2].

In order to provide more precise information for stress detection, certain parts of the spectrum can be combined to form reflectance indices (*RIs*) [2,7,9,10]. In this way, the spectral differences detected are amplified while the resulting *RIs* are used directly as a metric to quantify different aspects of plant physiology response.

So far, numerous successful case studies related to RIs and their relationship with crop, climate, or soil data from different plants have been performed. The photochemical reflectance index ($PRI = (R_{531} - R_{570})/(R_{531} + R_{570})$) is one of the most widespread indicators estimating rapid changes in de-epoxidation of the xanthophyll cycle [11–14]. Bajwa et al. [15] used the Modified soil-adjusted vegetation index ($MSAVI = 1/2 \times (2 \times (R_{810} + 1) - (\sqrt{(2 \times R_{810} + 1) \times 2 - 8 \times (R_{810} - R_{690})}))$) to predict green biomass ($R^2 > 0.70$). Elvanidi et al. [1,7,13] also showed that Modified red normalised vegetation and simple ration indices ($mrNDVI = (R_{750} - R_{705})/(R_{750} + R_{705} - 2 \times R_{445})$; $mrSRI = (R_{750} - R_{445})/(R_{705} - R_{445})$) were the two most relevant and sensitive indices indicating water deficit stress in tomato plants. Additionally, the Transformed chlorophyll absorption in the reflectance index ($TCARI = 3 \times [(R_{700} - R_{670}) - 0.2 \times (R_{700} - R_{550}) \times (R_{700}/R_{670})]$) was strongly correlated with leaf chlorophyll content variation [13,16].

In addition to reflectance indices, many statistical and machine learning models such as classification tree analysis, support vector machine, artificial neural network, and other classification procedures have been developed to extract optimal information from remotely sensed data [1]. Researchers observed that the classification tree (*CT*) was a very useful model to analyze complex data sets by providing visual results [13,17–19].

Therefore, the object of this work was to develop a model based on the *CT* method to analyse complex *RIs* datasets in order to provide visual assessments of plant water and nitrogen deficit stress. For this reason, *RIs* that indicate different aspects of tomato crop physiology (such as photosynthetic rate (*As*, μmol m^{-1} s^{-1}), chlorophyll_a content (*Chl_a*, μg cm^{-2}), nitrogen content (N, %) and substrate water content (θ, %)) under a controlled environment, were classified to predict remotely the (i) plant chlorophyll content, (ii) plant water content status and to identify (iii) healthy, water- and nitrogen-deficit stressed plants. The applied objective of this work was to develop a model based on simplified reflectance indices that could be adapted by multisensory platform methodologies to predict future irrigation events.

2. Materials and Methods

2.1. Experimental Set-Up

The experiment was carried out during August of 2014 and April of 2016 in a controlled growth chamber located in Velestino, Central Greece, with a ground area of 28 m^2 (4 m $\times$ 7 m) and height of 3.2 m. Air temperature, relative humidity, light intensity and CO_2 concentration were automatically controlled using a climate control computer (Argos Electronics, Athens, Greece). The light intensity was controlled using 24 high-pressure sodium lamps, 600 W each (MASTER GreenPower 600 W EL 400 V Mogul 1SL, Philips, Eindhoven, The Netherlands), operated in four clusters with six lamps per cluster. The average irradiance when all 24 lamps were used was 240 W m^{-2} (about 350 μmol m^{-1} s^{-1}).

Tomato plants (*Solanum lycopersicum* cv. Elpida, provided by Spyrou SA, Athens, Greece) were grown in slabs filled with perlite (ISOCON Perloflor Hydro 1, ISOCON S.A., Athens, Greece), at different time periods. Two units comprised of two crop lines each (18 plants per line) were used. The precise cultivation practices followed are described in References [1,7,13].

The nutrient solution was supplied via a drip system and was controlled by a time-program irrigation controller (8 irrigation events per day, at 07:00, 10:00, 102:00, 14:00, 16:00, 18:00, 19:30, and 03:30, local time), with set-points for electrical conductivity (*EC*) at 2.4 dS m^{-1} and a pH of 5.6.

In order to quantify the plant physiological response to their environment by crop reflectance characteristics, the plants had about 10 leaves each, were about 1 m in height, had a leaf area index (*LAI*) of about 0.8, and were under the imposition of varying water and nitrogen regimes. A nutrient solution containing from 0 to 100% coverage of plant actual water and nitrogen needs was supplied to the root zone of the plants for several days, varying the chlorophyll_a, nitrogen and substrate water content from 44.62 to 36.04 μg cm^{-2}, 4.77–3.30% and 54.81–35%, respectively. The control plants were irrigated with a nutrient solution with 100% of plant water and nitrogen according to Reference [6] mineral nutrient list (irrigation dose of 120 mL per plant; nutrient solution of 12.9 mmol NO$_3$ L^{-1} and 1.0 mmol NH$_4$ L^{-1}; 8 events per day. The concentrations of the rest of the macronutrients in the control treatment were K 7.5 mmol L^{-1}, Ca 4.8 mmol L^{-1}, Mg 2.5 mmol L^{-1}, H$_2$PO$_4$ 1.5 mmol L^{-1}).

To create a series of plant physiological dataset groups, reflectance measurements (*r*) along with measurements of plant physiology such as θ (%), *Chl_a* (μg cm^{-2}), and N (%), values were obtained for the same set of plant water and nitrogen characteristics. The measurements were carried out in young and fully developed leaves between the 3rd and 6th branches of three adjacent tomato plants. The resulting correlations between the factors were further presented in References [1,7,13]. In total, 160 groups were performed throughout the period considered, under known conditions of water and nitrogen supply and environmental conditions.

2.2. Measurements

Air temperature (*T*, °C) and relative humidity (*RH*, %) were measured using two temperature-humidity sensors (model HD9008TR, Delta Ohm, Caselle di Selvazzano, Italy). Irradiance (*Rg,i*, W m^{-2}) inside the growth chamber was recorded using a solar pyranometer (model SKS 1110, Skye instruments, Powys, UK). The sensors were calibrated before the experimental period and placed 1.8 m above ground level. The data were automatically recorded in a data logger system (Zeno 3200, Coastal Environmental Systems Inc., Seattle, WA, USA). Substrate volumetric water content (θ, %) was estimated using capacitance sensors (model WCM-control, Rockwool B.V., Roermond, The Netherlands) placed horizontally in the middle (height and width) of the hydroponic slabs. Measurements were performed every 30 s and 10-min average values were recorded.

In plants with known θ values, leaf chlorophyll content measurements were recorded by means of an Opti-Science sensor performing measurements in contact with the leaf (CCM 200, Opti-Science, Hudson, NH, USA). The values recorded by means of the CCM 200 sensor were correlated with *Chl_a* values (μg cm^{-2}) obtained in the lab for the same set of leaves using the Reference [20] protocol. The resulted equation was presented in Reference [13].

The nitrogen content in the plant tissue (leaf dry matter sample of the entire tomato plant) was also analysed in the laboratory with the Total Kjeldahl Nitrogen method (TKN) using [21] protocol. N determination was done with an automatic flow injection analyser system (FIAstar 5000 analyser, Tecator, Foss, Hillerød, Denmark). The impact of varying nitrogen regimes in plant tissue was further examined by Reference [1].

The radiation reflected by the plants were recorded with two spectra sensors: (1) a portable spectroradiometer (model ASD FieldSpec Pro, Analytical Spectral Devices, Boulder, CO, USA) and (2) a hyperspectral camera Imspec V10 (Spectral Imaging Ltd., Oulu, Finland). The spectroradiometer measures the radiation reflected in the range between 350 and 2500 nm, while the camera operates in the visible and near-infrared (*VNIR*) spectrum region between 400 and 1000 nm. The camera system was placed on a moving cart so that images of the vertical canopy axis could be obtained to cover the canopy area of young, fully developed leaves between the 3rd and 6th branches of three adjacent tomato plants. For extra illumination of the target area (70×100 cm), four quartz-halogen illuminators (500 W each) were used.

The system calibration procedures of both spectral sensors along with the method concerned the camera's set up, image segmentation, and plant reflectance calculation was done as described in References [1,2,7,13,22]. The basic steps of the experimental setup are described in Figure 1.

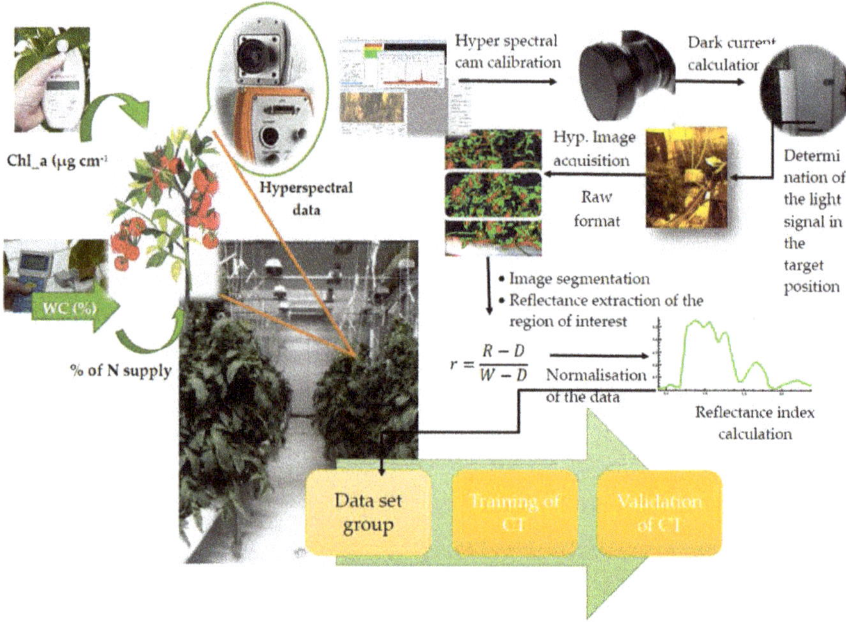

Figure 1. The experimental procedure for collecting 160 dataset groups to train and validate the system.

2.3. Calculations

Based on the available reflectance measurements, the following indices (based on the analysis performed in References [1,7,13]) were calculated and evaluated to train the models:

$$PRI = (R_{531} - R_{570})/(R_{531} + R_{570}) \tag{1}$$

$$TCARI = 3 \times [(R_{700} - R_{670}) - 0.2 \times (R_{700} - R_{550}) \times (R_{700}/R_{670})] \tag{2}$$

$$mrNDVI = (R_{750} - R_{705})/(R_{750} + R_{705} - 2 \times R_{445}) \tag{3}$$

$$mrSRI = (R_{750} - R_{445})/(R_{705} - R_{445}) \tag{4}$$

$$MSAVI = 1/2 \times (2 \times (R_{810} + 1) - (\sqrt{(2 \times R_{810} + 1)} \times 2 - 8 \times (R_{810} - R_{690}))] \tag{5}$$

$$OSAVI = (1 + 0.16) \times (R_{800} - R_{670})/(R_{800} + R_{670} + 0.16) \tag{6}$$

where R is the reflectance value performed in each band expressed in nm that is indicated by the subscript number, *mrSRI* is the Modified red simple ration index and *OSAVI* is the Optimised soil-adjusted vegetation index.

2.4. Statistical Analysis

A classification tree was performed using SPSS (Statistical Package for the Social Sciences, IBM, USA) to create a tree-based prediction model of future irrigation events. Based on *CT* methodology, three hypotheses based on RI prediction rules were used to predict (i) the plant chlorophyll_a content (*1st Hypothesis*); (ii) the substrate water content (*2nd Hypothesis*), and (iii) to identify healthy, water- and nitrogen-deficit stressed plants (*3rd Hypothesis*). Each hypothesis independently consisted of structure trees, in which different *RIs* were involved. The *CTs* developed were based on the classification regression tree (*CRT*) and the chi-squared automatic interaction detection (*CHAID*) method to control the number of *RIs* and the maximum number of levels of growth beneath the root. The *p*-value was computed each time by applying Bonferroni adjustments. The method followed was done as described in References [13,23].

The models developed were calibrated (using training data) and validated (on a different set of data). A simple-sample validation was performed by using random assignment. A total of 75% of the total (n = 140 data sets) number of datasets were used as the training sample sets and 25% were used as the testing sample sets. The training sets were used to build the classification models, which were subsequently applied to the test set, which consisted of records with unknown class labels. The system tries to decrease the training error by completely fitting all the training examples. Each partition of each hypothesis was marked as the class label. During the 1st and 2nd Hypothesis, the class label was marked as *Chl_a* and θ (the numerical variable) and was expressed in µg cm^{-2} and %, respectively, to predict the actual chorophyll_a and water status of the plant. To train the model, the measured *Chl_a* and θ values were considered as dependent variables.

During the 3rd Hypothesis, each partition was marked as either *C* (Control), *WS* (Water Stress), or *NS* (Nitrogen Stress) to answer the question if the crop is under water or nitrogen deficit stress. Thus, it could be considered that *C* referred to "no stressed plants/no irrigation or fertilization is needed", *WS* referred to "water stress plants/irrigation is needed", while *NS* referred to "nitrogen stress plants/nitrogen is needed". The model was built according to substrate water and chlorophyll_a content evolution, in which θ values lower than 39% have been defined as *WS*. However, when the θ values were higher than 39%, while the *Chl_a* values were lower than 40 µg cm^{-2}, then the plant was defined as *NS*. These set points were derived from the analyses in References [1,7,13]. The algorithm employed a greedy strategy to grow the decision tree by making a series of locally optimum decisions about which attribute to use for the partitioning data. Each node of the training or testing samples showed the predicted value, which was the mean value for the dependent variable at that node. The mean value along with the measurement variability (standard deviation, $\pm SD$) of the parameters measured are reported. The measure of the tree's predictive accuracy was calculated based on a risk estimation and its standard error, where the proportion of cases were incorrectly classified after adjustment for prior probabilities and misclassification costs. The letter "*n*" is used to designate the daily sample size of each parameter. The goal of the classification models was to predict the class label of the unknown records. The current methodology followed the steps performed by Morgan [19], Loh [23], Lewis [24] and IBM SPSS Statistics 21 guide [25].

3. Results

3.1. Automation for Plant Chlorophyll_a Content Measuring

During the model analysis of the *1st Hypothesis* where the *Chl_a* status of the tomato is predicted, two independently structured trees with different combinations of *RIs* (*CT1* and *CT2*) were developed (Figures 2 and 3). Both resulting structures consisted of eight nodes (two nodes with at least one child and four nodes without children). For each node, there is a table that provides the number (*n*) and the percentage (%) of *Chl_a* content cases in each reflectance index category set as a dependent variable.

According to *CT1*, it was calculated that when *TCARI* was ≤0.0949, then the *Chl_a* content value was more than 43.61 µg cm^{-2} (Figure 2). Additionally, when the *TCARI* ranged between 0.0949 and 0.1029, the *Chl_a* content varied close to 42.82 µg cm^{-2} (*SD* = ±1). Since there were no child nodes below it, this was considered as the terminal node.

On the other hand, if the *TCARI* varied between 0.1029 and 0.1386, the *PRI* readings (next best predictor) had to be taken into consideration to identify the plant chlorophyll status and that node three was to be omitted. In this case, when *PRI* readings were ≤0.0321, 23% of the sample returned *Chl_a* equal to 42.4 µg cm^{-2} (*SD* = ±0.8), otherwise (*PRI* > 0.03), 17% of the sample returned *Chl_a* equal to 41.4 µg cm^{-2} (*SD* = ±0.9). For *TCARI* readings between 0.14 and 0.15 (node 4) and >0.15 (node 5), *Chl_a* was equal to 39.8 µg cm^{-2} (*SD* = ±2.0) and 37.9 µg cm^{-2} (*SD* = ±1.5), respectively.

According to *CT2* (Figure 3), it was calculated that when *TCARI* was ≤0.0880 and between 0.0880 and 0.0984, then nodes 1 and 2 returned similar *Chl_a* content values to *CT1*. On the other hand, if the *TCARI* varied between 0.0984 and 0.1381, the *mrSRI* readings (next best predictor) had to be taken into consideration to identify the plant chlorophyll status and that node three was to be omitted. In this case, when *mrSRI* readings were ≤13.6523, 29% of the sample returned *Chl_a* equal to 42.67 µg cm^{-2} (*SD* = ±0.4), otherwise (*mrSRI* > 13.6523) 21.5% of the sample returned *Chl_a* equal to 41.73 µg cm^{-2} (*SD* = ±0.9).

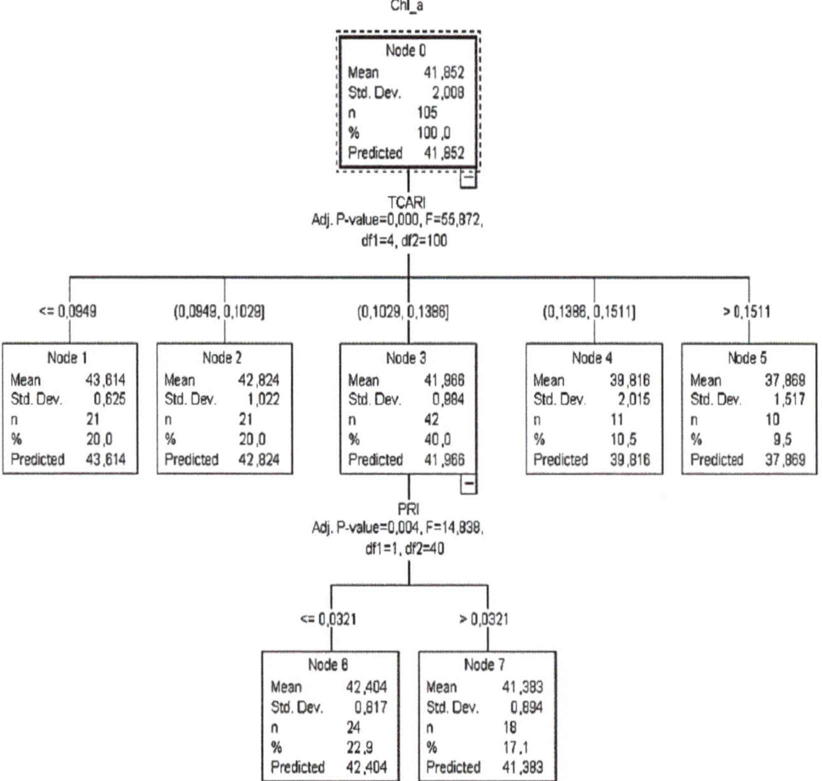

Figure 2. The classification tree (*CT1*) of the training sample for calibration to predict *Chl_a* concentration in the leaf area (µg cm^{-2}).

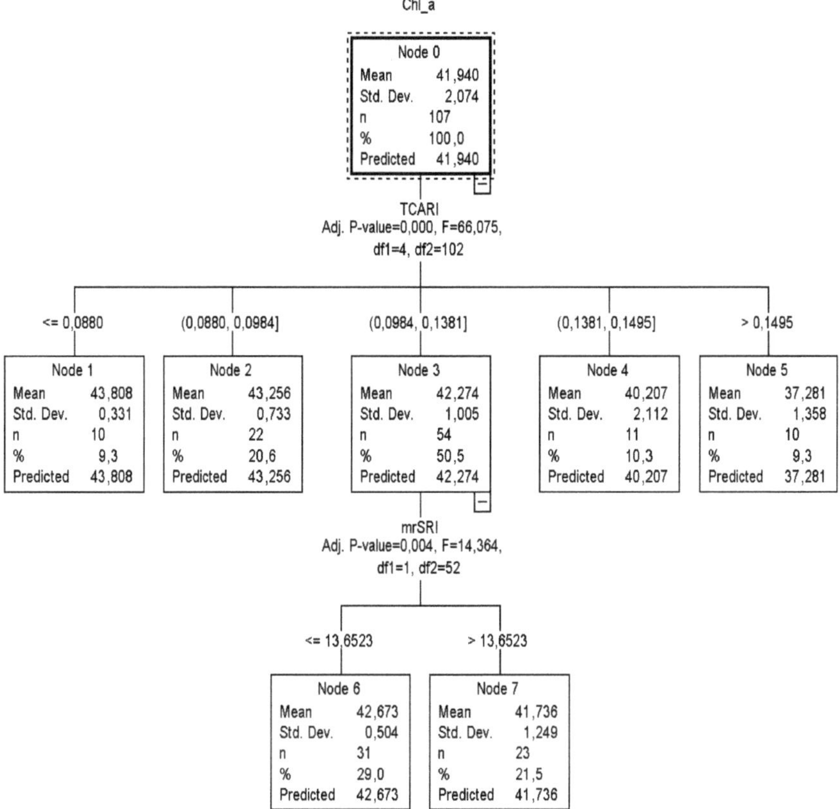

Figure 3. The *CT2* classification tree of the training sample for calibration to predict *Chl_a* concentration in the leaf area (μg cm^{-2}).

Both *CT1* and *CT2* structures did, however, reveal one potential problem with this model: for the plants that had a low level of *Chl_a* content, the standard deviation was high (>±1.0), which means that the data were widely spread and more than 20% of the predicted percentage inaccurately classified the *Chl_a* values lower than 39.8 μg cm^{-2}. This was also confirmed by the relevant node of the testing classification trees, where the *SD* was more than ±2. A comparison of the two structures, however, revealed that both *CT1* and *CT2* had a low estimation risk degree equal to 1.1. The trees resulting from the validation dataset provided an estimation risk degree equal to those of the training dataset (1.5 and 1.7, respectively).

3.2. Automation for Substrate Water Content Measuring

In the *2nd Hypothesis*, the tree (*CT3*) developed from both the training and testing samples had eleven nodes (five nodes with at least one child and six nodes without children) (Figure 4). In this tree, the predicted category was the estimation of the substrate water content (varied from 54.81 to 35%).

It was calculated that when *OSAVI* (a starter index) was ≤0.70, the tree returned the θ value as > 49.33%. Since there were no child nodes below it, this was considered the terminal node.

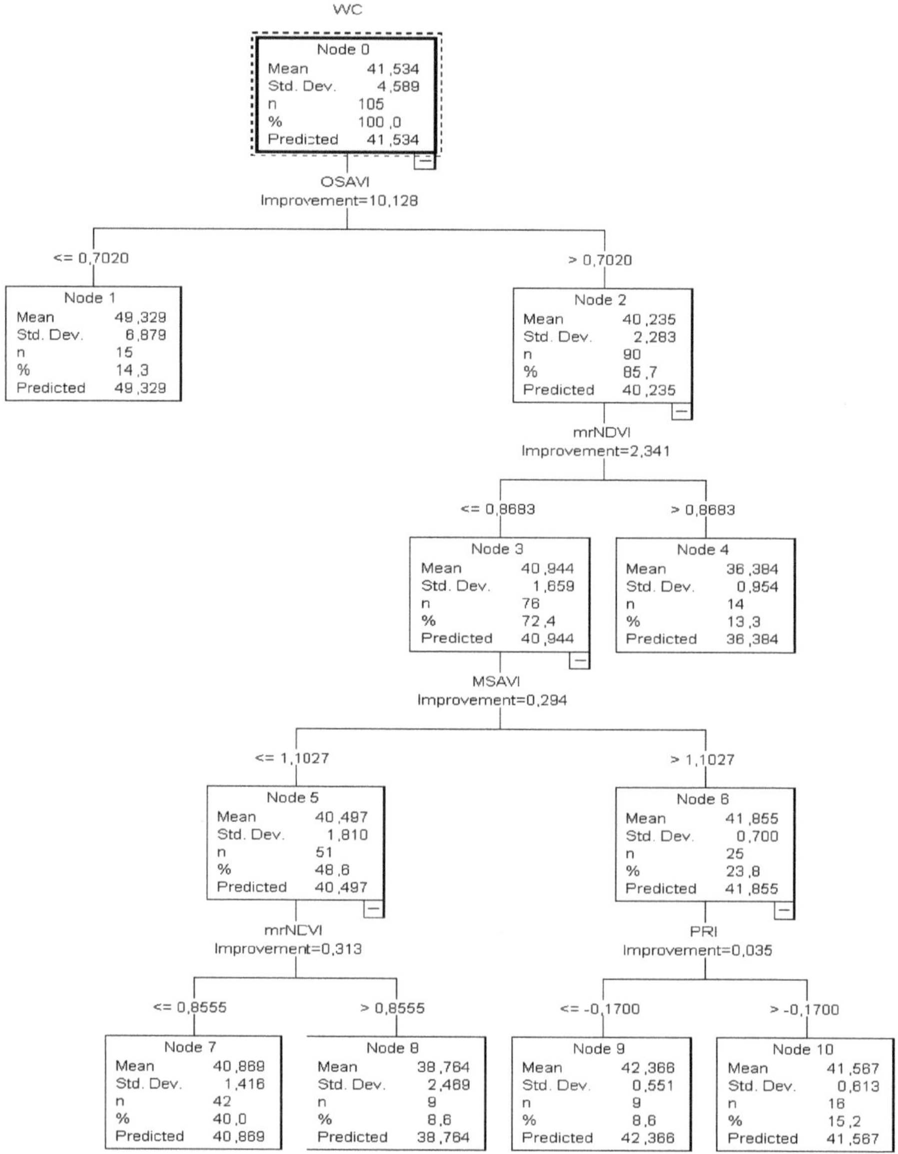

Figure 4. The *CT3* classification tree of the training sample for calibration to predict the plant θ values in the root zone (%).

On the other hand, if the *OSAVI* was higher than 0.7020, the *mrNDVI* readings (next best predictor) had to be taken into consideration to estimate the substrate water status and that node two was to be omitted. In this case, when *mrNDVI* readings were ≤0.8683, the sample of the plants returned θ as equal to "36.38%" (13.3%) and the node was considered terminal as no child node developed below. However, by contrast, when *mrNDVI* was less than 0.8683, a combination of *MSAVI* and *PRI* readings had to be taken into account to estimate the substrate water content. This was also confirmed by the relevant node of the testing classification tree. However, the classification tree developed from the training data revealed a high estimation risk degree (7.06), while the risk degree in the tree based on the validation dataset was high as well (5.4). The high values of the risk estimation indicated that the accuracy of this structure was uncertain due to the θ values' estimation and that more water content values should be included. The goal was to reduce the risk values to around 1.

3.3. Automation for Water and Nitrogen Deficit Stress Detection

During the model analysis of the *3rd Hypothesis*, two independently structured trees with different combinations of RIs (*CT4* and *CT5*) were developed (Figures 5 and 6). *CT4* consisted of nine nodes (four nodes with at least one child and five nodes without children) and *CT5* of seven nodes (three nodes with at least one child and four nodes without children). In those trees, the predicted category was the detection of water and nitrogen deficit stress (*C* or *WS* or *NS*). For each node, the mentioned table provides the number (*n*) and the percentage (%) of the *C* or *WS* or *NS* cases in each reflectance index category set as a dependent variable.

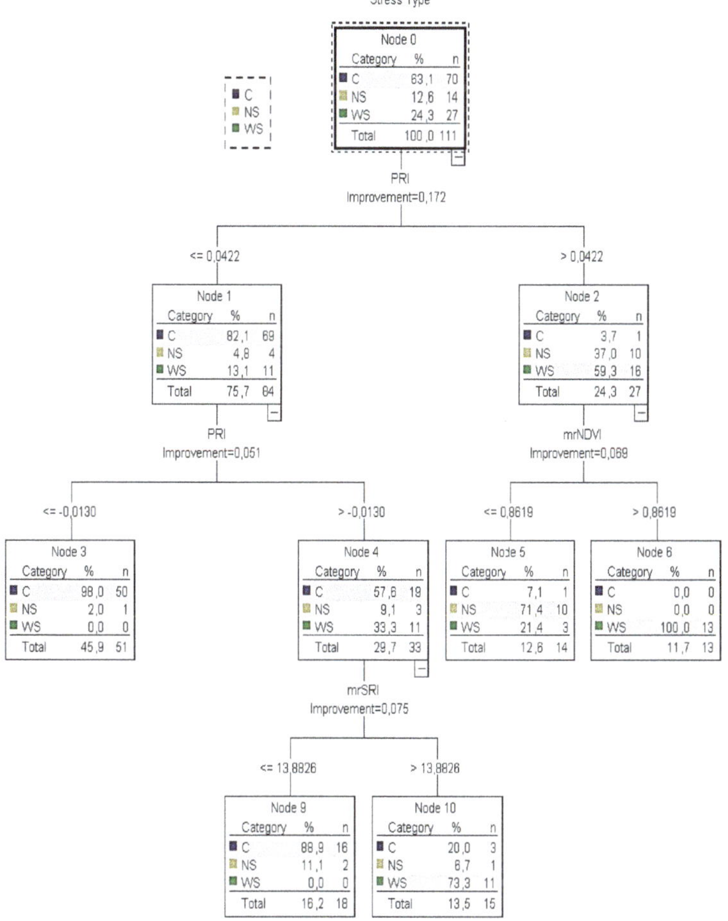

Figure 5. The *CT4* classification tree of the training sample for calibration to identify healthy, water- and nitrogen-deficit stressed plants. *C*: indication of control plants; *WS*: indication of water deficit stress plants; *NS*: indication of nitrogen deficit stress plants.

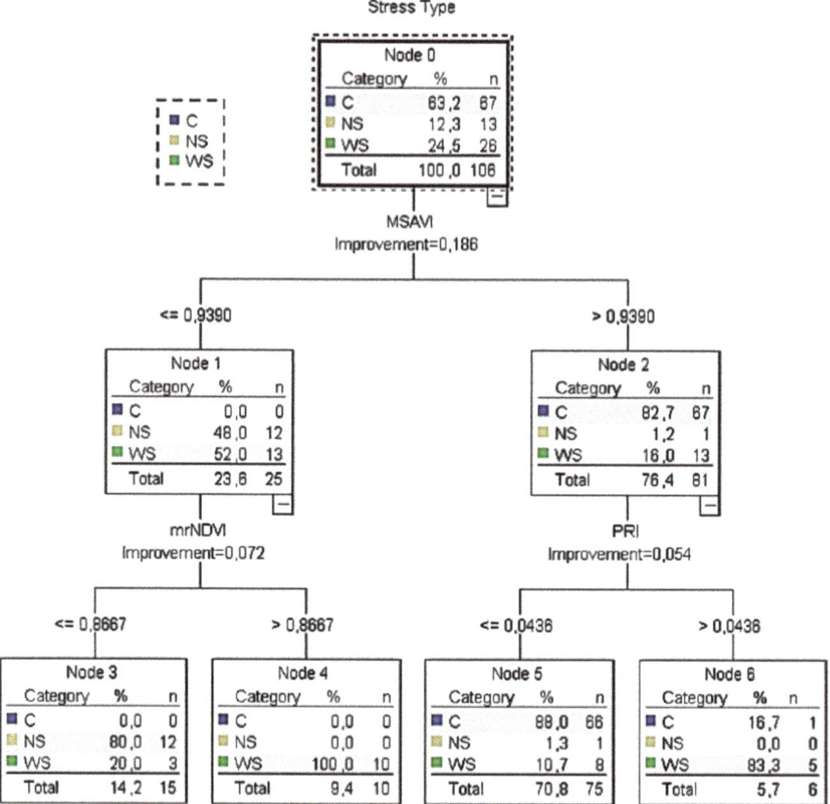

Figure 6. The *CT5* classification tree of the training sample for calibration to identify healthy, water- and nitrogen-deficit stressed plants. *C*: indication of control plants; *WS*: indication of water deficit stress plants; *NS*: indication of nitrogen deficit stress plants.

According to the training *CT4* (Figure 5), it was calculated that there was no water or nitrogen deficit stress when *PRI* was ≤ -0.0130 and *C* was returned (no irrigation is needed). On the other hand, if the *PRI* was between -0.01 and 0.04, the *mrSRI* readings (next best predictor) had to be taken into consideration to identify the water stressed plants and that node 4 was omitted. In this case, when *mrSRI* readings were ≤ 14, *C* was returned (no irrigation is needed). However, when *mrSRI* readings were >14, the majority of the sample (73%) returned *WS*. If the *PRI* was higher than 0.04, the *mrNDVI* readings (next best predictor) had to be taken into account. In this case, when *mrNDVI* readings were ≤ 0.86, *NS* was returned (nitrogen is needed). However, when *mrNDVI* readings were >0.86, the sample returned *WS* (irrigation is needed). However, contrary to node five of the training tree, in the testing tree, it was not clear when the plants were under water or nitrogen deficit stress because the sample of plants returning *NS* (60%) was very close to those returning *WC* (40%).

According to the training *CT5* (Figure 6), it was calculated that when the *MSAVI* was lower than 0.94, the *mrNDVI* readings (next best predictor) had to be taken into consideration. In this case, when the *mrNDVI* readings were ≤ 0.87, the majority of the sample (80%) returned *NS*. On the other hand, when *mrNDVI* was higher than 0.87, the total sample was characterised as water stressed (irrigation is needed). On the other hand, if *MSAVI* readings were higher than 0.94, the *PRI* readings (next best predictor) had to be taken into consideration. In this case, when *PRI* readings were ≤ 0.04 and *C* was returned (no irrigation is needed). However, when *PRI* readings were >0.04, the sample returned *WS* (irrigation is needed). Similar results were confirmed by the relevant node of the testing classification tree.

Table 1 presents the number of cases classified correctly and incorrectly for each category of the dependent variable. The predicted percent of the training sample was 90.1% in *CT4* and 89.6% in *CT5*, indicating that the model classified approximately 90.1% and 89.6%, respectively, of the sample correctly. Although the predicted percent of both *CT4* and *CT5* was similar, the *CT4* result revealed one potential problem with this model: for those plants cultivated under nitrogen stress, *NS* was predicted for 71.4% of them, which means that 28.6% of the stressed plants were inaccurately classified with *C* (non-stressed plants) or *WS* (water stressed plants). On the other hand, only 5.7% of non-stressed plants were inaccurately classified as *WS* or *NS* stressed plants. However, the final estimation risk of the misclassification of the training model was low, with small differences between the two models reflected at 9% and 10%, respectively.

Table 1. The classification accuracy of the training and testing sample for calibration and validation, respectively. *C*: control plants; *WS*: water deficit stress plants; *NS*: nitrogen deficit stress plants.

Classification Model	Sample	Observed	Predicted			
			C	*NS*	*WS*	Percent Correct
CT4	Training CT4	*C*	66	1	3	94.3%
		NS	3	10	1	71.4%
		WS	0	3	24	88.9.7%
		Overall Percentage	62.2%	12.6%	25.2%	90.1%
	Validation	*C*	23	2	4	83.3.3%
		NS	0	2	1	75.0%
		WS	1	1	3	62.5%
		Overall Percentage	64.9%	13.5%	21,6%	76.7%
CT5	Training	*C*	61	0	6	91.0%
		NS	1	12	0	92.3%
		WS	1	3	22	84.6%
		Overall Percentage	59.4%	14.2%	26.4%	89.6%
	Validation	*C*	20	0	1	95.2%
		NS	1	3	1	60.0%
		WS	0	0	9	100.0%
		Overall Percentage	60.0%	8.6%	23.5%	91.4%

4. Discussion

Despite many statistical and mathematical models such as principal component analysis [17], artificial neural network, and other classification procedures that have been developed to extract optimal information from remotely sensed data, Yohannes and Hoddinott [26] and Camdeviren et al. [27] believed that the *CT* method was a very useful model for analysing complex datasets by providing visual results. Goel et al. [18] applied a classification tree method to group hyperspectral data in order to identify weed stress and nitrogen status of corn and compared it with artificial neural networks. The advantages of tree-based classification are that it does not require the assumption of a probability distribution, specific interactions can be detected without previous inclusion in the model, non-homogeneity can be taken into account, mixed data types can be used, and dimension reduction of hyperspectral datasets is facilitated [28].

In the current study, in order to determine the water and nitrogen deficit stress severity, reflectance indices were investigated in the *CT* paths. Among the indices, the *CT* model selected (*MSAVI*) as a starter index to predict the water and nitrogen deficit stress. The *CT* analysis revealed that the classification accuracy for the training sample was 89.6% and the testing tree responded to the predicted expectation by approximately 91.4%.

The overall success rate of classification accuracy between the predicted and measured values of the stressed tomato indicated that the combination of *MSAVI*, *mrNDVI*, and *PRI* has the potential to determine water and nitrogen deficit stress and that classification tree algorithms have good potential in the classification of remotely sensed spectral data. Elvanidi et al. [13] also reveal *mrNDVI* and *PRI* in the *CT* path to determine water deficit stress in tomato plants, with classification accuracy values of 84.2% and 78.9% for the training and testing sample, respectively. Additionally, Genc et al. [17] also tested the ability of the classification tree algorithm to assess water stress in corn using hyperspectral reflectance spectra transformed into spectral vegetation indices. Their results demonstrated that water and nitrogen stress in corn was detectable through spectral reflectance analysis.

Generally, the use of machine learning models is not widespread in agriculture since they require a long time-series of datasets. The destructive methods or the complex sensors used in the last decades to quantify plant physiology did not allow for their progress. With the recent integration of a new age of computational intelligent sensors, more and more robust methodologies are being adapted. Up to now, less than 40 articles focused on machine learning models in agriculture. From those, 61% of the articles were related to different aspects of crop management [29].

However, the prediction performed by those models, in most cases, mostly correspond specifically to the area and conditions used in the training data, thereby trying to account for the otherwise invisible variations specific to that land and surroundings [30]. In the open field, for instance, the terrain in which the crop is cultivated affects the process of the crop water demand prediction. With the advent of the new era of computational intelligent sensors that can track various things that were previously not possible, more and more factors can be considered. Nevertheless, this new concept will ensure the development of more robust remote sensing approaches for monitoring plant physiology in order to train a decision support system with the aim of adjusting climate and irrigation control strategies within the greenhouse. Thus, in the future, the widespread usage of machine learning models is expected, allowing for the possibility of integrated and applicable tools. However, improvement in the performance of the decision tree classification approach with increases in the number of data sets further strengthens the belief that, by increasing the amount of data, model performance could probably be further improved.

5. Conclusions

In the current work, the ability to use a classification tree was tested to remotely predict leaf chlorophyll and substrate water content. Additionally, the classification tree was trained to assess different types of tomato stress such as water and nitrogen deficit stress. The model was trained by organizing, in the most effective way, the reflectance values measured by a hyperspectral camera. Among the reflectance indices, the classification tree model selected *TCARI* and *PRI* or *mrSRI* to predict leaf chlorophyll content. To estimate the substrate water content, on the other hand, the process was much more complex since more than four reflectance indices were involved in the procedure. Regarding the model trained to sense the actual plant water and nitrogen status, it was concluded that the combination of *MSAVI*, *mrNDVI*, and *PRI* involved in *CT5* made insufficient provision with a reasonable accuracy. These results are promising for designing a smart decision support system for better managing climate and irrigation control strategies. Additionally, less complicated reflectance sensors recording certain spectral bands could be adapted to monitor plant physiology in real-time in a cost-effective manner. Nevertheless, it has to be noted that the results presented are relevant to the conditions of the measurements and the specific crop studied. Therefore, it is expected that the use of machine learning models will be even more widespread in the near future.

Author Contributions: Conceptualization, methodology and data analysis A.E. and N.K.; Experimental measurements, A.E.; Writing-Original Draft Preparation, A.E.; Writing-Review & Editing, Supervision, N.K. and C.K.

Funding: This work has been co-financed by the European Union and Greek National Funds through the Operational Program "Education and Lifelong Learning" of the National Strategic Reference Framework (NSRF) – ARISTEIA "GreenSense" project, grant number 2632".

Conflicts of Interest: The authors declare no conflict of interest.

Nomenclature

As	Photosynthesis rate
C	Control: no stressed plants/no irrigation or fertilization is needed
CHAID	Chi-squared automatic interaction detection
Chl_a	Chlorophyll-a content
CT	Classification tree
CRT	Classification regression tree
EC	Electrical conductivity
LAI	Leaf area index
mrNDVI	Modified red edge normalised difference vegetation index
mrSRI	Modified red edge simple ratio index
MSAVI	Modified soil-adjusted vegetation index
n	Number of samples per day per treatment
NIR	Near-infrared region
nm	Nanometre
NS	Nitrogen stress: nitrogen stress plants/nitrogen is needed
OSAVI	Optimisation soil-adjusted vegetation
PRI	Photochemical reflectance index
R or r	Reflectance
R_{445}	Reflectance value in 445 nm band
R_{531}	Reflectance value in 531 nm band
R_{550}	Reflectance value in 550 nm band
R_{570}	Reflectance value in 570 nm band
R_{670}	Reflectance value in 670 nm band
R_{690}	Reflectance value in 690 nm band
R_{700}	Reflectance value in 700 nm band
R_{705}	Reflectance value in 705 nm band
R_{750}	Reflectance value in 750 nm band
R_{800}	Reflectance value in 800 nm band
R_{810}	Reflectance value in 810 nm band
$R_{g,i}$	Irradiance
RH	Relative humidity
RI	Reflectance index
rNDVI	Red edge normalised difference vegetation index
SD	Standard deviation
T	Air temperature
TCARI	Transformed chlorophyll absorption in reflectance index
TKN	Total Kjeldahl nitrogen
VIS	Visible spectrum
VNIR	Visible and near-infrared spectrum region
WS	Water stress: water stress plants/irrigation is needed
θ	Substrate water content

References

1. Elvanidi, A.; Katsoulas, N.; Augoustaki, D.; Loulou, I.; Kittas, C. Crop reflectance measurements for nitrogen deficiency detection in a soilless tomato crop. *Biosyst. Eng.* **2018**, *176*, 1–11. [CrossRef]
2. Katsoulas, N.; Elvanidi, A.; Ferentinos, K.P.; Kacira, M.; Bartzans, T.; Kittas, C. Crop reflectance monitoring as a tool for water stress detection in greenhouses: A review. *Biosyst. Eng.* **2016**, *151*, 374–398. [CrossRef]

3. Ray, S.S.; Das, G.; Singh, J.P.; Panigrahy, S. Evaluation of hyperspectral indices for LAI estimation and discrimination of potato crop under different irrigation treatments. *Int. J. Remote Sens.* **2006**, *27*, 5373–5387. [CrossRef]

4. Alchanatis, V.; Cohen, Y.; Cohen, S.; Moller, M.; Sprinstin, M.; Meron, M. Evaluation of different approaches for estimating and mapping crop water status in cotton with thermal imaging. *Precis. Agric.* **2010**, *11*, 27–41. [CrossRef]

5. Liaghat, S.; Balasundram, S.K. A review: The role of remote sensing in precision agriculture. *Am. J. Agric. Biol. Sci.* **2010**, *5*, 50–55. [CrossRef]

6. Sonneveld, C.; Voogt, W. *Plant Nutrition of Greenhouse Crops*; Springer: New York, NY, USA, 2009.

7. Elvanidi, A.; Katsoulas, N.; Bartzanas, T.; Ferentinos, K.P.; Kittas, C. Crop water status assessment in controlled environment using crop reflectance and temperature measurements. *Precis. Agric.* **2017**, *18*, 332–349. [CrossRef]

8. Story, D.; Kacira, M. Design and implementation of a computer vision-guided greenhouse crop diagnostics system. *Mach. Vis. Appl.* **2015**, *26*, 495–506. [CrossRef]

9. Kim, Y.; Glenn, D.M.; Park, J.; Ngugi, H.K.; Lehman, B.L. Hyperspectral image analysis for plant stress detection. *Am. Soc. Agric. Biol. Eng.* **2010**, 1009114. [CrossRef]

10. Amatya, S.; Manoj, K.; Alva, A.K. Identifying water stress in potatoes using leaf reflectance as an indicator of soil water content. *J. Agric. Eng.* **2014**, *1*, 52–61. [CrossRef]

11. Sarlikioti, V.; Driever, S.M.; Marcellis, L.F.M. Photochemical reflectance index as a mean of monitoring early water Stress. *Ann. Appl. Biol.* **2010**, *157*, 81–89. [CrossRef]

12. Zarco-Tejada, P.J.; González-Dugo, V.; Williams, L.E.; Suárez, L.; Berni, J.A.J.; Goldhamer, D.; Fereres, E.A. PRI-based water stress index combining structural and chlorophyll effects: Assessment using diurnal narrow-band airborne imagery and the CWSI thermal index. *Remote Sens. Environ.* **2013**, *138*, 38–50. [CrossRef]

13. Kacira, M.; Sase, S.; Okushima, L.; Ling, P.P. Plant response-based sensing for control strategies in sustainable greenhouse production. *J. Agric. Meteorol.* **2005**, *61*, 15–22. [CrossRef]

14. Magney, T.S.; Vierling, L.A.; Eitel, J.U.; Huggins, D.R.; Garrity, S.R. Response of high frequency Photochemical Reflectance Index (PRI) measurements to environmental conditions in wheat. *Remote Sens. Environ.* **2016**, *173*, 84–97. [CrossRef]

15. Bajwa, S.G.; Mishra, A.R.; Norman, R.J. Canopy reflectance response to plant nitrogen accumulation in rice. *Precis. Agric.* **2010**, *11*, 488–506. [CrossRef]

16. Lu, S.; Lu, L.; Zhao, W.; Liu, Y.; Wang, Z.; Omasa, K. Comparing vegetation indices for remote chlorophyll measurement of white poplar and Chinese elm leaves with different adaxial and abaxial surfaces. *J. Exp. Bot.* **2015**, *66*, 5625–5637. [CrossRef] [PubMed]

17. Genc, L.; Demirel, K.; Camoglu, G.; Asik, S.; Smith, S. Determination of plant water stress using spectral reflectance measurements in watermelon (Citrullus vulgaris). *Am. Eur. J. Agric. Environ. Sci.* **2011**, *11*, 296–304.

18. Goel, P.K.; Prasher, S.O.; Patel, R.M.; Landry, J.A.; Bonnell, R.B.; Viau, A.A. Classification of hyperspectral data by decision trees and artificial neural networks to identify weed stress and nitrogen status of corn. *Comput. Electron. Agric.* **2003**, *39*, 67–93. [CrossRef]

19. Morgan, J. *Classification and Regression Tree Analysis*; Technical report; University of Boston: Boston, MA, USA, 2014.

20. Lichtenthaler, H.K.; Wellburn, A.R. Determinations of total carotenoids and chlorophylls a and b of leaf extracts in different solvents. *Biochem. Soc. Trans.* **1983**, *11*, 591–592. [CrossRef]

21. Kjeldahl, J. A new method for the determination of nitrogen in organic matter. *Z. Anal. Chem.* **1883**, *22*, 366. [CrossRef]

22. Katsoulas, N.; Elvanidi, A.; Ferentinos, K.P.; Bartzanas, T.; Kittas, C. Calibration methodology of a hyperspectral imaging system for greenhouse plant water stress estimation. In Proceedings of the 6th Balkan Symposium on Vegetables and Potatoes, Zagreb, Croatia, 29 September–2 October 2014; Volume 1142, pp. 119–126.

23. Loh, W.Y. Classification and regression trees. In *Wiley Interdisciplinary Reviews: Data Mining and Knowledge Discovery*; John Wiley & Sons: Hoboken, NJ, USA, 2011; Volume 1, pp. 14–23.

24. Lewis, R. An introduction to classification and regression tree (CART) analysis. In Proceedings of the Annual Meeting of the Society of Academic Emergency Medicine, San Francisco, CA, USA, 22–25 May 2000; pp. 1–14.

25. IBM Corporation. *IBM SPSS Decision Trees 21 Guide*; IBM Corporation: Armonk, NY, USA, 2012.
26. Yohannes, Y.; Hoddinott, J. *Classification and Regression Trees: An Introduction*; International Food Policy Research Institute: Washington, DC, USA, 1999.
27. Çamdeviren, H.; Mendeş, M.; Ozkan, M.M.; Toros, F.; Şaşmaz, T.; ve Oner, S. Determination of depression risk factors in children and adolescents by regression tree methodology. *Acta Med. Okayama* **2005**, *59*, 19–26. [PubMed]
28. Delalieux, S.; Van Aardt, J.; Keulemans, W.; Schrevens, E.; Coppin, P. Detection of biotic stress (Venturia inaequalis) in apple trees using hyperspectral data: Nonparametric statistical approaches and physiological implications. *Eur. J. Agric.* **2007**, *27*, 130–143. [CrossRef]
29. Liakos, K.G.; Busato, P.; Moschou, D.; Pearson, S.; Bochtis, D. Machine learning in agriculture: A review. *Sensors* **2018**, *18*, 2674. [CrossRef] [PubMed]
30. Krupakar, H.; Akshay, J.; Dhivya, G. A review of intelligent practices for irrigation prediction. *arXiv*, 2016; arXiv:1612.02893.

 *horticulturae*

Article

Effect of LED Lighting and Gibberellic Acid Supplementation on Potted Ornamentals

Taylor Mills-Ibibofori [1], Bruce L. Dunn [1,*], Niels Maness [1] and Mark Payton [2]

[1]   Department of Horticulture & L.A., Oklahoma State University, Stillwater, OK 74078, USA
[2]   Department of Statistics, Oklahoma State University, Stillwater, OK 74078, USA
*   Correspondence: bruce.dunn@okstate.edu

Received: 16 April 2019; Accepted: 11 July 2019; Published: 15 July 2019

Abstract: Use of light emitting diode (LED) technology is beginning to replace traditional lighting in greenhouses. This research focused on the effects of LED lighting and gibberellic acid supplementation on growth and flowering of *Dahlia* spp. 'Karma Serena', *Liatris spicata* 'Kobold', and *Lilium asiatic* 'Yellow Cocotte'. Light treatments, used to extend photoperiod, included LED flowering lamps and halogen lamps that emitted a combination of red + far-red + white, red + white, and broad spectrum from late fall to early spring. Gibberellic acid treatments ranged from 40 to 340 mg L^{-1} for Asiatic lily 'Yellow Cocotte', 50 to 250 for gayfeather 'Kobold', and 50 to 150 for dahlia 'Karma Serena'. Results varied within species in response to light and gibberellic acid. A significant interaction of light with gibberellic acid influenced mean flower number and flowering percentage for dahlia 'Karma Serena', while flowering percentage and flower diameter were influenced for Asiatic lily 'Yellow Cocotte'. Effect of light was most significant on growth and flowering measurements, especially for gayfeather 'Kobold' and dahlia 'Karma Serena'. For gayfeather 'Kobold', flowering occurred two weeks earlier under sole LED lighting than under other light treatments and no supplemental light. Although flowering occurred the earliest for dahlia 'Karma Serena' under no supplemental light, plants under light treatments had greater height, width, and shoot weight. Significant effects of gibberellic acid on growth and flowering measurements for dahlia 'Karma Serena' and Asiatic lily 'Yellow Cocotte' were observed for height, width, and flower number.

Keywords: light emitting diodes; GA_3; extended photoperiod; greenhouse

---

1. Introduction

Light is the single most important variable with respect to plant growth and development and is often the most limiting factor in greenhouse production [1]. Therefore, using artificial lighting (AL) or grow lights (GL) in commercial greenhouses is beneficial for plants and growers. Altering photoperiod and increasing light levels are reasons for using these lights. The different lighting sources that growers can use include incandescent (INC) lamps, fluorescent lamps (FL), and high intensity discharge (HID) lamps. Light emitting diodes (LED) are fourth generation lighting sources and are the emerging technology in horticulture [2]. Before choosing a lighting device, several factors, such as efficiency, total energy emissions, life expectancy, and costs need to be considered. In addition, it is important to know the three most important light factors that affect plant growth, which are light quality, light intensity, and light duration [1]. LEDs have proven to be advantageous in all these factors when compared to traditional lighting sources [3].

Energy inputs range from 10% to 30% of total production costs for the greenhouse industry [4]. Thus, any new lighting technology that significantly reduces consumption of electricity for crop lighting, while maintaining or improving crop value is of great interest to growers. Light sources, such as fluorescent, metal halide, high pressure sodium, and incandescent lamps are generally used for plant growth under greenhouse conditions and have been around for half a century. However, these light sources have disadvantages, such as less suitable wavelength for plant growth and limited lifetime of operation. In addition, they require more electricity and produce heat that may injure plant leaves [5].

In the 1990s, light-emitting diodes (LEDs) were investigated for the first time for plant growth and were found to be efficient alternatives to traditional lamps used in lighting systems [6]. Compared with conventional lamps, LEDs are smaller in size and weight, have a long lifetime, low heat emissions, wavelength specificity, and much lower energy consumption [7]. In addition to changes in plant productivity, increased suppression of pathogens has been noted in tomato *(Solanum lycopersicum* L.) and cucumber *(Cucumis sativis* L.) [8]. Physiological and morphological effects of LEDs have been studied in several species, including potato *(Solanum tuberosum* L.), wheat *(Triticum aestivum* L.), lily *(Lilium candidum* L.), lettuce *(Lactuca sativa* L.), spinach *(Spinacia oleracea* L.), strawberry *(Fragaria* × *ananassa* Duchesne), marigold *(Tagetes erecta* L.), chrysanthemum *(Chrysanthemum indicum* L.), and salvia *(Salvia divinorum* Epling and Játiva) using various LED products [9].

Light-emitting diodes have the potential to shorten the crop time, reduce costs, and add new plants for specialty cut flower production during the winter [7]. This light source may also induce greater flowering for winter crops; however, research is limited to propagation, vegetables, and seedling production. Commercial LED fixtures for photoperiodic lighting have been recently developed for flowering applications and are alternatives to INC lamps. Craig and Runkle [10] quantified how red (R) to far-red (FR) ratio of photoperiodic lighting from LEDs influenced flowering and extended the growth of short-day plants. Kohyama [11] investigated the efficacy of commercial LED products developed for flowering applications on long-day plants. Meng and Runkle [12] coordinated grower trials to investigate the efficacy of R + white (W) + FR LEDs to regulate flowering of daylength-sensitive ornamental crops. For some plants, flowering is promoted with a combination of R and FR light [13,14].

Gibberellic acid (GA_3) is a hormone found in plants, which is produced in low amounts. Synthetic GA_3 is commonly used in commercial agriculture. This hormone is very influential and can control plant development, promote growth, and elongate cells. Gibberellic acid can also promote petal growth and enhance other flowering characteristics [15,16]. In certain plant species, GA_3 acts as a mobile signal transmitter for photoperiodic flowering stimulation [17]. For flower induction, soaking bulbs, rhizomes, corms, or spraying the foliage with a GA_3 solution are common applications [18–20]. There are limited but statistically valid interactions between light and GA_3. Both factors are known to have synergistic effects, but mainly on germination of seedlings [21,22]. In certain species, growth and flower initiation are affected by light and GA_3 application [23,24]. More current research needs to be conducted to assess the interaction of light and GA_3 further. Therefore, objectives of this study were to evaluate how gibberellic acid and different combinations of red and far-red light together from LED flowering lamps and halogen lamps, would influence growth and flowering of *Lilium* L., *Dahlia* Cav., and *Liatris* Gaertn. ex Schreb. species.

2. Materials and Methods

2.1. Plant Material and Culture

On 15 September 2015, bulbs of *Lilium asiatic* L. 'Yellow Cocotte' were graded at 16 to 19 cm. Cuttings of *Dahlia* spp. 'Karma Serena', which are short-day plants, arrived 14 October 2015. *Liatris spicata* (L.) Willd. 'Kobold' corms, which are long-day plants, arrived 12 November 2015 and were graded at 8 to 10 cm. Plant materials were obtained from a broker (Gloeckner and Company Incorporated, Harrison, NY, USA). Before transplanting, dahlia 'Karma Serena' cuttings were placed on a mist bench and Asiatic lily 'Yellow Cocotte' were placed in a cooler at 4 °C upon arrival for one month. Gayfeather 'Kobold' corms were immediately treated with GA_3 (Plant Hormones LLC, Auburn, WA, USA). All bulbs and corms were soaked in an aqueous solution of GA_3 for 30 min before being potted. Dahlia leaves were sprayed to glisten once with different rates of GA_3 solution after potting. Tween-20 (Sigma-Aldrich, St. Louis, MO, USA) was also added in the GA_3 solution as a surfactant at a concentration of 0.01%. The GA_3 treatment dates were 24 October 2015, 31 October 2015, and 13 November 2015 for 'Yellow Cocotte ', 'Karma Serena', and 'Kobold', respectively. Dahlia 'Karma Serena', Asiatic lily 'Yellow Cocotte', and liatris 'Kobold' were potted in standard 15 cm pots filled with Metro-Mix 360 media (Sun Gro Horticulture, Bellevue, WA, USA) and were placed in the greenhouses on 16 October 2015, 24 October 2015, and 12 November 2015, respectively.

2.2. Experimental Arrangement

The experiment was conducted at four research greenhouses of the Department of Horticulture and L.A. in Stillwater, OK. For each greenhouse, temperatures were set at 23 °C during the day and 18 °C during the night with a photosynthetic photon flux density (PPFD) between 600 to 1200 μmol m^2 s^{-1} and daily light integral of 10–15 mol m^2 d. One light treatment was established in each greenhouse. Light emitting diodes (Philips Green Power Flowering lamps, Amsterdam, The Netherlands) and standard halogen bulbs, which are broad spectrum across the photosynthetically active radiation region, were installed at 0.914 m above the bench area and 0.914 m apart. In the first light treatment, there were 19 14-watt LED R + W + FR flowering lamps (Phillips Lighting, Somerset, NJ, USA) with a spectrum from 420 to 780 nm and peaks at 660 (35%) and 740 (46%). The second light treatment had 11 15-watt LED R + W flowering lamps (Phillips Lighting, Somerset, NJ) with a spectrum from 420 to 720 nm and a peak at 660 (78%) and 12 40-watt halogen bulbs (Osram Sylvania, Wilmington, MA, USA) with a spectrum from 400 to 1200 nm with peaks at 600, 760, and 850 nm with lamps and bulbs installed alternatively. The third light treatment included 23 of the above mentioned 40-watt halogen bulbs, and the fourth treatment did not have lights (control). Plant species and GA_3 rates were randomized within light treatments. Plants were supplemented with seven hours of light after sunset. Before daylight savings time (8 November 2015), lighting was delivered from 1900 to 0200 HR. After daylight savings time, lighting was delivered between 1700 to 2400 HR using timers. A quantum sensor (Spectrum Technologies, Inc., Aurora, IL, USA) measured photosynthetic photon flux density (PPFD) of the LED lamps and halogen bulbs. In each greenhouse where the light was supplemented, measurements were randomly recorded across the bench area and were taken at pot level. The mean photon outputs were 10, 20, and 2 μmol m^{-2} s for LED emitting R + W + FR, LED emitting R + W, and halogen, respectively.

Gibberellic acid rates for gayfeather 'Kobold' were 50, 170, and 250 mg L^{-1} with 12 pots per rate per light treatment. Asiatic lily 'Yellow Cocotte' had rates of 40, 140, and 340 mg L^{-1} with 12 pots per rate per light source. Dahlia 'Karma Serena' rates were 50, 100, and 150 mg L^{-1} with 10 pots per rate per light source. All plants included a controlled rate in which water was used. Plants were watered with drip irrigation as needed. On 23 November 2015, a slow release fertilizer 16-9-12 (3–4 month, Osmocote® Plus, The Scotts Co., Marysville, OH, USA) at a rate of 10 g was added at time of potting and 200 mg L^{-1} 20-10-20 Peat-lite (Jacks, Allentown, PA, USA) water soluble fertilizer was supplemented after three weeks.

2.3. Harvesting and Measurements

Data collected from plants included the date of first flower (anthesis), which was only recorded when petals were fully opened. Flower diameter was recorded on 15 November 2015 for dahlia 'Karma Serena' and 22 December 2015 for Asiatic lily 'Yellow Cocotte' using a digital caliper (Tresna Instrument., LTD, Guilin, China). Flowering percent (flowering or not per pot), Number of flowers, plant height (from media surface to tallest flower or bud), and width (average of two perpendicular measurements) were recorded on 18 January 2016 for dahlia 'Karma Serena', 22 Feburary 2016 for Asiatic lily 'Yellow Cocotte', and 27 Feburary 2016 for gayfeather 'Kobold'. Shoot dry weight was recorded on 1 Feburary 2016 for dahlia 'Karma Serena', 29 Feburary 2016 for Asiatic lily 'Yellow Cocotte', and 7 March 2016 for gayfeather 'Kobold' by cutting the stems at the media level, and drying for 3 d at 54.4 °C.

2.4. Statistical Analysis

Pots were arranged in a completely randomized design with plant species, GA_3 and light treatments as the specified factors. Data were analyzed with SAS version 9.4 software (SAS Institute, Cary, NC, USA). An analysis of variance methods (PROC MIXED) was used with a two-factor factorial arrangement with light and GA_3 as the factors of interest. For percentage response variables, arcsine square root transformations were used to help normalize the data. Because the levels of the factors changed, separate analyses were conducted for each plant species. When interactions of light with GA_3 were significant, simple effects were reported. Mean separations were determined using protected Fisher-type comparisons (a DIFF option in an LSMEANS statement and a SLICE option when appropriate) and with 0.05, 0.01, 0.001, and 0.0001 levels of significance.

3. Results

3.1. Liatris spicata 'Kobold'

A main effect of light was found on all growth measurements, as well as on a number of terminal spikes and days to anthesis (Table 1). Plants under LEDs flowered the earliest, but were not different than halogen or LED + halogen (Table 2). The average number of spikes was greatest with natural light, which was not different than halogen. Plant height and width was greatest under LED and LED + halogen. Shoot dry weight was greatest with halogen lighting. Gibberellic acid rates had a significant effect on plant width, shoot dry weight, and mean spike number (Table 1). For width, plants receiving 0 mg L^{-1} GA_3 were greatest, but were not different from those treated at 50 and 170 mg L^{-1} GA_3 (Table 3). Shoot weight was greatest for 0 mg L^{-1} GA_3, but was not different from 50 and 250 mg L^{-1} GA_3. The average number of spikes was greatest at 250 mg L^{-1} GA_3, but was not different than 0 or 170 mg L^{-1} GA_3.

Table 1. Analysis of variance for growth and flowering measurements of *Liatris spicata* 'Kobold', *Dahlia* spp. 'Karma Serena', and *Lilium asiatica* 'Yellow Cocotte' grown with LED and halogen lights along with multiple rates of gibberellic acid.

Cultivar	Source	Height (cm)	Width (cm)	Shoot Dry Weight (g)	Flowers/Spikes Number [z]	Flower Diameter	Days to Anthesis	Flowering (%)
'Kobold'	Light	**** [y]	****	****	****	_ [x]	****	ns
	GA_3	ns	**	*	*	-	ns	ns
	Light × GA_3	ns	ns	ns	ns	-	ns	ns
'Karma Serena'	Light	****	****	****	****	ns	****	ns
	GA_3	****	ns	ns	**	*	ns	ns
	Light × GA_3	ns	ns	ns	*	ns	ns	*
'Yellow Cocotte'	Light	ns	ns	ns	ns	ns	ns	ns
	GA_3	ns	ns	ns	ns	ns	ns	ns
	Light × GA_3	ns	ns	ns	ns	*	ns	*

[z] Number of flowers for 'Karma Serena' and 'Yellow Cocotte', but the number of spikes for 'Kobold'. [y] NS, *, **, ***, **** indicate non-significant or significant at $p \leq 0.05$, 0.01, 0.001, 0.0001, respectively. [x] Data not taken.

Table 2. Growth and flowering measurements of *Liatris spicata* 'Kobold', *Dahlia* spp. 'Karma Serena', and *Lilium asiatica* 'Yellow Cocotte' affected by light averaged across GA$_3$.

Light Type	Height (cm)	Width (cm)	Shoot Dry Weight (g)	Flower Measurements [z]	Days to Anthesis	Flowering (%)
			'Kobold'			
Control	47.3b [y]	35.2c	13.9b	3.5a	88a	96a
LED	64.7a	49.4a	17.2b	2.3bc	70b	100a
Halogen	52.1b	40.9b	22.0a	3.1ab	73b	98a
LED + Halogen	65.9a	44.9ab	16.8b	1.8c	77ab	98a
			'Karma Serena'			
Control	58.9b	32.5c	9.1d	7.1a	46c	- [x]
LED	67.1b	43.9b	35.0c	7.1a	61b	-
Halogen	95.8a	46.7b	43.6b	8.5a	74a	-
LED + Halogen	85.9a	56.9a	52.9a	7.9a	80a	-
			'Yellow Cocotte'			
Control	45.5a	15.0a	4.0a	2.4a	54a	-
LED	44.5a	19.6a	3.5a	2.0a	47a	-
Halogen	38.4a	16.3a	4.2a	2.0a	43a	-
LED + Halogen	54.1a	19.8a	4.8a	2.1a	55a	-

[z] Mean number of flower spikes for 'Kobold', flower diameter (cm) for 'Karma Serena', and flower number for 'Yellow Cocotte'. [y] Means (n = 12 for 'Kobold' and 'Yellow Cocotte'; n = 10 for 'Karma Serena') with the same letter within the same column are not statistically significant ($p < 0.05$). [x] Interaction significant for plant measurements.

Table 3. Growth and flowering measurements of *Liatris spicata* 'Kobold', *Dahlia* spp. 'Karma Serena', and *Lilium asiatica* 'Yellow Cocotte' affected by GA$_3$ averaged across the light.

GA$_3$ Rate (mg L^{-1})	Height (cm)	Width (cm)	Shoot Dry Weight (g)	Flower Measurements [z]	Days to Anthesis	Flowering (%)
			'Kobold'			
0	59.7a [y]	47.4a	19.8a	2.4ab	77a	98a
50	59.5a	43.2ab	17.6ab	2.3b	76a	94a
170	54.6a	40.6ab	15.2b	2.6ab	78a	100a
250	56.3a	39.3b	17.3ab	3.5a	76a	100a
			'Karma Serena'			
0	65.0b	45.5a	30.5a	8.6a	67a	- [x]
50	81.0a	45.7a	35.8a	7.3ab	62a	-
100	81.3a	45.5a	38.5a	6.8b	64a	-
150	80.3a	43.4a	35.7a	7.8ab	69a	-
			'Yellow Cocotte'			
0	48.5a	19.6a	4.5a	2.1a	- [x]	- [x]
40	47.2a	16.8a	4.4a	2.4a	-	-
140	42.9a	17.3a	3.9a	2.0a	-	-
340	43.4a	17.0a	3.7a	2.0a	-	-

[z] Mean number of flower spikes for 'Kobold', flower diameter (cm) for 'Karma Serena', and flower number for 'Yellow Cocotte'. [y] Means (n = 12 for 'Kobold' and 'Yellow Cocotte'; n = 10 for 'Karma Serena') with the same letter within the same column are not statistically significant ($p < 0.05$). [x] Interaction significant for plant measurements.

3.2. Dahlia spp. 'Karma Serena'

There was a significant Light × GA$_3$ interaction for mean flower number and flowering percentage (Table 1). Flower number within the 50 mg L^{-1} GA$_3$ rate was greatest for plants under halogen, LED + halogen, and no supplemental light (Table 4). Plants treated with 100 mg L^{-1} GA$_3$ treatment, no supplemental light, LEDs, and halogen had the greatest number of flowers. The flowering percentage within the 50 and 150 mg L^{-1} GA$_3$ rates was greatest with no supplemental lighting, halogen, and

LED + halogen light. Plants treated with 100 mg L^{-1} GA$_3$ treatment, flowering was greatest with natural light, LED, and halogen lighting. The light had a significant effect on height, width, shoot dry weight, and days to anthesis (Table 1). Time to flower was longest under halogen and LED + halogen (Table 2). Height was greatest under halogen, which was not different than LED + halogen. Plant width and shoot dry weight were greatest under LED + halogen. Only height and flower diameter were significantly affected by GA$_3$ (Table 1). All GA$_3$ rates produced taller plants compared to no supplemental lighting. No supplemental lighting had the greatest number of flowers though 50 and 150 mg L^{-1} GA$_3$ were not different (Table 3).

Table 4. Mean flower number and flowering percent of *Dahlia* spp. 'Karma Serena' and *Lilium asiatica* 'Yellow Cocotte' affected by the interaction of light with GA$_3$.

Plant	Characteristic	Source	GA$_3$ (mg L^{-1})			
			0	50	100	150
'Karma Serena'	Flower number	Control	3.1c [z]	2.4b	2.3b	2.8a
		LED	2.2c	2.9b	3.1ab	3.1a
		Halogen	6.6a	5.4a	4.4a	3.7a
		LED + Halogen	5.1b	4.5a	4.4a	2.3a
	Flowering percent	Control	100a	100a	100a	100a
		LED	100a	89b	100a	80b
		Halogen	100a	100a	100a	100a
		LED + Halogen	100a	100a	80b	100a
			0	40	140	340
'Yellow Cocotte'	Flower diameter	Control	8.9b	9.2b	9.7a	9.7b
		LED	10.4a	9.8b	10.4a	10.1ab
		Halogen	9.8b	9.4b	10.1a	9.5b
		LED + Halogen	10.5a	10.7a	10.0a	10.9a
	Flowering percent	Control	58b	67b	58bc	75ab
		LED	100a	67b	75a	50bc
		Halogen	75ab	75a	33c	33c
		LED + Halogen	58b	75a	67ab	100a

[z] Means (n = 10 for 'Karma Serena'; n = 12 for 'Yellow Cocotte') with the same letter within the same column and within plant characteristic are not statistically significant ($p < 0.05$).

3.3. Lilium Asiatic 'Yellow Cocotte'

The interaction of Light × GA$_3$ was seen on flower diameter and flowering percentage (Table 1). Plants treated with 0 mg L^{-1} GA$_3$ rate, LED and LED + halogen had the greatest flower diameter (Table 4). Plants treated with 40 mg L^{-1} GA$_3$ rate had the greatest flower diameter under LED + halogen. Plants treated with 340 mg L^{-1} GA$_3$ rate, plants under LED and LED + halogen had the greatest flower diameters. The flowering percentage was greatest with halogen within the 0 mg L^{-1} GA$_3$ rate, but was not different from halogen. Plants treated with 40 mg L^{-1} GA$_3$ rate, plants with halogen and LED + halogen had the greatest flowering percentage. Plants treated with 140 mg L^{-1} GA$_3$ rate, plants with LED had the greatest flowering percentage, but were not different from LED + halogen. Plants treated with 340 mg L^{-1} GA$_3$ rate, plants under LED + halogen had the greatest flowering percentage, but were not different from natural lighting. No significant effects were seen by light or GA$_3$ as main effects on other growth and flowering measurements of 'Yellow Cocotte'.

4. Discussion

The use of LED, LED + halogen, and sole halogen lamps emitting R and FR light effectively promoted growth and flowering in gayfeather 'Kobold' and dahlia 'Karma Serena'. Red light is the most effective at inhibiting flowering in short-day plants (SDP). This was true for dahlia under LED, halogen, and LED + halogen (Table 2). Craig and Runkle [10] reported that flowering in SDPs, such as chrysanthemum (*Chrysanthemum indicum* L.) and dahlia was delayed under incandescent and LED lights. Inhibition of flowering by R light was also seen in cocklebur (*Xanthium strumarium* L.), chrysanthemum, and soybean (*Glycine max* L. Merr.) [25–27]. Delaying flowering in SDPs, such as dahlia especially during the winter months is ideal. During this season, the days are shorter, and the nights are longer. Therefore, SDPs will want to spend photosynthates in the production of reproductive organs, which will result in a lack of growth and development of vegetative parts. Extended growth and greater biomass are promoted under R light, and this was seen for liatris and dahlia under LED flowering lamps and halogen lamps (Table 2). Miyashita et al., [28] noted that R light from LEDs increased shoot length of potato (*Solanum tuberosum* L.) plantlets. Height was also greatest under either LED flowering lamps emitting R + W or R + W + FR, as well as incandescent lamps in ageratum (*Ageratum houstonianum* L.), calibrachoa (*Calibrachoa* x *hybrida* Cerv.), dianthus (*Dianthus* L.), and petunia (*Petunia* x *hybrida* Juss.). Height and shoot dry weight were greatest for salvia (*Salvia splendens* Sellow ex J.A. Schultes) and tomato (*Solanum lyopersicum* L.) under LEDs emitting red [29]. Meng and Runkle [12] reported that the stem length of verbena (*Verbena* x *hybrid* L.) increased under incandescent and LED flowering lamps compared to the control. Dry weight and plant width increased in poinsettia (*Euphorbia pulcherrima* Willd. ex Klotzsch) when grown under supplemental LED lighting emitting R and blue [30]. An increase in all these growth parameters is beneficial for cut flowers.

A combination of R + FR is effective for promoting flowering in long-day plants (LDP). This was true for liatris that were under sole LED lighting emitting R + W + FR (Table 2). Meng and Runkle [12] have also reported that photoperiodic lighting with a mixture of R and FR light from LEDs and incandescent lamps was most effective at promoting flowering in LDPs. The flowering of *Gypsophila paniculata* (L.) 'Baby's Breath' and *Eustoma grandiflorum* (Salisb.) 'Lisianthus' was also promoted under a combination of R and FR light [13,14]. The presence of FR in LED lamps shortened the flowering time and increased number of flowers in petunia. Hastening of flowering, while maintaining plant quality, will decrease the costs of labor and inputs, as well as assure an early market season. Neither R nor FR light from the lamps influenced flowering in Asiatic lily 'Yellow Cocotte'. Bieleski et al. [31] also reported that the use of R light as a night-break was not effective for increasing anthesis or flower bud opening in multiple cultivars of Asiatic lilies. It was also noted that flowering in lilies was more influenced by variations in day-length and not night interruption with supplemental lighting.

Gibberellic acid (GA$_3$) effectively promoted growth and flowering measurements in gayfeather 'Kobold', lily 'Karma Serena', and Asiatic lily 'Yellow Cocotte'. Previous research has noted the presence and influence of GA$_3$ in growing tissues, shoot apices, leaves, and flowers [32]. Cell division and expansion are stimulated by GA$_3$, especially in response to light or darkness [33]. Flower initiation, development, sex expression, and number are also regulated by GA$_3$ [34]. Bulyalert [35] reported that exogenous applications of GA$_3$ increased width and height, as well as the flowering percentage in liatris. The significant effect of GA$_3$ on flower diameter and height in three cultivars of dahlia was not analyzed, but an increase in these features was observed and reported [36]. Flower diameter was also increased in Asiatic hybrid cut lily flowers when treated with GA$_3$ and a standard preservative [37]. The following studies have reported similar results in other cut flowers. Application of GA$_3$ promoted shoot elongation in different cultivars of chrysanthemums [38,39]. Foliar application of GA$_3$ increased stem length in a variety of cut flower cultivars that were field-grown [40]. Bultynck and Lambers [41] reported that the addition of exogenous GA$_3$ promoted leaf elongation and increased shoot biomass in *Aegilops caudata* (L.) and *Aegilops tauschii* (L.). Pobudkiewicz and Nowak [42] found that flowering size of gerbera (*Gerbera jamesonni* Hooker f.) was enhanced when GA$_3$ was applied at 200 mg L^{-1}. Mean flower number was increased in philodendron (*Philodendron* Schott) 'Black Cardinal' as GA$_3$ concentrations increased [43]. Dobrowolska and Janicka [44] also reported that application of GA$_3$ at a concentration of 10 mg dm^{-3} increased flower number in *Impatiens hawkeri* (L.) 'Riviera Pink'.

Interaction of light with GA$_3$ effectively promoted growth and flowering measurements of dahlia 'Karma Serena' and Asiatic lily 'Yellow Cocotte'. Yamaguchi and Kamiya [45] have concluded that light and GA$_3$ are highly interactive and are involved in the same pathways that regulate germination and dormancy. Light and GA$_3$ are likely interacting with similar pathways regulating growth and flowering. A study reported that cell expansion was promoted in the leaves of dwarf bean (*Phaseolus vulgaris* L.)

and stem elongation was increased in garden peas (*Pisum sativa* L.) when exposed to FR light and saturated with GA_3 [46]. In Kentucky bluegrass (*Poa pratensis* L.), shoot elongation was increased when endogenous levels of GA_3 interacted with light [47]. Williams and Morgan [48] noted that the exposure of GA_3 to FR light hastened flowering in sorghum (*Sorghum bicolor* L.). White et al. [49] reported that although potted greenhouse plants *Aquilegia* × *hybrida* (L.) 'Bluebird' and 'Robin' all flowered when treated with 100 mg L^{-1} exogenous GA_3, there was no synergistic effect with the supplemental lights emitting R and FR. An increase in flower number was also observed, but not due to an interaction of light with GA_3. Another study reported that GA_3 should be applied to plants before cold temperature exposure and light treatments should be applied after cold temperature exposure to improve floral development. There could be even more of an effect between light and GA_3 on lily bulbs based on exposure to cold temperatures before or after as Asiatic lily 'Yellow Cocotte' were the only plants exposed to a cold treatment before applications of GA_3 and light treatments. Possibly, the exposure to cold temperatures before GA_3 treatment contributed to the lack of growth and flowering rates.

5. Conclusions

Light emitting diode flowering lamps are equally effective as halogen lamps at regulating growth and flowering. Although the LED flowering lamps and halogen bulbs have similar light intensity, the energy consumption of LEDs was 14 to 15 watts per lamp, whereas halogen bulbs use considerably more watts per bulb. Not only was there an improvement in energy use, but the quality of plants was maintained and improved with the use of LED flowering lamps. Results of this study and that of many others show that GA_3 also plays an important role in flowering stimulation, as well as plant growth. In addition, light and GA_3 have a synergistic relationship with each other regarding plant and flower development of plants. More research needs to be conducted using an array of LED flowering lamps with different spectrums, and in combination with the plant hormone GA_3 to control plant growth and flowering, as affects are species dependent.

Author Contributions: Conceptualization, B.L.D.; T.M.-I.; methodology, T.M.-I.; formal analysis, M.P.; investigation, T.M.-I.; writing—original draft preparation, T.M.-I.; writing—review and editing, B.L.D and N.M.; supervision, B.L.D.

Funding: This research was funded by ODAFF Specialty Block grant #G00010135.

Acknowledgments: We thank Stephen Stanphill for helping with data collection and greenhouse management.

Conflicts of Interest: The authors declare no conflict of interest.

References

1. Nelson, P. *Greenhouse Operation and Management*, 7th ed.; Pearson: Boston, MA, USA, 2012.
2. Morrow, R. LED lighting in horticulture. *HortScience* **2008**, *43*, 1947–1950. [CrossRef]
3. Bourget, C. An introduction to light-emitting diodes. *HortScience* **2008**, *43*, 1944–1946. [CrossRef]
4. Bessho, M.; Shimizu, K. Latest trends in LED lighting. *Electron. Commun. Jpn.* **2012**, *95*, 315–320. [CrossRef]
5. Singh, D.; Basu, C.; Meinhardt-Wollweber, M.; Roth, B. LEDs for Energy Efficient Greenhouse Lighting. Available online: https://arxiv.org/abs/1406.3016 (accessed on 9 October 2017).
6. Briggs, W.R.; Christie, J.M. Phototropin 1 and phototropin 2: Two versatile plant blue-light receptors. *Trends Plant Sci.* **2002**, *7*, 204–209. [CrossRef]
7. Massa, G.D.; Kim, H.H.; Wheeler, R.M.; Mitchell, C.A. Plant productivity in response to LED lighting. *HortScience* **2008**, *43*, 1951–1956. [CrossRef]
8. Kim, H.H.; Wheeler, R.M.; Sager, J.C.; Yorio, N.C.; Goins, G.D. Light-emitting diodes as an illumination source for plants: A review of research at Kennedy Space Center. *Habitation (Elmsford)* **2005**, *10*, 71–78. [CrossRef] [PubMed]
9. Heo, J.; Lee, C.; Chakrabarty, D.; Paek, K. Growth responses of marigold and salvia bedding plants as affected by monochromic or mixture radiation provided by a light-emitting diode (LED). *Plant Growth Regul.* **2002**, *38*, 225–230. [CrossRef]

10. Craig, D.S.; Runkle, E.S. A moderate to high red to far-red light ratio from light-emitting diodes controls flowering of short-day plants. *J. Am. Soc. Hortic. Sci.* **2013**, *138*, 167–172. [CrossRef]

11. Kohyama, F.; Whitman, C.; Runkle, E.S. Comparing flowering responses of long-day plants under incandescent and two commercial light-emitting diode lamps. *HortTechnology* **2014**, *24*, 490–495. [CrossRef]

12. Meng, Q.; Runkle, E.S. Controlling flowering of photoperiodic ornamental crops with light-emitting diode lamps: A coordinated grower trial. *HortTechnology* **2014**, *24*, 702–711. [CrossRef]

13. Nishidate, K.; Kanayama, Y.; Nishiyama, M.; Yamamoto, T.; Hamaguchi, Y.; Kanahama, K. Far-red light supplemented with weak red light promotes flowering of Gypsophila paniculata. *J. Jpn. Soc. Hortic. Sci.* **2012**, *81*, 198–203. [CrossRef]

14. Yamada, A.; Tanigawa, T.; Suyama, T.; Matsuno, T.; Kunitake, T. Red:Far-red light ratio and far-red light integral promote or retard growth and flowering in Eustoma grandiflorum (Raf.) Shinn. *Sci. Hortic.* **2009**, *120*, 101–106. [CrossRef]

15. Hu, J.; Mitchum, M.G.; Barnaby, N.; Ayele, B.T.; Ogawa, M.; Nam, E.; Lai, W.C.; Hanada, A.; Alonso, J.M.; Ecker, J.R.; et al. Potential sites of bioactive gibberellin production during reproductive growth in Arabidopsis. *Plant Cell* **2008**, *20*, 320–336. [CrossRef] [PubMed]

16. Gupta, R.; Chakrabarty, S.K. Gibberellic acid in plant: Still a mystery unresolved. *Plant Signal. Behav.* **2013**, *8*, 1–5. [CrossRef] [PubMed]

17. Kobayashi, Y.; Weigel, D. Move on up, its time for change: Mobile signals controlling photoperiod-dependent flowering. *Genes Dev.* **2007**, *21*, 2371–2384. [CrossRef] [PubMed]

18. Dennis, D.J.; Doreen, J.; Ohteki, T. Effect of gibberellic acid 'quick-dip' and storage on the yield and quality of blooms from hybrid Zantedeschia tubers. *Sci. Hortic.* **1994**, *57*, 133–142. [CrossRef]

19. Delvadia, D.V.; Ahlawat, T.R.; Meena, B.J. Effect of different GA_3 concentration and frequency on growth, flowering and yield in Gaillardia (Gaillardia pulchella Foug.) cv. Lorenziana. *J. Hortic. Sci.* **2009**, *4*, 81–84.

20. Ranwala, A.P.; Legnani, G.; Reitmeier, M.; Stewart, B.B.; Miller, W.B. Efficacy of plant growth retardants as preplant bulb dips for height control in LA and oriental hybrid lilies. *HortTechnology* **2002**, *12*, 426–431. [CrossRef]

21. Dissanayake, P.; George, D.L.; Gupta, M.L. Effect of light, gibberellic acid and abscisic acid on germination of guayule (Parthenium argentatum Gray) seed. *Ind. Crop Prod.* **2010**, *32*, 111–117. [CrossRef]

22. Toyomasue, T.; Tsuji, H.; Yamane, H.; Nakayama, M.; Yamaguchi, I.; Murofushi, N.; Takahasi, N.; Inoue, Y. Light effect on endogenous levels of gibberellins in photoblastic lettuce seeds. *J. Plant Growth Regul.* **1993**, *12*, 85–90. [CrossRef]

23. Lona, F.; Bocchi, A. Luterferenza dell'acido gibberellieo nell'effecto della lute rossa e rosso-estrema sull'allungamento dell fusto di Perilla ocy~noides L. *L'ateneo Parmense* **1956**, *7*, 645–649.

24. Lockhart, J. A reversal of the light inhibition of pea stem growth by the gibberellins. *Proc. Natl. Acad. Sci. USA* **1956**, *42*, 841–848. [CrossRef]

25. Borthwick, H.A.; Hendricks, S.B.; Parker, M.W. The reaction controlling floral initiation. *Proc. Natl. Acad. Sci. USA* **1952**, *38*, 929–934. [CrossRef]

26. Cathey, H.M.; Borthwick, H.A. Photoreversibility of floral initiation in Chrysanthemum. *Bot. Gaz.* **1957**, *119*, 71–76. [CrossRef]

27. Downs, R.J.; Borthwick, H.A.; Piringer, A.A. Comparison of incandescent and fluorescent lamps for lengthening photoperiods. *Proc. Am. Soc. Hortic. Sci.* **1958**, *71*, 568–578.

28. Miyashita, Y.; Kitaya, Y.; Kubota, C.; Kozai, T.; Kimura, T. Effects of red and far-red light on the growth and morphology of potato plantlets in-vitro: Using light emitting diodes as a light source for micropropagation. *Acta Hortic.* **1995**, *393*, 189–194. [CrossRef]

29. Wollaeger, H.M.; Runkle, E.S. Growth and acclimation of impatiens, salvia, petunia, and tomato seedlings to blue and red light. *HortScience* **2015**, *50*, 522–529. [CrossRef]

30. Bergstrand, K.J.; Asp, H.; Larsson-Jonsson, E.H.; Schussler, H.K. Plant developmental consequences of lighting from above or below in the production of poinsettia. *Eur. J. Hortic. Sci.* **2015**, *80*, 51–55. [CrossRef]

31. Bieleski, R.; Elgar, J.; Heyes, J.; Woolf, A. Flower opening in Asiatic lily is a rapid process controlled by dark-light cycling. *Ann. Bot.* **2000**, *86*, 1169–1174. [CrossRef]

32. Jones, R.L.; Phillips, I.D. Organs of gibberellin synthesis in light-grown sunflower plants. *Plant Physiol.* **1966**, *41*, 1381–1386. [CrossRef]

33. Feng, S.; Martinez, C.; Gusmaroli, G.; Wang, Y.; Zhou, J.; Wang, F. Coordinated regulation of Arabidopsis thaliana development by light and gibberellins. *Nature* **2008**, *451*, 475–479. [CrossRef]

34. Griffiths, J.; Murase, K.; Rieu, I. Genetic characterization and functional analysis of the GID1 gibberellin receptors in Arabidopsis. *Plant Cell* **2006**, *18*, 3399–3414. [CrossRef]

35. Bulyalert, O. Effect of Gibberellic Acid on Growth and Flowering of Liatris Corm (*Liatris spicata*) c.v. Florist Violet Propagated from Seed (Abstract). Available online: http://agris.fao.org/agris-search/search.do?recordID=TH9220028 (accessed on 9 October 2017).

36. Pudelska, K.; Podgajna, E. Decorative value of three dahlia cultivars (Dahlia cultorum Thorsr. et Reis) treated with gibberellin. *Mod. Phytomorphol.* **2013**, *4*, 83–86.

37. Rabiza-Swider, J.; Skutnik, E.; Jedrzejuk, A.; Lukaszewska, A.; Lewandowska, K. The effect of GA$_3$ and the standard preservative on keeping qualities of cut LA hybrid lily Richmond. *Acta Sci. Pol. Hortorum Cultus.* **2015**, *14*, 51–64.

38. Schmidt, C.; Bellé, A.B.; Nardi, C.; Toledo, A.K. The gibberellic acid (GA$_3$) in the cut chrysanthemum (*Dedranthema grandiflora* Tzevelev.) Viking: Planting summer/autumn. *Ciência Rural* **2003**, *33*, 267–274. [CrossRef]

39. Zalewska, M.; Żabicka, A.; Wojciechowska, I. The influence of gibberellic acid on the growth and flowering of cascade chrysanthemum cultivars in outside glasshouse. *Zesz. Probl. Post. Nauk. Roln.* **2008**, *525*, 525–533.

40. Bergmann, B.A.; Dole, J.M.; McCall, I. Gibberellic acid shows promise for promoting flower stem length in four field-grown cut flowers. *HortTechnology* **2016**, *26*, 287–292.

41. Bultynck, L.; Lambers, H. Effects of applied gibberellic acid and paclobbutraol on leaf expansion and biomass allocation in two Aegilops species with contrasting leaf elongation rates. *Physiol. Plant.* **2004**, *122*, 143–151. [CrossRef]

42. Pobudkiewicz, A.; Nowak, J. The effect of gibberellic acid on growth and flowering of Gerbera jamesonii Bolus. *Folia Hortic.* **1992**, *4*, 35–42.

43. Chen, J.; Henny, R.J.; McConnell, D.B.; Caldwell, R.D. Gibberellic acid affects growth and flowering of *Philodendron* Black Cardinal. *J. Plant Growth Regul.* **2003**, *41*, 1–6. [CrossRef]

44. Dobrowolska, A.; Janicka, D. The effect of growth regulators on flowering and decorative value of Impatiens hawkeri W. Bull belonging to Riviera group. *Rocz. AR Pozn. Ogrodn.* **2007**, *41*, 35–39.

45. Yamaguchi, S.; Kamiya, Y. Gibberellins and light-simulated seed germination. *J. Plant Growth Regul.* **2001**, *20*, 369–376. [CrossRef]

46. Vince, D. Gibberellic acid and light inhibition of stem elongation. *Planta (Berlin)* **1967**, *75*, 291–308. [CrossRef]

47. Tan, Z.G.; Qian, Y.L. Light intensity affects gibberellic acid content in Kentucky bluegrass. *HortScience* **2003**, *38*, 113–116. [CrossRef]

48. Williams, E.A.; Morgan, P.W. Floral initiation in sorghum hastened by gibberellic acid and far-red light. *Planta* **1979**, *145*, 269–272. [CrossRef]

49. White, J.W.; Chen, H.; Beattie, D.J. Gibberellin, light, and low temperature effects on flowering of Aquilegia. *HortScience* **1990**, *25*, 1422–1424.

 *horticulturae*

Article

Anaerobically-Digested Brewery Wastewater as a Nutrient Solution for Substrate-Based Food Production

Ignasi Riera-Vila, Neil O. Anderson, Claire Flavin Hodge and Mary Rogers *

Department of Horticultural Sciences, University of Minnesota, Saint Paul, MN 55108, USA;
i.riera.vila@gmail.com (I.R.-V.); ander044@umn.edu (N.O.A.); flavi010@umn.edu (C.F.H.)
* Correspondence: roge0168@umn.edu

Received: 30 January 2019; Accepted: 10 May 2019; Published: 2 June 2019

Abstract: Urban agriculture, due to its location, can play a key role in recycling urban waste streams, promoting nutrient recycling, and increasing sustainability of food systems. This research investigated the integration of brewery wastewater treatment through anaerobic digestion with substrate-based soilless agriculture. An experiment was conducted to study the performance of three different crops (mustard greens (*Brassica juncea*), basil (*Ocimum basilicum*), and lettuce (*Lactuca sativa*) grown with digested and raw brewery wastewater as fertilizer treatments. Mustard greens and lettuce grown in digested wastewater produced similar yields as the inorganic fertilizer control treatment, while basil had slightly lower yields. In all cases, crops in the digested wastewater treatments produced higher yields than raw wastewater or the no fertilizer control, indicating that nutrients in the brewery wastewater can be recovered for food production and diverted from typical urban waste treatment facilities.

Keywords: urban agriculture; reclaimed wastewater; controlled environment agriculture; soilless production; brewery; *Brassica juncea*; *Lactuca sativa*; *Ocimum basilicum*

---

1. Introduction

Urban agriculture is experiencing a resurgence in popularity in many parts of the world. Beyond the social benefits urban agriculture can provide, such as creating opportunities for community building, jobs, and education, as well as increased access to healthy food [1], urban agriculture is also gaining interest due to the possible environmental benefits it can provide to municipalities. Some of those benefits include increased green spaces, reduction of food imports, and nutrient recycling [2–5]; the latter is the focus of this paper. Modern cities function largely as nutrient sinks, with nutrients shipped in from rural locations through food and transformed to waste once consumed. Historically, and presently in the Global South, these nutrient-rich waste streams were and are prized for fertility in both rural and urban agriculture [5–7]. However, social stigma, concerns over sanitation, and potential presence of pathogens, nitrates, heavy metals, or pharmaceuticals [8] in urban wastewater make this integration a challenge today. Still, advocates argue that finding ways to recycle the organic fraction of waste streams to agricultural production in urban areas will not only reduce soil and water pollution, but also prove central to both urban waste management and agricultural production [9].

Integrating agricultural production with industries that produce wastewater suitable for irrigation may increase sustainability in both sectors [9,10] and reduce the challenge of reusing wastewater [11]. The brewing industry can serve as a model for wastewater and urban agriculture integration. The brewing process creates large amounts of wastewater that, if treated aerobically like most of the wastewater in the US, requires large amounts of energy for water treatment. This usually translates into substantial surcharges to the brewery [11]. The large energy requirement for treatment is due to processing the high number of organic compounds in the water. These organic compounds have the potential to produce hydrogen or methane energy if treated anaerobically, and the resulting nutrients could be a source of fertility for agricultural production [12–14].

Traditional soil-based agriculture may not be a feasible solution in urban environments due to soil contamination and the lack of available land. Soilless production, such as substrate-based, aquaponics (hydroponics) could offer a flexible solution adaptable to different settings [15]. Research has explored how to recycle wastewater for soilless production, mostly for hydroponic production, and found that if the wastewater or organic waste is mineralized and the optimal nutrient content is achieved, comparable yields to plants grown using inorganic fertilizer may be obtained. This has been shown using brewery wastewater [12]. In all cases, researchers promoted nutrient mineralization through either algae ponds [12], bioreactors [16], or introduction of desired bacteria in the system [17]. Part of the rationale of using organic substrates in this work is to enhance mineralizing and heterotroph bacteria and fungi populations [18] in order to promote organic matter breakdown and mineralization [19,20].

The objective of this research was to test the performance of both digested and raw brewery wastewater as a fertility source for substrate-based vegetable production. The use of both digested and raw wastewater seeks to explore the feasibility of using brewery wastewater for substrate production as well as to determine the potential benefits of digesting the wastewater. Three genetically diverse crops were used to study the adaptability of the brewery wastewater to grow high value crops suitable for controlled environment production in soilless media. Mustard greens (*Brassica juncea*) can be easily grown in soilless production as a leafy vegetable and are increasing in consumer demand, particularly in fresh salad mixes [21,22]. Lettuce (*Lactuca sativa*) is widely grown and studied in soilless production [23–26] and basil (*Ocimum basilicum*) is a popular herb that is also widely cultivated and studied in soilless systems [27,28].

Our hypothesis was that if the treatments with wastewater have similar nutrient profiles to the inorganic control treatment, crops should produce similar yields. We hypothesized that differences between the digested and raw wastewater treatments would be due to the higher nitrogen and lower organic carbon content of the digested wastewater as opposed to the raw wastewater treatment.

2. Materials and Methods

This research is part of a larger project that seeks to create an anaerobic wastewater treatment process to reduce carbon load of wastewater combined with urban food production via soilless agriculture. Specifically, we are interested in modeling a decentralized approach to this integration that could happen at various locations in an urban environment, unlike a centralized wastewater treatment facility. For this project, the anaerobic digestion process addresses two objectives: (1) to reduce organic load of the wastewater while producing hydrogen and methane energy and (2) to create a final water solution more suitable for plant uptake in soilless vegetable production, thereby closing a water usage loop.

Three experiments corresponding to three different crops—'Green Wave' mustard greens, 'Nufar' basil, and 'Salanova Green Butter' lettuce (Johnny's Selected Seeds, Waterville, ME)—were conducted between May and September 2018 at the University of Minnesota Plant Growth Facilities in St. Paul, MN, USA (44°59′17.8″ N lat., -93°10′51.6″ W long.). The experiments were the same in methods, only differing in crop type. Each experiment was set up as a completely randomized design, with $n = 5$ replicates of each of the four fertility treatments, for a total of 20 plants in each experiment.

Four different fertility treatments were evaluated: (1) an unfertilized control (i.e., only water); (2) an industry standard inorganic hydroponic fertilizer with the following concentrations: 150 ppm N, 52 ppm P, 215 ppm K, 116 ppm Ca, 53 ppm Mg, 246 ppm SO_4, 3 ppm Fe, 0.5 ppm Mn, 0.15 ppm Zn, 0.15 ppm Cu, 0.5 ppm B, and 0.1 ppm Mo, obtained by mixing 4 L of water with 2.56 g of $CaNO_3$ and 3.88 g of 5–12–26 (Jack's hydroponics fertilizers, JR Peters Inc., Allentown, PA, USA); (3) raw wastewater from a local brewery (Fulton Beer, Minneapolis, MN, USA); and (4) digested wastewater from the same brewery, diluted at 50% with deionized water after digestion (Dr. Paige Novak's laboratory, Dept. of Civil, Environmental and Geo-engineering, University of Minnesota, Minneapolis, MN, USA). All of the plants were watered with 100 mL of well water every morning, along with 50 mL of their respective fertility treatments six afternoons per week (Sunday to Friday), with rates based on previous experience; one day per week had only water applied. Fertility treatments were at the same greenhouse temperature when irrigating.

All raw and digested wastewater was collected before the experiments began and stored in 26.5 L polyethylene containers (Aqua-Tainer, Reliance Products, Winnepeg, Manitoba, Canada) in a cooler at 5 °C (Vollrath Inc., Sheboygan, WI, USA). A subsample of the digested and raw wastewater was submitted for analytical testing (Research Analytical Lab, University of Minnesota, St. Paul, MN, USA) for ammonium-N [29], nitrate/nitrite-N [30], and total phosphorus [31] at the beginning of the experiment. In the early stages of the project, a low nitrogen concentration in the digestate but a high electrical conductivity [32] was detected. Therefore, 150 mg/L of ammonium hydroxide was used for pH adjustment of the brewery wastewater instead of calcium carbonate as a means to obtain a digestate with higher nitrogen content but lower conductivity.

The three different crops—mustard greens, basil, and lettuce—were grown for six weeks in plastic containers with a 10.16 cm diameter and an 8.5 cm depth (Belden Plastics, Saint Paul, MN, USA) filled with a peat-based substrate (Professional Growing Mix #8, Sun Gro Horticulture, MA, USA). This substrate was chosen based on performance in preliminary experiments. Plants were started in a 128 plug tray (TO Plastics, Clearwater, MN, USA) in a peat-based propagating mix (Sunshine Propagation Mix, Sungro, MA, USA), and kept in a mist chamber, misted at intervals of 5 min for 7 days to encourage germination. The mean temperature was 25.78 ± 2.14 °C with 66.94 ± 16.29% relative humidity (RH). After germination, seedlings were moved to the greenhouse and placed on top of a flat 25.4 × 50.8 cm tray filled with 5 cm of water for 7–10 days. The photoperiod was set for long days (16:8 h day:night) with supplemental lighting supplied by high pressure sodium high intensity discharge (HID) lamps at a maxima of 1377 μmol m^{-2} s^{-1}. Mean temperature ±SD was 24.39 ± 4.24 °C. All environmental settings were controlled via an Argus Control Systems Ltd. computer (Surrey, BC, Canada).

Plant biomass, or final yield, was calculated for each plant at the end of the six weeks. Mustard green leaves longer than 10 cm were harvested weekly. Fresh weight of the leaves was recorded and then leaves were dried for 72 h at 70 °C in a hot air oven (Hatchpack, PI, USA) and dry weights were measured. The total above-ground biomass of mustard greens was calculated by adding the cumulative harvested yield to the dry mass of the above-ground plant material at the termination of this experiment. In the case of lettuce and basil, total above-ground biomass present at the end of the experiment was used as the measurement of yield.

Plant growth measurements were recorded weekly. The number of leaves on the terminal stem and stem width (mm) were measured on each experimental unit, or each plant, while chlorophyll level was measured using a soil-plant analyses development (SPAD) meter (SPAD 502, Spectrum technologies, IL, USA) for each leaf on each plant. The SPAD meter provides an output value between −9.9 to 199.9 in SPAD units; the higher the value, the greater the chlorophyll content [33]. The meter measures relative light absorbance at two different wavelengths (650 nm and 940 nm) [34].

Water quality of the leachate (collected after watering with 100 mL of well water) was measured weekly for pH, EC, infiltration, and nitrates. Three out of the five replications of each treatment were sampled, for a total of 12 samples from each crop, which were randomly selected at the beginning of the experiment. Infiltration time was measured as the duration of time from pouring the water until there was no standing water on the substrate surface. Electrical conductivity (EC) and pH of the collected leachate was measured using an electronic pH and EC meter (Milwaukee mw802, Milwaukee Electronics, WI, USA). Nitrate levels of the leachate were measured using a LAQUA twin NO3-11C pocket nitrate reader (Horiba scientific, Minami-ku Kyoto, Japan).

Statistical analyses were conducted using R software (The R foundation, v.3.4.1) to determine significance using Analysis of Variance (ANOVA). If the null hypothesis of no significant differences was rejected, mean separations were accomplished using Tukey's honestly significant difference (HSD) test at a significance level of $\alpha = 0.05$. Plant chlorophyll content and growth parameters were also tested for correlation with yield, using Pearson's correlation test at a significance level of $\alpha = 0.05$.

3. Results

3.1. Biomass and Yield

Fertility treatment had a significant effect on the number of harvested leaves (data not shown) and total above-ground biomass of mustard greens (Table 1). Fertility treatment also had a significant effect on total above ground biomass of lettuce and basil (Table 1). Mustard greens and lettuce plants grown using digested wastewater produced a similar amount of total above ground biomass when compared to plants in the inorganic fertilizer treatment, while raw wastewater plants performed at the same level as those in the no fertilizer treatment (Table 1). For basil, the highest biomass was observed in the inorganic fertilizer followed by digested wastewater treatments (Table 1). Mustard greens in both the inorganic fertilizer and digested wastewater treatments reached harvestable size the second week of the experiment and produced until week 6, while mustard greens grown in the raw wastewater only were harvestable on week 6, and no plants reached maturity in the unfertilized control treatment.

Table 1. Mean ± SE total above-ground biomass, or per plant dry weight (g), for mustard greens, lettuce, and basil grown with different fertility treatments in a greenhouse in St. Paul, Minnesota in 2018. One-way ANOVA results for fertility treatments were performed on each parameter and crop. Values within the same column with different letters are significantly different (Tukey's HSD, $\alpha = 0.05$).

Fertility Treatment	Total Dry Weight Per Plant (g)		
	Mustard Greens	**Lettuce**	**Basil**
No fertilizer	0.19 ± 0.04 b	0.19 ± 0.03 b	0.46 ± 0.02 c
Inorganic fertilizer	0.58 ± 0.02 a	2.83 ± 0.11 a	3.57 ± 0.09 a
Raw wastewater	0.29 ± 0.08 b	0.46 ± 0.09 b	0.73 ± 0.13 c
Digested wastewater	0.56 ± 0.04 a	2.09 ± 0.46 a	1.95 ± 0.36 b
ANOVA	$F(3,16) = 16.33$ $p < 0.01$	$F(3,16) = 27.51$ $p < 0.01$	$F(3,16) = 50.71$ $p < 0.01$

3.2. Plant Growth

Due to the leaf removal from mustard green plants, plant growth was monitored using stem width at the substrate level. No significant fertility treatment effects were found for mustard stem width during any of the weeks (data not shown). Weekly stem width measurements were moderately correlated with weekly harvest ($r = 0.57$, $p < 0.01$).

For lettuce and basil, the number of leaves per week was used as a measurement of plant growth. Significantly different numbers of leaves per lettuce plant were counted in weeks 2–6 ($p < 0.05$) between treatments. In weeks 2–5, lettuce grown in inorganic fertilizer treatments had the highest number of leaves, while in week 6, lettuce from both inorganic fertilizer and digested wastewater treatments had a similar number of leaves. Total number of lettuce leaves at the end of the experiment was positively correlated to final above-ground biomass ($r = 0.89$, $p < 0.01$). Basil plants, too, had a similar number of leaves through the first week, but subsequent significant differences were found in weeks 2–6 ($p < 0.01$) between treatments. In later weeks, basil from the inorganic fertility treatment had the highest number of leaves. Similar to the case of lettuce, at the end of the experiment, the total number of basil leaves was highly correlated to final above-ground biomass ($r = 0.97$, $p < 0.01$). Overall, plants from both inorganic and digested wastewater treatments appeared marketable for all crops (Figures 1–3).

Figure 1. Image of 'Green Wave' mustard green plants (day 42) grown in University of Minnesota's Plant Growth Facilities greenhouse (St. Paul, MN). Plants were grouped by fertility treatment in rows; from left to right: digested wastewater, raw wastewater, inorganic fertilizer, and no fertilizer.

Figure 2. Image of 'Salanova Green Butter' lettuce plants (day 42) grown in University of Minnesota's Plant Growth Facilities greenhouse (St. Paul, MN). Plants were grouped by fertility treatment in rows; from left to right: digested wastewater, raw wastewater, inorganic fertilizer, and no fertilizer.

Figure 3. Image of 'Nufar' basil plants (day 42) grown in University of Minnesota's Plant Growth Facilities greenhouse (St. Paul, MN). Plants were grouped by fertility treatment in rows; from left to right: digested wastewater, raw wastewater, inorganic fertilizer, and no fertilizer.

3.3. Chlorophyll Content

In all crops, the chlorophyll content of the digested wastewater treatments stayed at a similar level to inorganic fertilizer treatments throughout the 6 weeks (Table 2). Significant differences in SPAD values of mustard greens were not found until weeks 5 and 6 ($p = 0.03$), where digested wastewater treatments had the highest chlorophyll content and raw wastewater the lowest (data not shown). SPAD values for mustard greens in were not correlated with weekly harvest ($r = 0.05, p < 0.56$).

Table 2. Mean chlorophyll content (SPAD values) averaged across 6 weeks for mustard greens, lettuce, and basil grown with different fertility treatments in a greenhouse in St. Paul, Minnesota in 2018. Values within the same column with different letters are significantly different (Tukey's HSD, $\alpha = 0.05$).

Fertility Treatment	Chlorophyll Content, SPAD Units		
	Mustard Greens	**Lettuce**	**Basil**
No fertilizer	22.23 ab	13.33 c	17.79 c
Inorganic fertilizer	21.01 a	20.76 a	25.93 a
Raw wastewater	20.40 b	17.21 b	21.50 b
Digested wastewater	24.13 a	20.42 a	26.22 a
ANOVA			
	$F_{(3,133)} = 8.32$	$F_{(3,116)} = 22.72$	$F_{(3,116)} = 43.85$
	$p < 0.001$	$p < 0.001$	$p < 0.001$

Significant differences between SPAD measurements became apparent in the fourth week, where digested wastewater and inorganic fertilizer treatments had the highest chlorophyll content; the six-week average is shown in Table 2. SPAD values at the end of the experiment were correlated with lettuce total biomass ($r = 0.78, p < 0.05$).

We observed significant differences in the SPAD values of basil plants grown in the different fertility treatments as early as week 2 and continuing through week 6 ($p < 0.01$). In all cases, basil leaves from the digested wastewater and inorganic fertilizer treatments had the highest chlorophyll content, followed by raw wastewater treatment (Table 2). Basil SPAD values were correlated with total dry biomass at the end of the experiment ($r = 0.77, p < 0.01$).

3.4. Water Test

Comparing the total nitrogen ($T_N = NH_4 - N + NO_2 - N + NO_3 - N$) between the digested wastewater (50% diluted) and the raw wastewater, there was higher ammonium-N content in the digested wastewater, 171.50 ppm compared to 7.23 ppm (Table 3). The ammonium content of the digested wastewater was similar to that used in the inorganic fertilizer control and within the 100–250 ppm recommended range [35,36]. The only concerning factor was that the N ratio ($R_N = NH_4 - N/NO_3 - N$) [37] was extremely high: >1000, and since all the N was in the form of ammonium, this could generate problems of ammonia toxicity and yield reduction [38,39], though this was not observed. Phosphorus tests showed an orthophosphate concentration as low as 22 ppm in the pure digested wastewater (Table 3), a level similar to our previous hydroponic experiments [32] and lower than the raw wastewater, which can be explained by the 50% dilution of the digester leachate before use.

Table 3. Solution test results (pH, electrical conductivity (EC), ammonium-N, nitrate/nitrite N, total phosphorus) for each fertility treatment. Nutrient levels of digested and raw wastewater were obtained from laboratory testing while nutrient levels of inorganic fertilizer obtained from the fertilizer formulation (see Methods).

Fertility Treatment	pH	EC	Ammonium-N	Nitrate/Nitrite-N	Total Phosphorus
Digested wastewater	8.40	1.87	171.50	<0.1	22.00
Raw wastewater	4.40	0.84	7.23	<0.1	79.88
Inorganic fertilizer	6.10	1.72	0	150	52
No fertilizer [1]	7.90	0.24	-	-	-

[1] Tests for ammonium, nitrate/nitrite or phosphorus were not applicable for the unfertilized control.

3.5. Leachate Monitoring

Water infiltration time increased over time in raw wastewater treatments, while digested wastewater showed similar monitoring trends compared to the inorganic fertilizer treatments (data not shown). Nitrate, EC, and pH data of these leachate samples are reported in Table 4 averaged across weeks and crops, as fertility treatment effects were greater than crop effects. Overall, the highest nitrate levels were observed in the leachate from the beginning of the experiment (data not shown), and on average, N levels were highest in leachates from the digested wastewater and inorganic fertilizer treatments (Table 4). At the beginning of the experiment, the leachate was acidic (pH = 5.5–6), but turned more neutral through the experiment (data not shown); raw wastewater treatment leachate was the most basic (Table 4). Electrical conductivity also declined over time, with highest levels measured in the digested wastewater and inorganic fertilizer treatment leachate.

Table 4. Leachate measurements (Nitrate-N, EC, and pH) for each fertility treatment averaged across 6 weeks and crops (mustard greens, lettuce, and basil) grown with different fertility treatments in a greenhouse in St. Paul, MN in 2018. Values within the same column with different letters are significantly different (Tukey's HSD, $\alpha = 0.05$).

Fertility Treatment	Leachate Characteristics		
	Nitrate-N (ppm)	EC (DS/cm)	pH
No fertilizer	340 b	0.51 b	6.52 b
Inorganic fertilizer	604 a	0.90 a	6.36 bc
Raw wastewater	440 ab	0.65 ab	6.85 a
Digested wastewater	653 a	0.97 a	6.22 c
ANOVA			
	$F_{(3,248)} = 5.343$ $p < 0.01$	$F_{(3,248)} = 5.283$ $p < 0.01$	$F_{(3,248)} = 35.93$ $p < 0.001$

4. Discussion

Fertilizing with digested wastewater produced similar yields for lettuce and mustard greens compared to plants grown using commercial inorganic fertilizer. For basil, yields were lower when using the digested wastewater, but still significantly higher than the treatments that received only raw wastewater or well water. Therefore we suggest that digested brewery wastewater has the potential to provide plants with the required nutrients to obtain high yields, at least with certain crops. This is consistent with a similar study that grew tomato, *Solanum lycopersicum*, using digested brewery wastewater [12] and other research that successfully used nutrient solutions partially or totally made with wastewater or organic waste for soilless production [10,39–42].

Lettuce and basil plants that were only harvested once, compared to the multiple harvests of mustard leaves, exhibited the highest chlorophyll content in the inorganic fertilizer and digested wastewater treatments. Moreover, the higher chlorophyll contents at the end of the experiment were strongly positively correlated with yield. This reinforces the conclusion that digested brewery wastewater was able to provide enough nutrients, particularly N, for the plants to produce chlorophyll at the same level as the inorganic nitrogen of the synthetic fertilizer [23,28]. In contrast, SPAD values were not different among treatments for mustard greens. This could be due to the effect leaf removal had on plant health, since stress is a factor known to reduce SPAD measurements [33]. SPAD values for inorganic and digested wastewater were similar to those reported by other authors on basil grown in hydroponic and aquaponic systems [28]. Lettuce in inorganic and digested wastewater treatments obtained much higher SPAD values than those reported for plants hydroponically grown with reclaimed organic wastes [43], and similar to lettuce plants hydroponically grown with 150–200 ppm of total N [23]. The additional ammonium-N we added as a buffer and additional nutrient source is likely the reason for this finding.

Notably, most of the nitrogen present in the digested brewery wastewater came from the addition of ammonia in the digester. This addition may seem to contradict our research objective of growing horticultural crops with the digested wastewater, but due to the low nitrogen concentration found in the wastewater, increasing nitrogen concentration was needed to successfully grow plants, similar to what other studies found [39–42]. Ammonia can increase the risk of ammonium toxicity and nitrogen deficiency [44,45] in some crops, while in other species, like lettuce, it may be the preferred nitrogen source [17,38]. This highlights the need for further research on crop suitability for digested wastewater production in order to better understand which crops may produce acceptable yields, as well as the factors driving those differences and how to enhance nitrification for those crops where nitrate is preferentially taken up. Many factors affect nitrification rates including: pH of the substrate, substrate material [46], temperature [47], ammonia to nitrate ratio, carbon to nitrogen ratio [16], original bacteria population levels, and inoculation [19,37]. Additionally, future studies should test how to optimize this ammonia addition to minimize nitrogen concentration in the final effluent. A possible way to increase this nutrient removal could be through the recirculating of the nutrient solution in the system [48,49].

This study did not address the nutrient composition of the mature plants, nor of the wastewater nutrient solutions, besides nitrogen and phosphorus. No evident nutrient deficiencies were observed in any of the plants, though the plants used reached harvestable maturity relatively quickly and a different response might occur when growing longer maturing or fruiting crops [50]. Micronutrient deficiencies could also occur due to the alkaline pH of the digested wastewater [51]. A nutrient solution could also affect nutrient profile and taste of the produce [23], as the lack of certain elements like silicon or sodium do not lead to deficiencies but can effect quality and yield [25,52,53]. The high levels of sodium detected, around 130 ppm, did not exhibit salinity effects on the plants, but this should be considered in future analyses due to the issues with sodium accumulation in both the nutrient solutions and substrates [54,55]. Further understanding how the brewing process affects those nutrient fluctuations may help anticipate them.

5. Conclusions

This research shows that it is possible to use anaerobically digested brewery wastewater to grow different crops and obtain commercially acceptable yields. Anaerobic digestion is not only a possible way to produce energy [13,56], but also to adjust nutrient profiles and create a better nutrient solution for soilless production. Thus, it is possible to integrate brewery wastewater treatment with soilless urban agriculture. This integration, and energy generation through anaerobic digestion, could help reduce the high environmental footprint that many soilless urban farms have [57] and be a good way to increase food system sustainability by promoting nutrient reuse and reducing waste treatment energy requirements [5,10]. This integration of sectors opens the door to further synergies; for example, if the soilless production happened in protected environment, CO_2 from the brewing process could be used to enhance plant growth [58], the high temperature of the wastewater [14] could heat the growing space, or spent grains could be a large component of the substrate mix instead of peat [59]. Still, in order to promote the implementation of those solutions, there is a need to develop the knowledge and technologies. Important practical considerations will also need to be made, such as synchronizing digestate production and availability with crop needs, as well as scalability of this model.

The best method to integrate both soilless urban agriculture and brewery wastewater treatment still remains unclear. In this article, we have researched how to solve some of the agronomic and technological challenges of this integration, but additional complementary research is needed pertaining to the economic, legal, and social challenges of this decentralized urban system. Urban agriculture can provide food close to home, improve water use efficiency, and utilize locally available sources of nutrients, if other support systems exist. The economic viability of this integration would likely depend, in great measure, on successfully creating a marketing strategy that demonstrates value in sustainably managing urban waste streams and producing food locally.

Author Contributions: I.R.-V., N.O.A. and M.R. equally contributed to this work; C.F.H. provided review and edits.

Funding: Funding provided by the University of Minnesota's Grand Challenges Exploratory Research Grants.

Acknowledgments: We would like to acknowledge the researchers from the department of civil engineering (Kuang Zhu, Paige Novak and William Arnold) for all their help and knowledge as well as all the members of the Rogers and Anderson labs for their support.

Conflicts of Interest: The authors declare no conflict of interest. The founding sponsors had no role in the design of the study; in the collection, analyses, or interpretation of data; in the writing of the manuscript, and in the decision to publish the results.

References

1. DeLind, L.B. Where have all the houses (among other things) gone? Some critical reflections on urban agriculture. *Renew. Agr. Food Syst.* **2014**, *30*, 3–7. [CrossRef]
2. Mougeot, L.J.A. *Agropolis: The Social, Political and Environmental Dimensions of Urban Agriculture*; Mougeot, L.J.A., Ed.; Earthscan: London, UK, 2005; ISBN 1844072320.
3. Specht, K.; Siebert, R.; Hartmann, I.; Freisinger, U.B.; Sawicka, M.; Werner, A.; Thomaier, S.; Henckel, D.; Walk, H.; Dierich, A. Urban agriculture of the future: An overview of sustainability aspects of food production in and on buildings. *Agric. Hum. Values* **2014**, *31*, 33–51. [CrossRef]
4. Goldstein, B.; Hauschild, M.; Fernández, J.; Birkved, M. Urban versus conventional agriculture, taxonomy of resource profiles: A review. *Agron. Sustain. Dev.* **2016**, *36*, 1–19. [CrossRef]
5. Smit, J.; Nasr, J. Urban agriculture for sustainable cities: Using wastes and idle land and water bodies as resources. *Environ. Urban.* **1992**, *4*, 141–152. [CrossRef]
6. Metson, G.S.; Bennett, E.M. Phosphorus cycling in Montreal's food and urban agriculture systems. *PLoS ONE* **2015**, *10*, 1–18. [CrossRef] [PubMed]
7. Cohen, N. and Reynolds, K. Resource needs for a socially just and sustainable urban agriculture system: Lessons from New York City. *Renew. Agr. Food Syst.* **2014**, *30*, 103–114. [CrossRef]
8. Kretschmer, N.; Ribbe, L.; Gaese, H. Wastewater Reuse for Agriculture. *Technol. Resour. Dev.* **2000**, *2*, 37–64.

9. McClintock, N. Why farm the city? Theorizing urban agriculture through a lens of metabolic rift. *Camb. J. Reg. Econ. Soc.* **2010**, *3*, 191–207. [CrossRef]

10. Mohareb, E.; Heller, M.; Novak, P.; Goldstein, B.; Fonoll, X.; Raskin, L. Considerations for reducing food system energy demand while scaling up urban agriculture. *Environ. Res. Lett.* **2017**, *12*. [CrossRef]

11. Modic, W.; Kruger, P.; Mercer, J.; Webster, T.; Swersey, C.; Skypeck, C. *Brewers Association Wastewater Management Guidance Manual*; Brewers Association: Boulder, CO, USA, 2015; pp. 1–31.

12. Power, S.D.; Jones, C.L.W. Anaerobically digested brewery effluent as a medium for hydroponic crop production—The influence of algal ponds and pH. *J. Clean. Prod.* **2016**, *139*, 167–174. [CrossRef]

13. Weber, B.; Stadlbauer, E.A. Sustainable paths for managing solid and liquid waste from distilleries and breweries. *J. Clean. Prod.* **2017**, *149*, 38–48. [CrossRef]

14. Enitan, A.; Adeyemo, J.; Kumari, S. Characterization of Brewery Wastewater composition. *World Acad.* **2015**, *9*, 1043–1046.

15. Lind, O. *Organic Hydroponics—Efficient Hydroponic Production from Organic Waste Streams: Introductory Research Essay*; 2016; Available online: http://www.3rmovement.com/wp-content/uploads/2017/04/Hydroponic-produciton-organic-Olle-Lind.pdf (accessed on 30 January 2019).

16. Norström, A. Treatment of Domestic Wastewater Using Microbiological Processes and Hydroponics in Sweden. PhD Thesis, KTH Royal Institute of Technology, Stockholm, Sweden, 2005; 62p.

17. Shinohara, M.; Aoyama, C.; Fujiwara, K.; Watanabe, A.; Ohmori, H.; Uehara, Y.; Takano, M. Microbial mineralization of organic nitrogen into nitrate to allow the use of organic fertilizer in hydroponics. *Soil Sci. Plant Nutr.* **2011**, *57*, 190–203. [CrossRef]

18. Koohakan, P.; Ikeda, H.; Jeanaksorn, T.; Tojo, M.; Kusakari, S.I.; Okada, K.; Sato, S. Evaluation of the indigenous microorganisms in soilless culture: Occurrence and quantitative characteristics in the different growing systems. *Sci. Hortic. (Amsterdam)* **2004**, *101*, 179–188. [CrossRef]

19. Lang, H.J.; Elliott, G.C. Enumeration and inoculation of nitrifying bacteria in soilless potting media. *J. Amer. Soc. Hort. Sci.* **1997**, *122*, 709–714. [CrossRef]

20. Montagu, K.D.; Goh, K.M. Effects of forms and rates of organic and inorganic nitrogen fertilisers on the yield and some quality indices of tomatoes (*Lycopersicon esculentum* Miller). *N. Z. J. Crop Hortic. Sci.* **1990**, *18*, 31–37. [CrossRef]

21. Kansas State University Extension. *Mustard Greens*; Kansas State University: Manhattan, KS, USA, 2016.

22. Parkell, N.B.; Hochmuth, R.C.; Laughlin, W.L. *Leafy Greens in Hydroponics and Protected Culture for Florida*; University of Florida, IFAS Extension: Gainesville, FL, USA, 2016; pp. 1–7.

23. Mahlangu, R.I.S.; Maboko, M.M.; Sivakumar, D.; Soundy, P.; Jifon, J. Lettuce (*Lactuca sativa* L.) growth, yield and quality response to nitrogen fertilization in a non-circulating hydroponic system. *J. Plant Nutr.* **2016**, *39*, 1766–1775. [CrossRef]

24. Brechner, M.; Both, A.J. *Hydroponic Lettuce Handbook*; Cornell University CEA Program: Ithaca, NY, USA, 1996; p. 48.

25. Scuderi, D.; Restuccia, C.; Chisari, M.; Barbagallo, R.N.; Caggia, C.; Giuffrida, F. Salinity of nutrient solution influences the shelf-life of fresh-cut lettuce grown in floating system. *Postharvest Biol. Technol.* **2011**, *59*, 132–137. [CrossRef]

26. Samarakoon, U.C.; Weerasinghe, P.A.; Weerakkody, W.A.P. Effect of electrical conductivity [EC] of the nutrient solution on nutrient uptake, growth and yield of leaf lettuce (*Lactuca sativa* L.) in stationary culture. *Trop. Agric. Res.* **2006**, *18*, 13–21.

27. Treadwell, D.D.; Hochmuth, G.J.; Hochmuth, R.C.; Simonne, E.H.; Davis, L.L.; Laughlin, W.L.; Li, Y.; Olczyk, T.; Sprenkel, R.K.; Osborne, L.S. Nutrient management in organic greenhouse herb production: Where are we now? *Horttechnology* **2007**, *17*, 461–466. [CrossRef]

28. Saha, S.; Monroe, A.; Day, M.R. Growth, yield, plant quality and nutrition of basil (*Ocimum basilicum* L.) under soilless agricultural systems. *Ann. Agric. Sci.* **2016**, *61*, 181–186. [CrossRef]

29. RFA. *Methodology Ammonia Nitrogen A303–S171*; Astoria-Pacific Int.: Clackamas, OR, USA, 1986.

30. RFA. *Methodology Nitrate/Nitrite A303–S170*; Astoria-Pacific Int.: Clackamas, OR, USA, 1985.

31. RFA. *Methodology Total Phosphorus A303–S050*; Astoria-Pacific Int.: Clackamas, OR, USA, 1986.

32. Riera-Vila, I.; Anderson, N.O.; Rogers, M. Anaerobically digested brewery wastewater as a nutrient solution for non-circulating hydroponics. Unpublished work. 2018.

33. Ling, Q.; Huang, W.; Jarvis, P. Use of a SPAD-502 m to measure leaf chlorophyll concentration in *Arabidopsis thaliana*. *Photosynth. Res.* **2011**, *107*, 209–214. [CrossRef] [PubMed]

34. Spectrum technologies inc SPAD 502 plus chlorophyll meter product manual. *SPAD Man.* **2009**, 1–23, 32.

35. Liu, W.K.; Yang, Q.C.; du Lian, F.; Cheng, R.F.; Zhou, W.L. Nutrient supplementation increased growth and nitrate concentration of lettuce cultivated hydroponically with biogas slurry. *Acta Agric. Scand. Sect. B Soil Plant. Sci.* **2011**, *61*, 391–394. [CrossRef]

36. Resh, H.M. *Hydroponic Food Production*, 7th ed.; CRC Press: Boca Raton, FL, USA, 2013.

37. Trejo-tellez, L.; Gomez-Merino, F. Nutrient Solutions for Hydroponic Systems- A standard methodology for plant biological researchers. *Toshiki Asao IntechOpen* **2012**. [CrossRef]

38. Savvas, D.; Passam, H.C.; Olympios, C.; Nasi, E.; Moustaka, E.; Mantzos, N.; Barouchas, P. Effects of ammonium nitrogen on lettuce grown on pumice in a closed hydroponic system. *HortScience* **2006**, *41*, 1667–1673. [CrossRef]

39. Guo, S.; Brück, H.; Sattelmacher, B. Effects of supplied nitrogen form on growth and water uptake of French bean (*Phaseolus vulgaris* L.) plants: Nitrogen form and water uptake. *Plant. Soil.* **2002**, *239*, 267–275. [CrossRef]

40. Dos Santos, J.D.; Lopes da Silva, A.L.; da Luz Costa, J.; Scheidt, G.N.; Novak, A.C.; Sydney, E.B.; Soccol, C.R. Development of a vinasse nutritive solution for hydroponics. *J. Environ. Manag.* **2013**, *114*, 8–12. [CrossRef]

41. Kawamura-Aoyama, C.; Fujiwara, K.; Shinohara, M.; Takano, M. Study on the hydroponic culture of lettuce with microbially degraded solid food waste as a nitrate source. *Jpn. Agric. Res. Q.* **2014**, *48*, 71–76. [CrossRef]

42. Eregno, F.E.; Moges, M.E.; Heistad, A. Treated greywater reuse for hydroponic lettuce production in a green wall system: Quantitative health risk assessment. *Water* **2017**, *9*, 454. [CrossRef]

43. da Silva Cuba Carvalho, R.; Bastos, R.G.; Souza, C.F. Influence of the use of wastewater on nutrient absorption and production of lettuce grown in a hydroponic system. *Agric. Water Manag.* **2018**, *203*, 311–321. [CrossRef]

44. Silber, A.; Bar-Tal, A. *Nutrition of Substrate-Grown Plants*, 1st ed.; Elsevier Ltd.: Amsterdam, The Netherlands, 2008; ISBN 9780444529756.

45. Ikeda, H.; Tan, X. Urea as an organic nitrogen source for hydroponically grown tomatoes in comparison with inorganic nitrogen sources. *Soil Sci. Plant. Nutr.* **1998**, *44*, 609–615. [CrossRef]

46. Hashida, S.-N.; Johkan, M.; Kitazaki, K.; Shoji, K.; Goto, F.; Yoshihara, T. Management of nitrogen fertilizer application, rather than functional gene abundance, governs nitrous oxide fluxes in hydroponics with rockwool. *Plant. Soil.* **2014**, *374*, 715–725. [CrossRef]

47. Russell, C.A.; Fillery, I.R.P.; Bootsma, N.; McInnes, K.J. Effect of temperature and nitrogen source on nitrification in a sandy soil. *Commun. Soil Sci. Plant. Anal.* **2002**, *33*, 1975–1989. [CrossRef]

48. Bar-Yosef, B. *Fertigation Management and Crops Response to Solution Recycling in Semi-Closed Greenhouses*, 1st ed.; Elsevier Ltd.: Amsterdam, The Netherlands, 2008; ISBN 9780444529756.

49. Bugbee, B. Nutrient management in recirculating hydroponic culture. *Acta Hortic.* **2004**, *648*, 99–112. [CrossRef]

50. Wortman, S.E.; Douglass, M.S.; Kindhart, J.D. Cultivar, growing media, and nutrient source influence strawberry yield in a vertical, hydroponic, high tunnel system. *Horttechnology* **2016**, *26*, 466–473.

51. Tyson, R.V.; Simonne, E.H.; Treadwell, D.D.; Davis, M.; White, J.M. Effect of water pH on yield and nutritional status of greenhouse cucumber grown in recirculating hydroponics. *J. Plant. Nutr.* **2008**, *31*, 2018–2030. [CrossRef]

52. Manzocco, L.; Foschia, M.; Tomasi, N.; Maifreni, M.; Dalla Costa, L.; Marino, M.; Cortella, G.; Cesco, S. Influence of hydroponic and soil cultivation on quality and shelf life of ready-to-eat lamb's lettuce (*Valerianella locusta* L. Laterr). *J. Sci. Food Agric.* **2011**, *91*, 1373–1380. [CrossRef]

53. Petersen, K.K.; Willumsen, J.; Kaack, K. Composition and taste of tomatoes as affected by increased salinity and different salinity sources. *J. Hortic. Sci. Biotechnol.* **1998**, *73*, 205–215. [CrossRef]

54. Riera-Vila, I.; Anderson, N.O.; Rogers, M. Wastewater test of 4 local breweries. Unpublished work.

55. Raviv, M.; Lieth, J.H.; Bar-Tal, A.; Silber, A. *Growing Plants in Soilless Culture: Operational Conclusions*, 1st ed.; Elsevier Ltd.: Amsterdam, The Netherlands, 2008; ISBN 9780444529756.

56. Simate, G.S.; Cluett, J.; Iyuke, S.E.; Musapatika, E.T.; Ndlovu, S.; Walubita, L.F.; Alvarez, A.E. The treatment of brewery wastewater for reuse: State of the art. *Desalination* **2011**, *273*, 235–247. [CrossRef]

57. Goldstein, B.; Hauschild, M.; Fernández, J.; Birkved, M. Testing the environmental performance of urban agriculture as a food supply in northern climates. *J. Clean. Prod.* **2016**, *135*, 984–994. [CrossRef]

58. Tremblay, N.; Gosselin, A. Effect of carbon dioxide enrichment and light. *Horttechnology* **1998**, *8*, 524–528. [CrossRef]
59. NiChualain, D.; Prasad, M. Evaluation of three methods for determination of stability of composted material destined for use as a component of growing media. *Acta Hortic.* **2009**, *819*, 303–310. [CrossRef]

 *horticulturae* 

Article

Use of Diatomaceous Earth as a Silica Supplement on Potted Ornamentals

Taylor Mills-Ibibofori [1], Bruce Dunn [1,*], Niels Maness [1] and Mark Payton [2]

[1] Department of Horticulture & L.A., Oklahoma State University, Stillwater, OK 74078, USA; tsmills@okstate.edu (T.M.-I.); niels.maness@okstate.edu (N.M.)

[2] Department of Statistics, Oklahoma State University, Stillwater, OK 74078, USA; mark.payton@okstate.edu

* Correspondence: bruce.dunn@okstate.edu

Received: 28 January 2019; Accepted: 22 February 2019; Published: 1 March 2019

Abstract: The role of silica as a needed supplement in soilless media is gaining interest. This research studied the effects of diatomaceous earth as a supplement on growth and flower characteristics, physiology, and nutrient uptake in dahlia (*Dahlia Cav.* × *hybrida* 'Dahlinova Montana'), black-eyed Susan (*Rudbeckia hirta* L. 'Denver Daisy'), and daisy (*Gerbera jamesonii* L. 'Festival Light Eye White Shades'). Plants were either well-watered at 10 centibars or water-stressed at 20 centibars. Silicon treatments included top-dressed at 20, 40, 60, and 80 g, or incorporated at 50, 100, 150, and 200 g, in Metro-Mix 360 media without silica plus a control and one treatment of new Metro-Mix 360 with silica already incorporated. Significant effects were seen from diatomaceous earth supplementation, irrigation, and interaction in all plants; growth and flower characteristics, leaf nutrient content, and tolerance to stress were improved by application of diatomaceous earth. An increase in leaf N, P, K, Mg, and Ca was observed for dahlia 'Dahlinova Montana' and black-eyed Susan 'Denver Daisy'. Transpiration was maintained in all three species due to silica supplementation under water-stress. Metro-Mix with silica was similar to the Metro-mix without silica and equivalent to most treatments with supplemental silica for all three species.

Keywords: greenhouse; metro-mix; *Dahlia*; *Rudbeckia*; *Gerbera*

1. Introduction

Silicon is the second most abundant element on earth and is present in various forms, including silicon dioxide, also known as silica (Si). In plants, except for members of the family Equisetaceae, Si is a nonessential and beneficial element, meaning that plants can complete their life cycles without the mineral nutrient [1]. However, plants deficient in Si are often weaker structurally and more prone to abnormalities of growth, development, and reproduction. The benefits of Si are mostly evident when plants are under stress conditions [2]. Several studies have shown that plants benefit in many ways from supplemental soluble Si, including greater tolerance of environmental stresses, drought, salinity, mineral toxicity or deficiency, improved growth rates, and resistance to insects and fungi [3–5].

Common use of soilless substrates in greenhouse and nursery production limits the availability of Si to plants [6]. Plants grown with soilless media often appear weaker structurally compared to crops grown in the field [7]. Therefore, adding Si-related compounds as an amendment has been highly recommended. Miyake and Takahashi [8] brought interest to Si nutrition of horticultural crops when they observed Si deficient tomatoes (*Lycopersicon esculentum* Mill.). In the Netherlands, the use of Si supplementation in a hydroponic system was recommended for crops such as cucumber (*Cucumus sativus* L.) and roses (*Rosa hybrida* L.) [9,10]. Roses with Si added to the nutrient formula also showed a decrease in leaf and flower senescence [11]. The shelf life of *Chrysanthemum* L. cut flowers was also extended [12]. Hydroponically-produced gerbera plants supplemented with Si had improved overall crop and flower quality [13].

Other considerations such as solubility, availability, physical properties, and contaminants must be considered before choosing a Si source. Silica is available from natural resources, fertilizers (organic and inorganic), and industrial by-products. Most horticultural studies use Si from by-products such as liquid silicates, slag, and basalt dust [14–16]. Diatomaceous earth (DE) is a sedimentary rock formed from the deposition of silica-rich diatoms. The cell walls of diatoms contain amorphous silica ($SiO_2 \cdot H_2O$). There has been limited research focused on the effects of DE regarding growth and flower characteristics, as well as water-stress related issues in horticultural crops. Most studies utilizing DE focused on retention of water or circulation of oxygen in plant media [17]. However, supplementation of DE has been proven to improve plant growth, quality, and nutrient uptake in agronomic crops, such as rice (*Oryza sativa* L.) [18]. Use of DE to improve plant growth of ornamentals is limited, thus the objectives of this study were to determine the effects of DE as a Si supplement on three potted ornamentals under well-watered and water-stressed conditions.

2. Materials and Methods

2.1. Plant Material and Culture

On 8 May 2015, two 128 plug cell trays of black-eyed Susan (*Rudbeckia hirta* L. 'Denver Daisy'), five 51 plug cell trays of dahlia (*Dahlia Cav.* × *hybrida* 'Dahlinova Montana'), and two 128 plug cell trays of daisy (*Gerbera jamesonii* L. 'Festival Light Eye White Shades') were obtained from Park Seed (Greenwood, SC, USA). Before transplanting, all species were placed on a mist bench. Cuttings and plugs were transplanted into standard 15 cm pots, filled with media (Metro-Mix (MM) 360; Sun Gro Horticulture, Bellevue, WA, USA) that did not contain Si on 28 May 2015, and a single treatment of media (MM + Si; Sun Gro Horticulture, Bellevue, WA, USA) that contained 20 to 50 ppm soluble Si (RESiLIENCE™) derived from wollastonite [19]. A single plant was placed in each pot and plants were grown at the Department of Horticulture and L.A. research greenhouses in Stillwater, OK under natural photoperiods. Temperatures were set at 37 °C during the day and 26 °C during the night.

2.2. Experimental Arrangement

Ten Si treatments were established by adding diatomaceous earth (Perma-Guard, Inc., Kamas, UT, USA). Application of DE included top-dressed (TD) rates at 20, 40, 60, and 80 g, and incorporated (INC) rates at 50, 100, 150, and 200 g. An MM control (using media without DE) and the MM + Si treatments were also included. For each species, there were six pots per Si treatment per irrigation treatment, which served as single pot replicates. Plants were well-watered at 10 centibars or water-stressed at 20 centibars using one drip emitter per pot. Tensiometers (IRROMETER, Riverside, CA, USA) were used to control irrigation by placing a single tensiometer in the middle of a bench in a pot at a depth of 10 cm, which resulted in no leachate. Plant species and Si treatments were randomized within irrigation, which served as blocks.

2.3. Harvesting and Measurements

Data collected on plants included height from the media surface to the tallest opened flower, width (average of two perpendicular measurements), shoot dry weight, number of flowers, flower diameter, and transpiration. Shoot dry weight was determined by cutting the stems at media level and drying for 2 d at 52.2 °C. For foliar nutrient analyses, mature leaves from the middle to upper level of the plant were collected from five plants per Si and irrigation treatment of each species. Soil and leaf nutrient analysis was performed by the Soil, Water, and Forage Analytical Laboratory (SWFAL) at Oklahoma State University, using a LECO TruSpec Carbon and Nitrogen Analyzer (LECO Corporation, St. Joseph, MI, USA). Soil and leaf Si analysis was performed, using the 0.5 M ammonium acetate method [20]. Transpiration was recorded weekly using a LI-1600 Steady State Porometer (LI-COR Inc., Lincoln, NE, USA).

2.4. Statistical Analysis

Pots were arranged in a randomized block design with irrigation serving as the block. Analysis of variance methods (PROC MIXED) were used with a two-factor factorial arrangement, with irrigation and silicon treatment as the factors of interest. Separate analyses were conducted for each of the plant species. When interactions of irrigation and Si treatment were significant, simple effects were reported. Mean separations were determined using a DIFF option in an LSMEANS statement and a SLICE option (when appropriate) and with a 0.05 level of significance.

3. Results

3.1. Dahlia xhybrida 'Dahlinova Montana'

A significant interaction of Si treatment with irrigation was seen for transpiration and soil Si (Table 1). Under the well-watered condition, soil Si was greatest when supplemented with 60 and 80 g TD, as well as 100, 150, and 200 g INC (Table 2). Transpiration was greatest in plants under the control as well as 40 and 60 g TD. Under the water-stressed condition, soil Si was greatest for INC plants compared to TD plants and MM + Si plants. Plants treated with 100 g or less DE within the INC treatment had the greatest transpiration.

Table 1. Analysis of variance for growth, flowering, leaf nutrient content, soil silica, and physiology of *Dahlia* × *hybrida* 'Dahlinova Montana' after application of diatomaceous earth (DE) and irrigation controlled with a tensiometer.

Source	Height (cm)	Width (cm)	Shoot Dry Weight (g)	Stem Diameter (cm)	Mean Flower Number	Flower Diameter (cm)
DE Treatment	* z	ns	**	***	ns	ns
Irrigation	****	****	****	***	****	****
DE Treatment × Irrigation	ns	ns	ns	ns	ns	ns

Source	N (%)	P (%)	S (%)	K (%)	Mg (%)	Ca (%)	Na (%)	Si (ppm)	Zn (ppm)	Cu (ppm)	Fe (ppm)	Mn (ppm)	Ni (ppm)
DE Treatment	**	*	*	ns	***	****	ns	****	ns	***	**	****	ns
Irrigation	ns	ns	ns	ns	ns	ns	ns	ns	ns	ns	***	ns	ns
DE Treatment × Irrigation	ns	ns	ns	ns	ns	ns	ns	ns	ns	ns	ns	ns	ns

Source	Transpiration	Soil silica (ppm)
DE Treatment	ns	****
Irrigation	****	****
DE Treatment × Irrigation	****	**

z NS, *, **, ***, **** indicates non-significant or significant at $p \leq 0.05$, 0.01, and 0.001, respectively.

A significant effect of irrigation was seen for all growth and flowering characteristics (Table 1). Well-watered plants had greater height, width, shoot dry weight, mean flower number, stem diameter, and flower diameter, compared to water-stressed plants (Table 3). A significant effect of DE treatment was seen for height, shoot dry weight, and stem diameter (Table 1). Height was greatest for control plants, all TD plants, INC plants at 100 and 200 g, as well as MM + Si plants (Table 4). Shoot dry weight was greatest for control plants, all TD plants, and INC plants at 100 g. Stem diameter was greatest for all TD plants, as well as INC plants at 50 and 100 g.

Table 2. Soil silica (Si), transpiration, and leaf Si affected by interaction of diatomaceous earth treatment with irrigation in *Dahlia × hybrida* 'Dahlinova Montana', *Gerbera jamesonii* 'Festival Light Eye White Shades', and *Rudbeckia hirta* 'Denver Daisy'.

Cultivar	Application and Rate (g) [z]	Well-Watered (10 cb)			Water-Stressed (20 cb)		
		Soil Si (ppm)	Transpiration	Leaf Si	Soil Si (ppm)	Transpiration	Leaf Si
Dahlinova Montana	0	47.8 e [y]	8.9 a	x	53.5 bcd	4.8 b	x
	TD 20	56.1 de	6.8 bc	x	49.5 d	4.5 b	x
	TD 40	59.7 bcd	9.0 a	x	51.5 cd	2.9 cd	x
	TD 60	65.8 abc	8.9 a	x	47.8 d	4.2 bc	x
	TD 80	65.1 abc	6.8 bc	x	49.7 d	3.3 cd	x
	INC 50	58.8 cd	7.0 b	x	62.0 ab	4.9 b	x
	INC 100	67.9 ab	5.2 d	x	59.3 abc	6.5 a	x
	INC 150	70.1 a	7.3 b	x	63.1 a	3.2 cd	x
	INC 200	72.7 a	5.7 cd	x	63.6 a	2.9 d	x
	MM + Si	53.1 de	6.7 bc	x	50.3 d	4.6 b	x
Festival Light Eye White Shades	0	43.6 bc	9.0	x	32.7 cd	7.3 abc	x
	TD 20	41.3 c	9.6	x	30.4 d	8.2 ab	x
	TD 40	45.5 bc	9.1	x	35.8 cd	8.5 ab	x
	TD 60	43.0 bc	10.4	x	48.2 b	8.5 ab	x
	TD 80	47.2 bc	9.7	x	47.6 b	8.7 a	x
	INC 50	42.3 c	8.9	x	41.5 bc	6.9 bc	x
	INC 100	50.4 bc	8.3	x	59.5 a	6.1 c	x
	INC 150	52.1 ab	9.2	x	59.4 a	7.0 bc	x
	INC 200	56.9 a	9.4	x	45.3 b	5.8 c	x
	MM + Si	43.6 bc	9.6	x	36.3 cd	7.2 abc	x
Denver Daisy	0	x	x	276.6 bc	x	x	345.7 bc
	TD 20	x	x	281.5 abc	x	x	373.4 bc
	TD 40	x	x	260.2 bc	x	x	357.8 bc
	TD 60	x	x	313.8 abc	x	x	412.0 ab
	TD 80	x	x	261.9 bc	x	x	321.3 bcd
	INC 50	x	x	292.3 abc	x	x	346.3 bc
	INC 100	x	x	377.4 ab	x	x	261.1 cd
	INC 150	x	x	196.4 cd	x	x	208.9 d
	INC 200	x	x	91.4 c	x	x	201.6 d
	MM + Si	x	x	406.5 a	x	x	522.4 a

[z] Top-dressed (TD), Incorporated (INC), and Metro-Mix media (MM). [y] Means (n = 6) with the same letter within the same column and within cultivar are not statistically different at $p \leq 0.05$. [x] Main effects were significant for factors.

Table 3. Growth and flowering characteristics affected by irrigation, controlled by a tensiometer, averaged across diatomaceous earth treatments in *Dahlia × hybrida* 'Dahlinova Montana', *Gerbera jamesonii* 'Festival Light Eye White Shades', and *Rudbeckia hirta* 'Denver Daisy'.

Cultivar	Irrigation Rate (cb)	Height (cm)	Width (cm)	Shoot Dry Weight (g)	Stem Diameter (cm)	Mean Flower Number	Flower Diameter (cm)
Dahlinova Montana	10	27.4 a [z]	29.7 a	19.8 a	4.3 a	8.4 a	6.8 a
	20	22.0 b	24.2 b	11.2 b	3.8 b	4.8 b	4.1 b
Festival Light Eye White Shades	10	12.3	23.6 a	6.2	1.1	0.9	1.4
	20	11.2	20.8 b	5.3	0.9	0.6	1.1
Denver Daisy	10	36.8 a	29.9 a	25.0 a	4.6 a	12.6 a	7.9 a
	20	22.0 b	24.2 b	11.2 b	3.7 b	4.8 b	4.1 b

[z] Means (n = 6) with the same letter within the same column and within cultivar are not statistically significant at $p \leq 0.05$.

Table 4. Growth and flowering characteristics affected by diatomaceous earth treatment averaged across irrigation, controlled by a tensiometer, in *Dahlia* × *hybrida* 'Dahlinova Montana', *Gerbera jamesonii* 'Festival Light Eye White Shades', and *Rudbeckia hirta* 'Denver Daisy'.

Cultivar	Application and Rate (g) [z]	Height (cm)	Width (cm)	Shoot Dry Weight (g)	Stem Diameter (cm)	Mean Flower Number	Flower Diameter (cm)
	0	24.9 ab [y]	26.9	15.8 a–d	3.8 bc	7.1	5.7
	TD 20	24.1 ab	29.1	18.9 a	4.5 a	8.0	5.8
	TD 40	24.8 ab	27.9	19.0 a	4.5 a	8.5	5.4
	TD 60	26.9 a	29.3	17.9 ab	4.3 ab	5.7	4.9
Dahlinova	TD 80	26.8 a	26.9	16.2 abc	4.5 a	7.0	5.3
Montana	INC 50	22.2 b	24.3	14.5 bcd	3.9 ac	5.9	4.9
	INC 100	26.6 a	27.5	15.2 a–d	4.3 ab	6.3	6.3
	INC 150	22.1 b	25.9	11.8 d	3.4 c	5.4	5.3
	INC 200	23.9 ab	25.1	12.6 cd	3.4 c	5.5	5.4
	MM + Si	24.9 ab	27.4	13.1 cd	3.8 bc	6.4	5.4
	0	8.9 b [y]	20.8 b	3.9 c	0.9	0.4	0.6
	TD 20	11.1 b	21.7 ab	5.9 bc	1.2	1.1	2.1
	TD 40	11.2 b	22.5 ab	4.9 bc	0.7	0.5	1.1
Festival	TD 60	10.4 b	22.4 ab	5.4 bc	0.6	0.3	0.2
Light Eye	TD 80	11.8 ab	20.0 b	5.9 bc	0.3	0.8	1.2
White	INC 50	12.1 ab	21.6 ab	5.7 bc	1.3	0.6	1.0
Shades	INC 100	15.1 a	25.2 a	8.6 a	1.5	1.2	2.8
	INC 150	15.3 a	24.9 a	7.2 ab	1.5	1.0	2.1
	INC 200	10.8 b	23.1 ab	5.0 bc	1.5	1.0	1.6
	MM + Si	10.8 b	19.5 b	4.9 bc	0.8	0.8	0.6
	0	30.8 bc [y]	23.0 bc	12.8	4.3	6.4	6.2 b
	TD 20	31.1 bc	26.9 ab	17.6	4.4	10.5	5.4 bc
	TD 40	30.9 bc	26.9 ab	18.1	3.9	9.8	5.5 bc
	TD 60	34.8 abc	27.2 ab	23.4	4.7	11.1	5.9 bc
Denver	TD 80	30.9 bc	28.2 a	24.2	4.5	9.7	6.4 b
Daisy	INC 50	32.2 bc	26.1 abc	18.2	4.2	9.6	6.5 ab
	INC 100	46.4 a	27.8 a	19.9	3.8	8.9	5.9 bc
	INC 150	25.3 c	21.9 c	15.5	4.0	7.0	4.9 bc
	INC 200	27.4 bc	23.1 bc	16.1	3.7	8.5	4.5 a
	MM + Si	39.2 ab	29.2 a	20.2	4.3	11.1	8.1 a

[z] Top-dressed (TD), Incorporated (INC), and Metro-Mix media (MM). [y] Means (n = 6) with the same letter within the same column and within cultivar are not statistically different at $p \leq 0.05$.

A significant effect of DE treatment was seen for leaf nutrient content (Table 1). Total nitrogen (N) was greatest for all TD plants and INC plants with rates of 100, 150, and 200 g (Table 5). TD plants at 80 g and INC plants at 100, 150, and 200 g had the greatest values of phosphorus (P). Magnesium (Mg) was greatest for TD plants at 40 and 80 g, as well as INC plants at 50, 100, and 150 g. Calcium (Ca) was greatest for INC plants at 50, 100, and 150 g, as well as MM + Si plants. Sulfur (S) was greatest for all TD plants and INC plants at 100, 150, and 200 g (Table 6). Silica was greatest for control plants, TD plants at 20, 40, and 60 g, INC plants at 100 and 150 g, as well as MM + Si plants. For copper (Cu), the greatest values were seen for TD plants at 40 and 80 g. Iron (Fe) was greatest for TD plants at 40, 60, and 80 g, as well as INC plants at 100 and 150 g. Manganese (Mn) was greatest for INC plants at 150 and 200 g.

Table 5. Leaf macronutrient content affected by diatomaceous earth treatment across irrigation, controlled by a tensiometer, in *Dahlia × hybrida* 'Dahlinova Montana', *Gerbera jamesonii* 'Festival Light Eye White Shades', and *Rudbeckia hirta* 'Denver Daisy'.

Cultivar	Application and Rate (g) [z]	N (%)	P (%)	K (%)	Mg (%)	Ca (%)
Dahlinova Montana	0	3.55 bcd [y]	0.29 c	3.51	0.92 bc	1.74 de
	TD 20	3.88 ab	0.33 bc	3.33	0.91 bc	1.81 cd
	TD 40	4.11 ab	0.34 bc	3.19	0.97 ab	1.75 de
	TD 60	4.09 ab	0.33 bc	3.23	0.84 c	1.61 e
	TD 80	3.83 ab	0.34 abc	3.47	0.96 ab	1.85 bcd
	INC 50	3.21 cd	0.29 c	3.48	0.98 ab	1.93 abc
	INC 100	3.75 abcd	0.35 abc	3.54	1.03 a	1.99 ab
	INC 150	3.79 abc	0.38 ab	3.81	0.98 ab	2.00 ab
	INC 200	4.27 a	0.39 a	3.54	0.92 bc	1.88 bcd
	MM + Si	3.19 d	0.33 bc	3.44	0.82 c	2.07 a
Festival Light Eye White Shades	0	2.86 [y]	0.26	2.63	0.62	1.45
	TD 20	2.92	0.31	3.18	0.66	1.55
	TD 40	3.05	0.28	3.30	0.62	1.48
	TD 60	2.93	0.46	3.06	0.77	2.12
	TD 80	3.14	0.54	3.31	0.79	2.02
	INC 50	2.74	0.29	3.10	0.73	1.71
	INC 100	2.82	0.27	3.06	0.62	1.42
	INC 150	2.95	0.26	3.03	0.55	1.26
	INC 200	2.98	0.26	2.97	0.56	1.33
	MM + Si	2.99	0.37	3.10	0.70	1.88
Denver Daisy	0	2.40 de [y]	0.20 e	3.09 d	1.21 a	3.35 b
	TD 20	2.79 cde	0.22 cde	3.31 bcd	1.20 ab	3.33 b
	TD 40	2.98 bc	0.26 bcd	3.53 bcd	1.10 ab	2.89 bc
	TD 60	2.94 bc	0.23 cde	3.25 cd	1.13 ab	3.28 bc
	TD 80	2.87 cd	0.21 de	3.29 bcd	1.12 ab	3.17 bc
	INC 50	2.82 cde	0.24 b–e	3.27 bcd	1.15 ab	3.27 bc
	INC 100	2.96 bc	0.27 abc	3.74 ab	1.12 ab	3.26 bc
	INC 150	3.38 ab	0.28 ab	3.72 abc	1.09 bc	3.26 bc
	INC 200	3.51 a	0.31 a	4.13 a	0.97 cd	2.79 c
	MM + Si	2.37 e	0.21 de	3.14 d	0.96 d	4.01 a
Optimum levels [x]		2.50–4.50	0.20–0.75	1.50–5.50	0.25–1.00	1.00–4.00

[z] Top-dressed (TD), incorporated (INC), and Metro-Mix media (MM). [y] Means (n=6) with the same letter within the same column and within cultivars are not statistically significant at $p \leq 0.05$. [x] According to Kalra (26).

3.2. Gerbera jamesonii 'Festival Light Eye White Shades'

A significant interaction of DE with irrigation was seen for soil Si and transpiration (Table 7). Under the well-watered condition, soil Si was greatest when DE was supplemented at 150 and 200 g INC (Table 2). Under the water-stressed condition, soil Si was greatest for 100 and 150 g INC plants. Transpiration was greatest for control plants, TD plants, and MM + Si plants. A significant effect of irrigation was seen for width and leaf nutrient content (Table 7). Well-watered plants had greater widths compared to water-stressed plants (Table 3). Potassium, Ca, sodium (Na), and Mn levels were greater in water-stressed plants compared to well-watered plants (Table 8).

Table 6. Leaf micronutrient and trace element content affected by diatomaceous earth treatment across irrigation, controlled by a tensiometer, in *Dahlia* × *hybrida* 'Dahlinova Montana', *Gerbera jamesonii* 'Festival Light Eye White Shades', and *Rudbeckia hirta* 'Denver Daisy'.

Cultivar	Application and Rate (g) [z]	S (%)	Na (%)	Si (ppm)	Zn (ppm)	Cu (ppm)	Mn (ppm)	Ni (ppm)	Fe (ppm)
Dahlinova Montana	0	0.33 bc [y]	0.02	84.1 a	38.5 a	12.0 cd	151.6 d	0.0	92.2 d
	TD 20	0.39 ab	0.03	69.1 abc	46.6 a	14.4 cd	149.3 d	0.0	106.6 cd
	TD 40	0.39 ab	0.03	77.3 ab	53.6 a	18.2 a	163.9 cd	0.0	232.7 a
	TD 60	0.35 abc	0.02	69.7 abc	39.5 a	14.4 cd	150.2 d	0.0	161.7 a–d
	TD 80	0.41 a	0.02	38.1 c	44.9 a	17.5 ab	188.8 bc	0.0	171.5 abc
	INC 50	0.34 bc	0.02	47.9 bc	45.2 a	14.8 bcd	189.9 bc	0.0	133.8 bcd
	INC 100	0.39 ab	0.03	82.9 a	57.9	15.1 bc	204.6 b	0.0	188.6 ab
	INC 150	0.37 ab	0.03	67.5 abc	42.2	14.1 cd	272.1 a	0.1	163.1 a–d
	INC 200	0.37 ab	0.02	45.8 bc	39.9	13.3 cd	250.4 a	0.0	104.5 cd
	MM + Si	0.31 c	0.03	94.3 a	42.5	11.9 d	194.7 b	0.3	97.1 d
Festival Light Eye White Shades	0	0.40 [y]	0.07	107.6	46.6	24.8	145.4	2.29 a	591.1
	TD 20	0.48	0.08	253.4	56.6	31.5	129.8	0.128 b	392.1
	TD 40	0.39	0.07	185.2	52.8	20.9	129.2	0.002 b	235.6
	TD 60	1.01	0.16	308.0	84.1	133.8	156.9	0.191 b	390.6
	TD 80	1.03	0.14	264.7	115.2	171.8	190.8	0.066 b	566.7
	INC 50	0.42	0.07	223.7	57.9	14.4	165.3	2.12 a	649.9
	INC 100	0.38	0.12	263.2	51.8	20.3	161.9	2.23 a	736.4
	INC 150	0.32	0.07	172.1	43.9	11.6	161.8	0.103 b	305.9
	INC 200	0.31	0.06	163.9	43.6	10.5	219.1	0.131 b	291.1
	MM + Si	0.62	0.08	308.1	79.2	67.9	198.8	1.48 ab	742.7
Denver Daisy	0	0.41 cd	0.02	x	39.7	6.03 de	137.6 bc	0.002	97.9 c
	TD 20	0.49 abc	0.02	x	40.4	8.0 b–e	131.7 bcd	0.002	130.9 bc
	TD 40	0.48 abc	0.02	x	42.7	8.9 b–e	114.4 d	0.002	134.5 bc
	TD 60	0.52 ab	0.04	x	39.8	8.6 a–d	125.7 cd	0.003	132.6 bc
	TD 80	0.44 bcd	0.04	x	33.1	7.5 cde	129.5 cd	0.462	91.7 c
	INC 50	0.55 a	0.02	x	44.2	10.9 ab	137.1 bc	0.533	209.2 ab
	INC 100	0.53 ab	0.03	x	36.8	10.0 abc	152.1 b	0.308	168.8 bc
	INC 150	0.48 abc	0.03	x	37.6	7.6 cde	208.7 a	0.145	139.5 bc
	INC 200	0.50 abc	0.03	x	43.6	11.4 a	203.8 a	0.575	285.2 a
	MM + Si	0.34 d	0.01	x	37.1	5.1e	140.3 bc	0.002	105.9 bc
Optimum levels		0.2–0.8	w	w	27–100	5.0–30.0	20–300	0–5	100–500

[z] Top-dressed (TD) or incorporated (INC). [y] Means (n = 6) with the same letter within the same column and within cultivars are not statistically significant at $p \leq 0.05$. [x] Significant interactions between diatomaceous earth and irrigation reported in another table. [w] Optimum levels not reported.

Table 7. Analysis of variance for growth, flowering, leaf nutrient content, soil silica, and physiology of *Gerbera jamesonii* 'Festival Light Eye White Shades' with application of diatomaceous earth (DE) and irrigation controlled with a tensiometer.

Source	Height (cm)	Width (cm)	Shoot Dry Weight (g)	Stem Diameter (cm)	Mean Flower Number	Flower Diameter (cm)
DE Treatment	* [z]	*	*	ns	ns	ns
Irrigation	ns	**	ns	ns	ns	ns
DE Treatment × Irrigation	ns	Ns	ns	ns	ns	ns

Source	N (%)	P (%)	S (%)	K (%)	Mg (%)	Ca (%)	Na (%)	Si (ppm)	Zn (ppm)	Cu (ppm)	Fe (ppm)
DE Treatment	ns	ns	ns	ns	ns	ns	ns	ns	ns	ns	ns
Irrigation	ns	ns	ns	ns	*	*	**	*	ns	ns	ns
DE Treatment × Irrigation	ns	ns	ns	ns	ns	ns	ns	ns	ns	ns	ns

Source	Transpiration	Soil silica (ppm)
DE Treatment	ns	****
Irrigation	****	*
DE Treatment × Irrigation	**	**

[z] NS, *, **, ***, **** indicates non-significant or significant at $p \leq 0.05$, 0.01, and 0.001, respectively.

Table 8. Leaf nutrient content affected by irrigation, controlled by a tensiometer, across diatomaceous earth treatments in *Dahlia* × *hybrida* 'Dahlinova Montana', *Gerbera jamesonii* 'Festival Light Eye White Shades', and *Rudbeckia hirta* 'Denver Daisy'.

Cultivar	Application and Rate (g) [z]	N (%)	P (%)	S (%)	K (%)	Mg (%)	Ca (%)	Na (%)	Si (ppm)	Zn (ppm)
Dahlinova	10	3.73 [z]	0.35	0.37	3.39	0.93	1.83	0.03	68.1	46.1
Montana	20	3.79	0.33	0.35	3.52	0.94	1.89	0.03	67.2	44.1
Festival Light Eye	10	2.89 [z]	0.27	0.41	3.01 b	0.61	1.45 b	0.07 b	184.2	51.5
White Shades	20	2.99	0.24	0.66	3.14 a	0.72	1.79 a	0.12 a	265.6	74.8
Denver Daisy	10	2.88 [z]	0.25	0.43 b	3.48	1.03 b	2.82 b	0.03	y	37.5
	20	2.92	0.24	0.52 a	3.42	1.18 a	3.70 a	0.22	y	41.5
Optimum levels [x]		2.50–4.50	0.20–0.75	0.25–1.00	1.50–5.50	0.25–1.00	1.00–4.00	w	w	27.0–100.0

[z] Means (n = 6) with the same letter within the same column and within cultivar are not statistically significant $p \leq 0.05$. [y] Significant interaction between diatomaceous earth treatment and irrigation. [x] According to Kalra (26). [w] Optimum levels not reported.

A main effect of DE was seen for height, width, shoot dry weight, and nickel (Ni) content in the leaf tissue (Table 7). Height was greatest for 80 g TD plants, as well as 50, 100, and 150 g INC plants (Table 4). Width was greatest for TD plants at 20, 40, and 60 g, and all INC plants. Shoot dry weight was greatest for INC plants at 100 and 150 g. Nickel was greatest for control plants, INC plants at 50 and 100 g, as well as MM + Si plants (Table 6).

3.3. Rudbeckia hirta 'Denver Daisy'

A significant interaction of DE with irrigation was seen for leaf Si content (Table 9). Under the well-watered condition, silica in the leaf was greatest for 20 and 60 g TD plants, 50 and 100 g INC plants, as well as MM + Si plants (Table 2). Under the water-stressed condition, Si in the leaf was greatest for 60 g TD plants and MM + Si plants.

Table 9. Analysis of variance for growth, flowering, leaf nutrient content, soil silica, and physiology of *Rudbeckia hirta* 'Denver Daisy' after application of diatomaceous earth (DE) and irrigation controlled with a tensiometer.

Source	Height (cm)	Width (cm)	Shoot Dry Weight (g)	Stem Diameter (cm)	Mean Flower Number	Flower Diameter (cm)
DE Treatment	* [z]	*	ns	ns	ns	**
Irrigation	**	****	****	***	****	****
DE Treatment × Irrigation	ns	ns	ns	ns	ns	ns

Source	N (%)	P (%)	S (%)	K (%)	Mg (%)	Ca (%)	Na (%)	Si (ppm)	Zn (ppm)	Cu (ppm)	Fe (ppm)	Mn (ppm)	Ni (ppm)
DE Treatment	****	****	***	***	***	**	ns	****	ns	ns	*	****	ns
Irrigation	ns	ns	ns	ns	****	****	ns	****	ns	**	ns	*	ns
DE Treatment × Irrigation	ns	ns	ns	ns	ns	ns	ns	***	ns	ns	ns	ns	ns

Source	Transpiration	Soil silica (ppm)
DE Treatment	ns	****
Irrigation	****	****
DE Treatment × Irrigation	ns	ns

[z] NS, *, **, ***, **** indicates non-significant or significant at $p \leq 0.05$, 0.01, and 0.001, respectively.

A significant effect of irrigation was seen for all growth and flower characteristics, as well as leaf nutrient content, transpiration, and soil Si (Table 9). Plants that were well-watered grew taller and wider, and had greater shoot dry weight, stem diameter, flower number, and flower diameter (Table 3). The nutrients S, Mg, Ca, and Mn were greater in the leaf tissue of water-stressed plants, compared to well-watered plants (Table 8). Soil Si and transpiration were greater in plants under the water-stressed condition, compared to those that were well-watered (Table 10). A main effect of DE treatment was seen for height, width, and flower diameter (Table 9). Height was greatest for 60 g TD plants, 100 g INC plants, and MM + Si plants (Table 4). Width was greatest for all TD plants, INC plants at 50 and 100 g, and MM + Si plants. Flower diameter was greatest for 50 g INC plants and MM + Si plants.

Table 10. Soil silica and transpiration of *Rudbeckia hirta* 'Denver Daisy' affected by irrigation, controlled by a tensiometer, averaged across silica treatments.

Source	Irrigation Rate (cb)	Soil Si (ppm)	Transpiration
Well-watered	10	35.8 b [z]	4.56 b
Water-stressed	20	47.7 a	8.10 a

[z] Means (n = 6) with the same letter within the same column are not statistically significant at $p \leq 0.05$.

4. Discussion

Amending the soilless substrate with varying rates of DE by top-dressing or by incorporating into the substrate increased plant height, width, shoot dry weight, stem diameter, and flower diameter in dahlia 'Dahlinova Montana', daisy 'Festival Light Eye White Shades', and black-eyed Susan 'Denver Daisy', in this study. Several other studies have reported similar benefits of supplemental Si on growth and flowering characteristics. Hwang et al. [21] reported that adding 200 mg L^{-1} of potassium (K) metasilicate increased plant height and shoot dry weight in cut roses. Stem quality was also improved in cut roses when Si was added to a recirculated nutrient solution in a closed hydroponic system [22]. Flower diameter of calibrachoa (*Calibrachoa* × *hybrida* Cerv.), fuchsia (*Fuchsia hybrid* hort. Ex Siebold & Voss), and petunia (*Petunia* × *hybrida* Vilm.) increased when supplemented with a weekly drench of K silicate at 100 mg L^{-1} [23]. Silica supplementation improved growth of two cultivars of French marigolds (*Tagetes patula* L.) by increasing stem diameter, shoots, and dry weights [24]. Growth and biomass parameters were increased in begonia (*Begonia semperflorens* Link et Otto) and pansy (*Viola* × *wittrockiana* Hort.) grown in vitro when supplemented with K silicate [25]. Savvas et al. [13] reported a greater percentage of flowers in hydroponically-grown gerbera (*Gerbera jamesonii*) supplemented with Si.

Amending the soilless substrate with DE increased nutrient content, despite being inert. Based on the analysis of Kalra [26], most nutrients were within the optimum range adequate for plant growth and levels greater than the maximum range were not considered excess or toxic. Nickel concentrations in dahlia 'Dahlinova Montana' were less than the minimum range (Table 4). However, these levels were not considered insufficient because often there are no symptoms to accurately determine Ni deficiency [27]. Epstein [1] has noted that the presence of Si does, in fact, affect absorption and translocation of several macro-nutrients and micro-nutrients. Early studies conducted by Fisher [28] reported that the addition of Si made P more available in barley (*Hordeum vulgare* L.). Mali and Aery [29] found that, in wheat (*Tritium aestivum* L.), potassium uptake was improved even at low concentrations of Si by H-ATPase being activated. Phosphorus and K are essential nutrients for flowering characteristics. Friedman et al. [30] conducted a study on sunflower (*Helianthus annuus* L.) and celosia (*Celosia argentea* L.), and reported that growth and flower parameters were increased when supplemented with an effluent containing high amounts of N, P, and other nutrients. Kamenidou et al. [7–9] also found an increase in N for sunflowers and gerbera, but most of the levels exceeded the optimum range. Nitrogen metabolism is a major factor in stem and leaf growth and too much can delay or prevent flowering. Calcium is part of the structure of cell walls and is necessary for cell growth and division. Ma and Takahashi [31] reported that there was an antagonistic effect between Si and Ca in rice, in which one can decrease the amount of the other. However, our study found the opposite effect, in which MM + Si increased Ca content in dahlia 'Dahlinova Montana' and 'Denver Daisy'. Kamenidou et al. [9] and Savvas et al. [13] also reported that supplemental Si increased Ca within gerbera. There was an increase in metals such as Cu, Fe, and Mn in dahlia 'Dahlinova Montana' and daisy 'Festival Light Eye White Shades', due to DE having trace amounts of these elements [18]. Silica levels in the leaf tissue and media for all the plants were observed in low amounts. Potentially, the plants could be classified as non-accumulators of Si (<0.5%) which has been reported for gerbera [32].

Amending the soilless substrate with varying methods and rates of DE showed mixed results on transpiration in all three species. Improvements in this physiological trait were mostly seen when Si (DE) supplementation interacted with irrigation. Kamenidou et al. [33] found that a foliar spray of Na silicate at 100 mg L^{-1} decreased transpiration in zinnias (*Zinnia* L.). Yoshida and Kitagishi [34] noted that the effects are related to Si being deposited in the cuticular layers of leaves, serving as a barrier which reduces the loss of water. A decrease in transpiration can benefit the floricultural market by improving quality and shelf life of cut flowers [35]. Considering the effect of irrigation, plants under the well-watered condition had greater growth and flowering, which was expected. However, Si is known to maintain the growth and flowering characteristics as well as nutrient levels in water-stressed plants. In Kentucky bluegrass (*Poa pratensis* L.), drought stress hindered physiological and quality attributes, but application of Si alleviated the adverse effects [36].

5. Conclusions

Several growth and flowering characteristics were improved, depending on the rate and application method, by application of DE. Benefits of DE included increased height, width, shoot dry weight, stem, and flower diameter. An increase in nutrients, such as N, P, K, Mg, and Ca, was seen mostly for dahlia 'Dahlinova Montana' and black-eyed Susan 'Denver Daisy'. The adverse effects that typically occur under water-stressed conditions were alleviated and plant quality, as well as transpiration, was maintained in all three plants due to Si supplementation. Silicon is known to play an important role in cell membrane integrity, in which osmosis, photosynthesis, and transpiration all occur. Diatomaceous earth as supplemental Si was beneficial for plant growth, flowering, and nutrient content under both well-watered and water-stressed conditions. For growth and flower characteristics, MM + Si was similar to the control (MM) with no added silica, and equivalent to most treatments with supplemental silica. To conclude, this research supports that DE, one of the many Si sources, is beneficial to plants; however, this is dependent upon species, Si rate, and the method of application. Benefits of DE include an increase in growth parameters, leaf nutrient content, and tolerance to stress, in which plant quality can be maintained. Future studies should further assess the use of DE on other crops and stress conditions.

Author Contributions: Conceptualization, B.D.; methodology, T.M.-I.; formal analysis, M.P.; investigation, T.M.-I.; writing—original draft preparation, T.M.-I.; writing—review and editing, B.D and N.M.; supervision, B.D.

Funding: This research was funded by ODAFF Specialty Block grant #G00010135.

Acknowledgments: We thank Stephen Stanphill for helping with data collection and greenhouse management.

Conflicts of Interest: The authors declare no conflict of interest.

References

1. Epstein, E. The anomaly of silicon in plant biology. *Proc. Natl. Acad. Sci. USA* **1994**, *91*, 11–17. [CrossRef] [PubMed]
2. Ma, J.F.; Yamaji, N. Silicon uptake and accumulation in higher plants. *Plant Sci.* **2006**, *11*, 392–397. [CrossRef] [PubMed]
3. Chérif, M.; Asselin, A.; Bélanger, R.R. Defense responses induced by soluble silicon in cucumber roots infected by *Pythium* spp. *Phytopathology* **1994**, *84*, 236–242. [CrossRef]
4. Ma, J.F. Role of silicon in enhancing the resistance of plants to biotic and abiotic stresses. *Soil Sci. Plant Nutr.* **2004**, *50*, 11–18. [CrossRef]
5. Liang, Y.; Sun, W.; Zhu, Y.G.; Christie, P. Mechanisms of silicon-mediated alleviation of abiotic stresses in higher plants: A review. *Environ. Pollut.* **2006**, *147*, 422–428. [CrossRef] [PubMed]
6. Voogt, W.; Sonneveld, C. *Silicon in Agriculture*; Elsevier: Amsterdam, The Netherlands, 2011.

7. Kamenidou, S.; Cavins, T.J.; Marek, S. Silicon supplements affect horticultural traits of greenhouse-produced ornamental sunflowers. *HortScience* **2008**, *43*, 236–239. [CrossRef]

8. Miyake, Y.; Takahashi, E. Silicon deficiency of tomato plant. *Soil Sci. Plant Nutr.* **1987**, *24*, 175–189. [CrossRef]

9. Kamenidou, S.; Cavins, T.J.; Marek, S. Silicon supplements affect floricultural quality traits and elemental nutrient concentrations of greenhouse produced gerbera. *Sci. Hortic.* **2010**, *123*, 390–394. [CrossRef]

10. De Kreij, C.; Voogt, W.; Baas, R. Nutrient solutions and water quality for soilless cultures. In *Research Station for Floriculture and Glasshouse Vegetables Brochure*; Naaldwijk Office: Naaldwijk, The Netherlands, 1999.

11. Reezi, S.; Babalar, M.; Kalantari, S. Silicon alleviates salt stress, decreases malondialdehyde content and affects petal color of salt stressed cut rose (*Rosa xhybrida* L.) Hot Lady. *Afr. J. Biotechnol.* **2009**, *8*, 1502–1508.

12. Carvalho-Zanao, M.P.; LAZ, J.; Barbosa, J.G.; Grossi, J.A.S.; Ávila, V.T. Yield and shelf life of *Chrysanthemum* in response to the silicon application. *Hortic. Bras.* **2012**, *30*, 403–408. [CrossRef]

13. Savvas, D.; Manos, G.; Kotsiras, A.; Souvaliotis, S. Effects of silicon and nutrient-induced salinity on yield, flower quality and nutrient uptake of gerbera grown in a closed hydroponic system. *J. Appl. Bot. Food Qual.* **2002**, *76*, 153–158.

14. Berthelsen, S.; Noble, A.D.; Kingston, G.; Hurney, A.; Rudd, A.; Garside, A. Improving Yield and ccs in Sugarcane through the Application of Silicon-Based Amendments; Final Report on SRDC Project CLW009. 2003. Available online: http://hdl.handle.net/11079/12957 (accessed on 9 January 2019).

15. Muir, S. Plant-Available Silicon (Si) as A Protectant Against Fungal Diseases in Soil-Less Potting Media. Available online: https://www.ngia.com.au/Story?Action=View&Story_id=1782 (accessed on 9 January 2019).

16. Savant, N.K.; Snyder, G.H.; Datnoff, L.E. Silicon management and sustainable rice production. *Adv. Agron.* **1996**, *58*, 151–199.

17. Meerow, A.W.; Broschat, T.K. Growth of *Hibiscus* in media amended with a ceramic diatomaceous earth granule and treated with a kelp extract. *HortTechnology* **1996**, *6*, 70–73. [CrossRef]

18. Pati, S.; Pal, B.; Badole, S.; Hazra, G.C.; Mandal, B. Effect of silicon fertilization on growth, yield, and nutrient uptake of rice. *Commun. Soil Sci. Plant Anal.* **2016**, *47*, 284–290. [CrossRef]

19. King, P.A.; Reddy, S. Soilless Growth Medium Including Soluble Silicon. Available online: https://patents.google.com/patent/US6074988A/en (accessed on 9 January 2019).

20. Jim, J.W.; Dolda, S.K.; Henderson, R.E. Soil silicon extractability with seven selected extractants in relation to colorimetric and ICP determination. *Soil Sci.* **2004**, *169*, 861–870.

21. Hwang, S.J.; Park, H.M.; Jeong, B.R. Effects of potassium silicate on the growth of miniature rose Pinnochio grown on rockwool and its cut flower quality. *J. Jpn. Soc. Hortic. Sci.* **2005**, *74*, 242–247. [CrossRef]

22. Ehret, D.L.; Menzies, J.G.; Helmer, T. Production and quality of greenhouse roses in recirculating nutrient systems. *Sci. Hortic.* **2005**, *106*, 103–113. [CrossRef]

23. Mattson, N.S.; Leatherwood, W.R. Potassium silicate drenches increase leaf silicon content and affect morphological traits of several floriculture crops grown in a peat-based substrate. *HortScience* **2010**, *45*, 43–47. [CrossRef]

24. Sivanesan, I.; Son, M.S.; Lee, J.P.; Jeong, B.R. Effects of silicon growth of *Tagetes patula* L. 'Boy Orange' and 'Yellow Boy' seedlings cultured in an environment-controlled chamber. *Propag. Ornam. Plants* **2010**, *10*, 136–140.

25. Lim, M.Y.; Lee, E.J.; Jana, S.; Sivanesan, I.; Jeong, B.R. Effect of potassium silicate on growth and leaf epidermal characteristics of Begonia and Pansy grown in vitro. *Hortic. Sci. Technol.* **2012**, *30*, 579–585. [CrossRef]

26. Kalra, Y. *Handbook of Reference Methods for Plant Analysis*; CRC Press: Boston, MA, USA, 1998.

27. Buechel, T. Role of Nickel in Plant Culture. Available online: http://www.pthorticulture.com/en/training-center/role-of-nickel-in-plant-culture/ (accessed on 9 January 2019).

28. Fisher, R.A. A preliminary note on the effect of sodium silicate in increasing the yield of barley. *J. Agric. Sci.* **1929**, *19*, 132–139. [CrossRef]

29. Mali, M.; Avery, N.C. Influence of silicon on growth, relative water contents and uptake of silicon, calcium and potassium in wheat grown in nutrient solution. *J. Plant Nutr.* **2008**, *31*, 1867–1876. [CrossRef]

30. Friedman, H.; Bernstein, N.; Bruner, M.; Rot, I.; Ben-Noon, Z.; Zuriel, A.; Zuriel, R.; Finklestein, S.; Umiel, N.; Hagiladi, A. Application of secondary-treated effluents for cultivation of sunflower (*Helianthus annus* L.) and celosia (*Celosia argentea* L.) as cut flowers. *Sci. Hortic.* **2007**, *115*, 62–69. [CrossRef]

31. Ma, J.F.; Takahashi, E. Interaction between calcium and silicon in water-cultured rice plants. *Plant Soil* **1993**, *148*, 107–113. [CrossRef]

32. Bloodnick, E. Role of Silicon in Plant Culture. Available online: http://www.pthorticulture.com/en/training-center/role-of-silicon-in-plant-culture/ (accessed on 9 January 2019).

33. Kamenidou, S.; Cavins, T.J.; Marek, S. Evaluation of silicon as a nutritional supplement for greenhouse zinnia production. *Sci. Hortic.* **2009**, *119*, 297–301. [CrossRef]

34. Yoshida, S.; Ohnishi, Y.; Kitagishi, K. Chemical forms, mobility and deposition of silicon in rice plant. *Soil Sci. Plant Nutr.* **1962**, *8*, 15–21. [CrossRef]

35. Jana, S.; Jeong, B.R. Silicon: The most under-appreciated element in horticultural crops. *Hortic. Res.* **2014**, *4*, 1–19.

36. Saud, S.; Li, X.; Chen, Y.; Zhang, L.; Fahad, S.; Hussain, S.; Sadiq, A.; Chen, Y. Silicon application increases drought tolerance of Kentucky bluegrass by improving plant water relations and morphophysiological function. *Sci. World J.* **2014**, *23*, 17647–17655. [CrossRef] [PubMed]

 *horticulturae*

Review

Irrigation of Greenhouse Crops

Georgios Nikolaou [1],*, Damianos Neocleous [2], Nikolaos Katsoulas [1] and Constantinos Kittas [1]

[1]   Department of Agriculture Crop Production and Rural Environment, School of Agricultural Sciences, University of Thessaly, Fytokou Str., 38446 Volos, Greece; nkatsoul@gmail.com (N.K.); ckittas@uth.gr (C.K.)

[2]   Department of Natural Resources and Environment, Agricultural Research Institute, 1516 Nicosia, Cyprus; d.neocleous@ari.gov.cy

*   Correspondence: gnicolaounic@gmail.com; Tel.: +30-24-2109-3249

Received: 29 November 2018; Accepted: 29 December 2018; Published: 15 January 2019

Abstract: Precision agricultural greenhouse systems indicate considerable scope for improvement of irrigation management practices, since growers typically irrigate crops based on their personal experience. Soil-based greenhouse crop irrigation management requires estimation on a daily basis, whereas soilless systems must be estimated on an hourly or even shorter interval schedule. Historically, irrigation scheduling methods have been based on soil or substrate monitoring, dependent on climate or time with each having both strengths and weaknesses. Recently, plant-based monitoring or plant reflectance-derived indices have been developed, yet their potential is limited for estimating the irrigation rate in order to apply proper irrigation scheduling. Optimization of irrigation practices imposes different irrigation approaches, based on prevailing greenhouse environments, considering plant-water-soil relationships. This article presents a comprehensive review of the literature, which deals with irrigation scheduling approaches applied for soil and soilless greenhouse production systems. Irrigation decisions are categorized according to whether or not an automatic irrigation control has the ability to support a feedback irrigation decision system. The need for further development of neural networks systems is required.

Keywords: accumulated radiation method; feedback irrigation system; fuzzy control system; irrigation dose; precision irrigation; phyto-sensing; soilless culture; transpiration; water use efficiency

---

1. Introduction

The concept of "precision agriculture" is used to define technologies that support customized agricultural practices aimed at higher efficiency and a lower impact on the environment [1]. Greenhouse production systems decrease crop water requirements by as much as 20% to 40% compared to open field cultivation; however, growers routinely apply more irrigation water than the estimated water consumption [2–4]. Irrigation practices are generally based on the personal perspective of the grower; i.e., irrigation without monitoring the soil or plant water status [5]. Considering the number of different plant species grown in prevailing greenhouse environments, the types of substrate and container sizes, field and soil characteristics, and the different irrigation systems, it becomes obvious why irrigation scheduling becomes complex if it is to be achieved with any level of precision [6–8]. Therefore, an accurate short term estimation of crop water requirements in protected cultivation are a prerequisite for optimal irrigation scheduling; as evapotranspiration (ET_C) could occur so rapidly that water loss can cause plant damage before wilting symptoms become visible [9,10]. Irrigation management is typically expected to achieve maximum water supply for plant growth and production, with soil or substrate water content being maintained close to field capacity [11].

Even in soilless cultivation systems, irrigation represents a very large and potentially important loss of nutrients and a source of environmental pollution (i.e., drain to waste hydroponics systems) as a surplus of 20% to 50% of the plant's water uptake in each irrigation cycle is often recommended [12–16].

Indeed, annual use of irrigation water ranges from 150 to 200 mm (e.g., leafy vegetable) in soil-based greenhouse crops to 1000 to 1500 mm in soilless-grown (e.g., Solanaceae, cucurbits) [11]. For container nursery production, as cited by Fulcher et al. [17], those values could be as high as 2900 mm.

Considering the scarcity of water resources combined with the operational energy irrigation costs, maintaining the sustainable use of water is a major water-climate policy challenge since excessive irrigation results in low water use efficiency, increases in runoff and contributes to higher CO_2 emissions [18–20]. Several institutions have worked to improve water use in irrigation, developed various models of water efficiency, reducing the environmental problems associated with irrigation in order to mitigate severe structural water deficits, yet these models are not commercialized [4,21].

This paper presents a review of the literature dealing with irrigation of greenhouse crops. The necessity, the advantages and the limitations of each irrigation approach used are discussed in relation to different greenhouse types and the ability of an irrigation controller unit to support a feedback irrigation decision system.

2. Background

The exact time and volume of irrigation are probably the most important factors for efficient irrigation management and saving water, and these in turn also improve the productivity and quality of crops grown in the greenhouse [22,23]. This is especially true as the high potential efficiency of fertigation (i.e., irrigation combined with fertilization) has become a routine cultural practice, therefore the terms "irrigation" and "fertigation" are often used interchangeably [24–27]. Yet, irrigation management of substrate-based greenhouse crops still requires much more accurate control than for the same crop grown in soil, taking into account that substrates have very little nutrient buffering capacity [28].

Soilless growth systems in readily made artificial media commonly use organic (i.e., coconut coir, peat moss, pine bark) or inert substrates (i.e., perlite, rockwool, vermiculite); with substrate volume at approximately 10 to 40 L m^{-2} as is the case of rockwool or perlite slabs [29,30]. Horticultural production has historically been increasingly based on those ready-made substrates produced on an industrial scale with unique characteristics such as a limited cation exchange and low buffering capacity, good water permeability and adequate aeration [31]. Compared to soil cultivation systems, soilless growth systems are superior for plant growth as less energy is required by plants to extract water at field capacity, therefore experiencing a lower risk of oxygen deficiency [32,33]. In the same manner, all containerized production systems can be considered as hydroponic (i.e., soilless growth system) since they consist of an artificial root zone aimed at optimizing water and nutrient availability [34]. However, the restricted root volume may negatively affect the supply of nutrients to the plants as the water in the substrate may be rapidly decreased [32,35–37]. In addition, changes are induced in air and water retention characteristics of organic and inorganic substrates when they are used for longer periods than one growing season [38,39]. Therefore, according to Deepagoda et al. [29], a porous media should preferably be inert to prevent chemical and biological interactions.

Irrespective of the type of greenhouse cultivation system used (i.e., soil or soilless), irrigation scheduling should be managed (I) to supply plants with the volume of water equal to the volume of transpired water for maintaining crop productivity, (II) to overcome the differences in water discharge achieving high water uniformity (III) to move excessive salts towards the rooting system, avoiding soil salination [30,40]. Even in the latter case, for greenhouse cultivation systems there is always a risk of erroneous choices in the matching irrigation supply to crop evapotranspiration, as it may be affected by sudden changes in outside weather conditions or the use of climate control systems such as heating and ventilation [41]. That is another reason why for open hydroponic systems the main irrigation strategy is to supply nutrient solutions, with a surplus of 30% to 50% of the water uptake by the plants [14].

The leaching requirement in greenhouse soil-grown cultivation can be estimated based on irrigation water salinity and crop salt tolerance following FAO [42] as below:

$$LR = \frac{EC_{iw}}{5EC_e - EC_{iw}},\qquad(1)$$

where LR is the minimum leaching requirement needed to control salts within the tolerance EC_e of the crop; EC_{iw} is the electrical conductivity of the irrigation water applied (dS m^{-1}); and, EC_e the average soil salinity tolerated by the crop as measured on a soil saturation extract (dS m^{-1}).

However, as cited by Ben-Gal et al. [43], traditional guidelines for the calculation of the crop-specific leaching requirement is imprecise due to failure to consider soil type, climate, or salinity-induced reduction in plant transpiration. Such omissions could possibly result in underestimating actual leaching and over-estimation of leaching requirements.

Micro-irrigation is often promoted as a technology that can increase the application efficiency of water, and improve crop production and quality. The sub irrigation system also applies for the production of many ornamental hydroponic crops. However, the tendency for salts to build up in the upper portion of the root zone represents a drawback [44,45]. Harmanto et al. [46] working with soil-based greenhouse tomatoes (*Solanum lycopersicum*) in a tropical environment indicated that by applying drip irrigation, the water savings inside the greenhouse could be as much as 20% to 25% higher compared to an open field drip irrigated farming system.

For scheduling irrigation in soil or soilless greenhouses, it is essential to estimate the crop evapotranspiration and, according to the soil or substrate, the irrigation dose. In addition, as cited in Incrocci et al. [16], the irrigation dose of container growing medium could be estimated based on water potential or volumetric water content, with the use of soil moisture sensors. In the meantime, the adoption of soil moisture monitoring in vegetables has been restricted by means of sensor accuracy and price as well as labor required for installation, removal, and collection of readings [7]. A recent review by Bianchi et al. [47] summarized the four macro-groups of soil water potential devices and their operational characteristics.

According to Cahn and Johnson [7], an advantage of tension thresholds is the lesser influence by soil texture in comparison to volumetric moisture thresholds. Even so, as cited by Nikolaou et al. [48], sensors that estimate dielectric capacitance or dielectric permittivity of substrates (e.g., time domain reflectometry, frequency domain) have a propensity to be more reliable for soilless culture systems, as opposed to sensors measuring water availability through the matric potential such as the tensiometers.

According to Baille [49], in the short-term, decision level irrigation can be triggered based either on greenhouse microclimate or on soil/substrate moisture status. Irrigation scheduling based on direct or indirect measurement of plant water status and plant physiological responses to drought by using plant-based methods was comprehensively reviewed by several authors [50,51]. The different methods of irrigation scheduling in greenhouses is summarized below (Table 1).

Table 1. Greenhouse irrigation methods for soil and soilless greenhouse cultivation systems.

Scheduling Irrigation	Based on	Method/Device Use	Decisions Made	Reference
Time clock based	Time	Irrigation controllers	Irrigation frequency	[52,53]
Climate monitoring	Evapotranspiration	Lysimeters	Determine evapotranspiration (ET_C)	[54–56]
		Class A Pan	Determine reference evapotranspiration (ET_O)	[57,58]
		Reduce Class A Pan	Determine reference evapotranspiration (ET_O)	[2,59]
		Atmometer	Determine reference evapotranspiration (ET_O)	[15]
		Evapotranspiration models	Crop water used	[9,41]
	Solar radiation	Pyranometer	Irrigation frequency	[60,61]
Soil or substrate monitoring	Water potential	Tensiometer	Irrigation frequency/dose mainly for soil cultivations	[62]
		Electrical resistance sensor (e.g., gypsum blocks)	Irrigation frequency for soil	[62]
	Volumetric water content	Dielectric sensor (e.g., time domain reflectometry, frequency domain)	Irrigation frequency for soilless and soil cultivations	[62–64]
	Electrical conductivity	Electrical conductivity sensor	Irrigation frequency for soilless cultivation	[65–67]
	Physical properties	Mathematic formula	Irrigation dose/frequency for soilless and soil cultivations	[23,51,65,68]
	Percentage of drainage	Mathematic formula, weighting devices	Irrigation volume and frequency based on trial and error for soilless	[69,70]
Phyto-sensing	Leaf water potential	Pressure chamber	Irrigation timing	[33]
	Stomata resistance	Diffusion porometer	Irrigation timing	[33]
	Canopy temperature	Infrared thermometry	Irrigation timing	[33,71,72]
	Flow on water in the stem	Heat balance sap flow sensor	Irrigation timing/detect water shortages	[33,73,74]
	Changes in stem diameter	Dentrometer	Irrigation timing	[33]
	Crop reflectance	Sensing system equipment and plant reflectance indices (e.g., photochemical reflectance index, normalized difference vegetation index)	Detect water stress	[51,75]

2.1. Monitoring Irrigation in Greenhouse Crops

Irrigation scheduling may have an impact on crop water productivity, affecting fruit yield and quality as well [76–78]. However, the targeted performance of a crop is largely situational; as irrigation might also be used as a tool for increasing water use efficiency, for maximizing yield or economic return [79].

For soil-based greenhouse cucumber (*Cucumis sativus*), Alomran [77] indicated that applying deficit irrigation at specific crop stages with 80% ET_C (i.e., decrease irrigation water up to 40% ET_C) is the most appropriate irrigation strategy for high crop water productivity and yield. For greenhouse tomatoes, partial root drying resulted in a water savings of 50%, but negatively affected the total fruit and total dry mass. However, the considerable savings of water could make partial root drying feasible in areas where water is scarce and expensive [80].

For soilless greenhouse cucumber, between transplanting and flowering, irrigation should be scheduled so as to induce slight water stress and increase root growth, while tomatoes should be stressed for a longer period (i.e., about three weeks) in order to set the first and second trusses [81]. In addition, several authors [53,60,82–84] indicated that increasing the irrigation intervals in soilless culture with the same daily amount of water applied positively influenced crop growth and production and minimized the outflow of water and nutrients from the greenhouse into the environment. However, that is not always the case, because results are often crop and substrate specific, and are also dependent on the experimental conditions and the limiting growth factor(s) [85].

A more rational approach for optimizing irrigation is through automatic irrigation controllers. Therefore, irrigation management approaches may be categorized according to the ability of a controller unit to support a feedback system [86]. Irrigation operations are often automated by using timers, specialized controllers, or computer control [87]. In the simplest form of automation in an "open loop irrigation control system", no measurements of the system outputs are used to modify the inputs and irrigation is based on preset time intervals (i.e., time clock scheduling) [86,88,89]. In a "feedback based irrigation closed-loop control system", the system provides growers with output data in real time (i.e., percentage of drainage, plant water status) which are evaluated in order to reschedule or perform irrigation. In a "feed forward irrigation control system" water uptake is predicted by using growth and transpiration models [14,71]. In addition, computerized-controlled irrigation systems can utilize a range of data to achieve accurate delivery of water according to crop requirements [71]. These systems are often mentioned as a fuzzy-logic control system, artificial intelligent system or multicriterion decision-making system. They are gaining importance because of their inherent ability to judge alternative scenarios for the selection of the best alternative which may be further analyzed before implementation [90].

2.2. The Soil/Substrate Physical Properties and the Irrigation Dose

Evapotranspiration rates depend on greenhouse environmental conditions, and are also affected by the water supply to the roots [91]. For scheduling irrigation, hydraulic properties and water content dependence on substrate suction must be known as they influence the water movement and retention in the substrate [91–93]. Water retention curves or moisture characteristic curves relate the water content in a specific substrate to the matric potential at a given tension or height [94]. Different kinds of substrates, as Fields et al. [95] indicated, have different water retention curves (Figure 1).

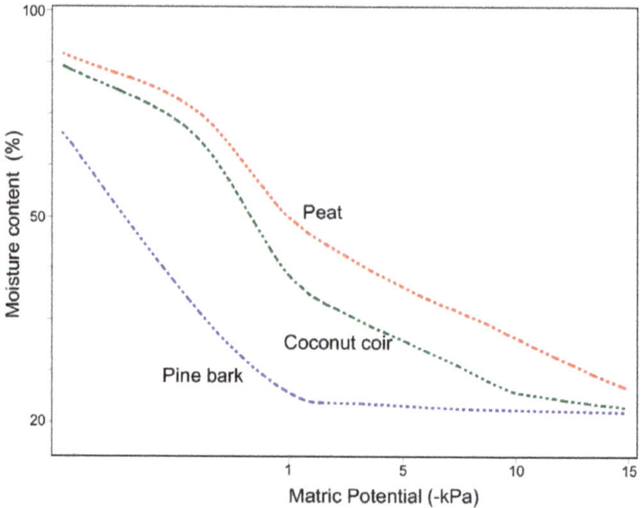

Figure 1. Moisture retention curves of peat, coconut coir and aged pine bark substrate components. Data adapted from [95].

From Figure 1, we can observe that the easily available water content (i.e., water released between 1 and 5 kPa) in coir is higher in comparison to pine bark; therefore, the crop may absorb more available water, reducing the need for applying a high frequency irrigation program. In addition, a taller container proportionately holds less water, as a percentage of water content by volume. Irrigation should take into account indices related to substrate availability of water, to container geometry and to specific substrate characteristics [64]. In general, water held by tensions higher than 10 kPa are considered unavailable to the crop, and water held between 5 to 10 kPa tensions are referred to as the substrate water buffer capacity. The available water in the container can be estimated according to Baudoin as follows [96]:

$$AWcont. = +0.64AW + 0.30P - 67h + 4.1,\qquad(2)$$

where AW is the water that is available in a specific substrate as obtained from the water release curve (%); P represents the substrate porosity; h is the height of the container (m).

However, according to Raviv [97], water and nutrient availability to plants depends on the actual moisture flux from the medium to the plant roots rather than on the water volume in the container. By measuring water contents at different pressure heads, the soil water retention function can be determined. However, the unsaturated hydraulic conductivity measurement is often difficult as it may require expensive equipment [91,94,98]. The reason is because substrates containing predominantly organic components decompose during crop production cycles resulted in changes in air to water ratios. Additionally, shrinkage and compaction of substrates generates problems with watering, hydration, and generally leads to worsening the air to water relationship [99,100]. In the same way, hydraulic conductivity of rockwool and similar substrates is high when well-watered, but declines drastically as it dries out and plants experience a water deficit [101].

Mavrogianopoulos [65] proposed a simple equation for the estimation of the irrigation dose based on substrate characteristics as below:

$$Q = \frac{Y \times W_W \times N}{(1 - dr)},\qquad(3)$$

where Q is the irrigation dose (L slab^{-1}); Y is the water holding capacity of the substrate inside the slab (L slab^{-1}); W_w is the percentage of the water holding capacity that is easily available water (%); N

represents a critical percentage of the easily available water that when reached, irrigation should start (i.e., typical values between 5% to 35%); *dr* is the percentage of drainage (%).

According to the same author, the substrate water holding capacity could be estimated by weighing it when dry, then filling it up with water for up to 24 h to complete the wetting process, draining up to 12 h, and reweighing it. The difference in weight is the water holding capacity, in kg slab^{-1} or L slab^{-1}; however, this procedure should be repeated at different stages throughout plant growth.

In addition, another equation for estimating the irrigation dose was proposed by Katsoulas [82], taking into account the crop transpiration rate plus an extra amount of water for leaching purposes. This method simply reflects the substrate influence on crop water uptake, but requires frequent measurements of crop's transpiration rates.

$$E = \frac{T_r}{(1-D)},$$ (4)

$$Tr = \zeta RG_o,$$ (5)

$$\zeta = \frac{K_c \tau \alpha}{\lambda},$$ (6)

where E is the amount of water applied in (Kg m^{-2}); Tr is the crop transpiration (kg m^{-2}); K_c is the species-specific crop coefficient; τ is the greenhouse radiation transmission coefficient; RG_o is the solar radiation measured outside the greenhouse (Kj m^{-2}); a is the evaporation coefficient; λ is the latent heat of vapourization of water (kJ kg^{-1}); and D is the drainage rate equal to 50% of irrigation water apply.

For greenhouse soil-based cultivation the amount of water which is "available" for root water uptake is defined as the amount of soil water between field capacity (i.e., soil matric pressure -10 or -33 kPa) and the permanent wilting point (i.e., -1500 kPa) expressed in m of water per m of soil depth. However, besides its use in irrigation management, field capacity is not an adequate soil physical quantity to assess soil water availability to crops, as a considerable (10% to 50%) fraction of transpired water is acquired from the soil at water contents above field capacity [102].

The most common values for typical soil texture classes are cited by Snider [103] in Table 2.

Table 2. Average values of available water holding capacity of the main soil texture groups (cm of water per cm of soil). Data adapted from [103].

Common Name	Field Capacity	Wilting Point	Available Water
Sandy soils	0.06–0.20	0.02–0.08	0.04–0.12
Loamy soils	0.23–0.27	0.10–0.12	0.13–0.15
Clayey soils	0.28–0.40	0.13–0.25	0.15–0.18

When the available soil moisture within the rooting zone has attained a predefined level of available water (i.e., the management allowable deficit—MAD), irrigation is triggered. The estimation of MAD is difficult because it depends on plant species and the evaporative conditions [104]. In general, the MAD can be calculated as a percentage of the available water, usually 30% to 50% in soil and 10% in soilless cropping systems. Then, the irrigation dose can be calculated by multiplying the MAD with a coefficient with typical values from 1.15 to 2, to account for water application uniformity and salinity. Typically, the frequency of irrigation can be estimated when the accumulated daily ET_C for the periods between irrigations approaches the MAD [96]. In line with this, the irrigation frequency of greenhouse soil cultivations can be estimated by dividing the readily available water with crop evapotranspiration [68].

Zeng et al. [23], working with drip-irrigated greenhouse soil cultivation of muskmelon, defined the irrigation dose by measuring the soil water content daily. When the water content was reduced to the irrigation start point, then the irrigation amount was decided:

$$Ir = \frac{\gamma \times h \times \theta_{fc} \times (g1 - g2)}{IE} \times 10,$$ (7)

$$SWC = \frac{FW - DW}{DW} \times 100,$$ (8)

where *Ir* represents the water amount by the drip irrigation system (mm); γ (gamma) is the soil bulk density (1.36 g·cm^{-3}); h is the depth of the soil which is irrigated in accordance with the vegetative stage; θ_{fc} is the water field capacity (32.9%); g_1 is the irrigation application rate (%); g_2 is the irrigation start point (60%); *IE* is the irrigation efficiency; *SWC* is the soil water content (%); *FW* is the fresh weight; and, *DW* is the dry weight.

3. Open and Feed Forward Irrigation Control System

3.1. Time Clock Scheduling and the Accumulated Radiation Method

In greenhouses with no feedback system control (i.e., open loop system), irrigation scheduling is determined according to the grower's perspective. Usually growers use a standard irrigation dose and change the frequency of irrigation; thus, they automate irrigation only on the basis of time [8,88]. For soil-based greenhouse crops the irrigation frequency is usually on a daily basis under warm and sunny conditions, and every 3–4 days under cooler and cloudy conditions [28]. In soilless systems, irrigation usually starts one hour after sunrise and stops one hour before sunset, with hourly or even shorter irrigation intervals during a day under high radiation conditions. For some substrates (i.e., rockwool) night irrigation is also recommended, avoiding drying to account for crop transpiration [105,106]. In line with this, according to Schröder and Lieth [81], irrigation at night is advised if the moisture content of the substrate has fallen below 8% to 10% from the previous morning. However, those rules-of-thumb obviously do not apply when the weather conditions are changing quickly from day to day [52,107]. Lizarraga et al. [52] evaluated the efficiency of timed scheduling, and concluded that this method does not actually meet the irrigation requirements of hydroponic tomatoes resulting in over and under irrigation during the morning and in the afternoon, respectively. Similarly, Incrocci et al. [16], working with several species of ornamentals in container nursery crops, reported an increase of the water use with timed irrigation scheduling by 20% to 40% and nutrient emissions of 39% to 74% in comparison with model-based irrigation.

A more rational approach for irrigation scheduling is the accumulated radiation method, allowing more closely matched water supply to the *ET* rate, which is primarily a day time phenomenon depending strongly on solar radiation [14,69,89,108,109]. However, Shin et al. [110] indicated that the transpiration rate of soilless paprika (*Capsicum annuum*) plants did not proportionally increase with an increase in light intensity, especially in high light conditions.

For estimating how much energy the crop has received, a light sensor (e.g., pyranometer) is used to measure incident solar radiation. Once this has been intercepted, a threshold value of light energy, an irrigation event, is triggered. Schröder and Lieth [81] suggested light sums inside greenhouses between 0.4 and 0.6 MJ m^{-2} in closed and 1.4 and 1.8 MJ m^{-2} in open hydroponic cultures with drainage volume factors of 30% and 15%, respectively. However, for rockwool substrate in a free drainage system, Lee [111] recommended accumulated values of 0.8 MJ m^{-2} with a minimum rest time set (i.e., not irrigated if the target value is reached) between 20 min in bright and 50 min on dark days according to the crop growth phase. Additionally, for bell pepper in container growth with peat mix, perlite, and pine bark media, Jovicich et al. [61] indicated that the first-quality fruit weight was enhanced at solar radiation integral levels of 0.34 MJ m^{-2}, while Lizarraga et al. [52] proposed indoor values of 0.81 MJ m^{-2} for tomatoes grown in perlite in bags of 40 L. In addition, Nikolaou et al. [60], working with cucumber in rockwool, indicated a 9% lower drainage amount between high and low irrigation frequency treatments (i.e., accumulated radiation outside greenhouse 1.3 MJ m^{-2} as opposed to 3 MJ m^{-2}), with no negative impact on production.

Despite the fact that the frequency of irrigation can be calculated based on the accumulated radiation method, the threshold values of light energy requires frequent evaluation, as affected by changes of crop coefficient values and cultivation techniques (e.g., defoliation). In any case, the solar radiation method was used only in soilless systems.

3.2. Crop Evapotranspiration and the Water Balance Method

Crop evapotranspiration assessment is necessary to correctly quantify crop irrigation water needs, playing a crucial role in cooling greenhouse crop canopies [112]. In order to evaluate crop evapotranspiration (ET_C), environmental conditions and physical, morphological and physiological features of soil-plant systems have to be considered [113].

A lot of research has been conducted in the field of estimating crop water irrigation needs, in real time, similar to the initial Penman–Monteith evapotranspiration models, which were initially developed for open field cultivations. However, the majority of these studies indicated a drastic effect of different greenhouse types to the transpiration rate and the need for model recalibration in prevailing greenhouse environments [72,114]. A brief summary of the most common evapotranspiration models in different greenhouse types, from literature cited by Fazlil Ilahi [115].

Class A evaporation pans (Figure 2) are considered one of the most widely used systems for climatic measurements in the estimation of the evapotranspiration rate for open field and protected cultivation, because of their simplicity and low cost [2,48]. The pan has proven its practical value and has been used successfully to estimate reference evapotranspiration by observing the evaporation loss from a water surface and applying empirical coefficients to relate pan evaporation to reference evapotranspiration (ET_O) [116].

Figure 2. Class A evaporation pan.

The estimation of crop evapotranspiration with the use of a class A evaporation pan can be calculated according to Allen et al. [117].

$$ET_C = E_P \times K_P \times K_C,$$ (9)

where ET_C is the maximum daily crop evapotranspiration measured (mm); E_P is the daily evaporation from class A Pan (mm); K_P is the pan coefficient;and, K_C is the crop coefficient.

However, there is difficulty in obtaining accurate field measurements with the use of pan A for herbaceous plants, because the crop coefficient (K_C value) is constantly changing throughout the growth, pruning and harvesting phases [48]. Abdel-Razzak et al. [118], working with cherry tomatoes, verified crop coefficients between 0.4 and 1.1, depending on the growth stage, while Gallardo et al. [28] indicated higher K_C values for supporter melon (*Cucumis* spp.) crops in relation to non supporter types. Regarding the pan A coefficient, a constant value of 0.79 provides a good estimation of reference evapotranspiration (ET_O) rates in plastic greenhouses under Mediterranean conditions [119], while Çakir et al. [57], working with cucumber in a solar greenhouse covered with netting material, indicated a plant-pan coefficient between 1.25 to 1.50. In line with this, for several crops in Cyprus, the evapotranspiration rate was calculated from reference evapotranspiration based on pan evaporation data, following the methodology proposed by Allen et al. [118] as indicated in Table 3 derived from Markou and Papadavid and Christou et al. [120,121]:

Table 3. Monthly and yearly estimated evapotranspiration requirements for several crops in protected cultivation (mm). Data adapted from [122,123].

Crop	J	F	M	A	M	J	J	A	S	O	N	D	Total
Tomato	42	60	85	120	180	168				12	40	36	743
Cucumber	42	48	72	120	208						40	36	566
French bean	42	48	84	140	70						24	28	436
Aubergines	12	24	40	60	76	100	68						380
Pepper	12	24	40	60	76	100	112						424
Watermelon	10	20	32	48	84	28							222
Sweet melon	10	20	32	48	84	28							222
Zucchini	12	24	50	78	136	88							388

The estimation of reference evapotranspiration is common in China and in Japan, the use of a reduced-size 20 cm diameter pan, eliminating the disadvantage of the large area occupied by a class A pan, (i.e., 121 cm diameter) [2,109,122]. Zeng et al. [23] working in soil-based greenhouse cultivation, indicated that K_{CP} values of cucumber equal to one can be recommended for the most appropriate irrigation scheduling from a standard 0.2 m diameter pan.

Commercially available atmometers can be used as an alternative approach to estimate evapotranspiration rate [7]. The estimated evapotranspiration values using a Piche atmometer (evaporated surface of paper disc), a modified atmometer (evaporated surface of a porous-porcelain plate) and a reduced evaporation pan were compared with the Penman–Monteith evapotranspiration method. As results indicated, atmometers had the best performance for estimating crop evapotranspiration in a greenhouse and could be used advantageously in relation to the evaporation pans [123].

According to several authors, devices that measure actual plant–soil evapotranspiration confined within containers (i.e., lysimeters) provide the only direct measure of water flux from a vegetative surface (i.e., can detect losses as small as 0.01 mm of water) and as such, they provide a standard against which other methods can be tested and calibrated [54,117]. Weighing lysimeters, measuring ET_C directly through changes in mass, while drainage lysimeters calculate ET_C through water budgets, where excess water removed by drainage or vacuum is subtracted from a known water volume applied to the soil surface [54]. In addition, Shin and Son [69] used load cells for the direct estimation of irrigation and drainage water amounts in soilless systems. Measurement practices, as cited by Sabeh [124], have ranged from weighting lysimeters measuring output every 10 min to calculating a 60 min average of 1 min measurements. However, the expense of these lysimeters limits their use to research and plants grown in containers (i.e., soilless culture systems; Figure 3) [7].

Figure 3. Mounting a lysimeter in a greenhouse structure (left), weighing device S-Beam load cell (centered) and plants grown in a lysimeter (right).

In soil-based systems simple portable scales must be replaced by expensive lysimeters. In this case sampling and representativeness problems become serious [125]. In order to eliminate this problem for soil-based greenhouse systems, it is preferable to use the water balance method, although is not very accurate, by monitoring all additions to and losses from a field's water [103]. In low volume/high frequency irrigation systems, the method has generally been sufficiently robust under a wide range of conditions [50]. Çakir et al. [57] and Mao et al. [126] estimated cucumber greenhouse crop evapotranspiration as follows:

$$ET = I \pm \Delta\Sigma - D - R, \tag{10}$$

where ET is crop evapotranspiration (mm); I is the total irrigation amount applied (mm); $\Delta\Sigma$ is change in soil water storage (mm); R and D are run-off and water loss, respectively, through deep percolation (mm) which is assumed to be zero since the amount of irrigation water is controlled and the precipitation or discharge rate of the irrigation system is below the soil infiltration rate.

4. Feedback Irrigation System

4.1. Soil/Substrate Monitoring

The frequency of irrigation could be monitored in soilless systems by measuring the change in salt concentration inside the cultivation slab. In this case, irrigation starts when the substrate electrical conductivity increased in relation to the nutrient solution electrical conductivity to a certain limit (e.g., 0.3–1 m·S·dm^{-1}) [65]. Using sensors for monitoring the EC, the pH and the amount of drainage could also be used as a tool for evaluating irrigation scheduling, taking into account seasonal transpiration differences [74]. As cited by Lizarraga et al. [52], the EC of the drainage solution should not be higher than 1 m·S·cm^{-1} compared to the EC of the irrigation solution. In addition, the percentage of the drainage amount could be tuned for irrigation control in greenhouses using a trial-end-error approach (e.g., the percentage of drainage should not be higher than 30% of the irrigation applied). Although the irrigation control system considers drainage amount as a single variable, it could not calculate the exact water amount used by the plant [69].

Greenhouse soil cultivation thresholds of soil potential have been used by many authors as a tool for irrigation management; even though soil matric potential values have been used, they appear to be based on experience [127]. For example, the irrigation of tomatoes and cucumbers growing in clay soils, with water potential set-points of −40 and −30 kPa, resulted in water savings of 35% and 46%, respectively, compared with irrigation set points at −10 kPa [128]. In line with this, for zucchini grown in artificial sand-mulched soil, a threshold soil matric potential of −25 kPa favored production and water savings in comparison with irrigation at −10 or −40 kPa [129]. On the other hand, for soilless crops, Depardieu et al. [130] indicated that plant growth and fruit production for strawberries (*Fragaria X ananassa*) grown in organic substrate (i.e., peat-sawdust mixture, aged bark, coconut fiber) were enhanced if irrigation started at −1.0 to −1.5 kPa, instead of −1.5 to −2.5 kPa.

4.2. Plant Monitoring

Plant phyto-sensing (e.g., leaf water potential, canopy temperature, crop reflectance) has been developed for an early, quantitative detection of plant responses to actual soil water availability, in order to define in real time, irrigation strategies to maximize plant growth [131]. However, a significant limitation is the fact that they do not provide a direct measure of the irrigation volume required. Hence, plant based sensing (Figure 4) is commonly used in conjunction with other irrigation techniques such as soil moisture measurement and the water balance approach [68].

Figure 4. Stem micro-variation sensor (left) and a contact leaf temperature sensor (right).

In general, the use of plant phyto-sensing indicators as a tool for irrigation scheduling requires the estimation of reference or threshold values [50]. For example, Seelig et al. [132] used leaf thickness as an input parameter for automated irrigation control of container soil greenhouse cowpea (*Vigna unguiculata*) plants; indicating that 25% to 45% of irrigation water could be conserved compared with a typical timed irrigation schedule. Similarly, Nikolaou et al. [72] indicated a good correlation between leaf temperature of soilless cucumber with transpiration, and established a relationship between transpiration and leaf temperature by modified the simplified Penman–Monteith equation.

Although remote plant phyto-sensing and crop reflectance indices have been applied with great success in open fields, in greenhouses it has not yet been fully tested, as there are problems associated with greenhouse cover and structure shading [133].

The sensors and approaches used for crop reflectance measurements, and the indices used for crop water and nutrient status detection in greenhouse crops, have been reviewed by Katsoulas et al. [51]

5. Artificial Neural Networks and Fuzzy-Logic Control Systems

Artificial neural networks are analogue computer systems, which are made up of a large number of highly interconnected processing units which encompass computer algorithms that can solve several types of problems, based on different input units [134]. The use of artificial neural networks in agricultural systems is supported, as the plants' responses to their environment can be considered chaotic [135]. Until now, those systems have been applied mainly for open field cultivation, in the estimation of soil moisture content based on various soil and environmental parameters and for irrigation planning [90,136,137]. Pérez-Castro et al. [138] indicated that the water requirements within a greenhouse (i.e., evapotranspiration) can be calculated based on virtual sensors by monitoring external greenhouse climatic conditions. In line with this, virtual water sensors for soilless greenhouse tomato based on crop growth, substrate water and crop transpiration rate were also used by Sánchez-Molina et al. [139]. On the other hand Ben Ali et al. [140], developed a fuzzy logic control system in order to promote a suitable microclimate by activating the appropriate actuators installed inside the greenhouse with the appropriate rate.

Agriculture in developed countries seems to be in a transition, with increasing use of ICT (Information and Communications Technology) within the agricultural ecosystems [141]. Additionally, virtual plants have already been used to develop a case study for the irrigation processes of a greenhouse [142].

6. Concluding Remarks

This paper presents a review of irrigation management in soil and soilless crop production in greenhouses where irrigation scheduling should match the diurnal course of evapotranspiration as affected by the prevailing greenhouse environment through soil/substrate and crop characteristics. The majority of irrigation methods used in the past implement a feed forward or a feedback irrigation decision support system, and, in addition, water stress indices were developed based on plant-sensing. However, a gap in commercialized solutions exists despite the significant research work in the field of precision irrigation of greenhouse crops. It is important that a large margin of progress in greenhouse water and fertilizer use efficiency is managed by farmers [3]. The information presented reveals a need for the development of a commercial irrigation controller unit, in order to model and monitor the soil-plant-atmosphere utilizing artificial intelligence analyses.

Author Contributions: G.N. and D.N. conducted the literature review and produced final drafts of the manuscript. Commentary and review of manuscript drafts were conducted by N.K. and C.K.

Funding: This research received no external funding.

Conflicts of Interest: The authors declare no conflict of interest

References

1. Kittas, C.; Elvanidi, A.; Katsoulas, N.; Ferentinos, K.P.; Bartzanas, T. Reflectance indices for the detection of water stress in greenhouse tomato (*Solanum lycopersicum*). *Acta Hortic.* **2016**, *1112*, 63–70. [CrossRef]
2. Fernandes, C.; Corá, J.; Araújo, J. Reference evapotranspiration estimation inside greenhouses. *Sci. Agric.* **2003**, *60*, 591–594. [CrossRef]
3. Kitta, E.; Bartzanas, T.; Katsoulas, N.; Kittas, C. Benchmark irrigated under cover agriculture crops. *Agric. Agric. Sci. Procedia* **2015**, *4*, 348–355. [CrossRef]
4. Levidow, L.; Zaccaria, D.; Maia, R.; Vivas, E.; Todorovic, M.; Scardigno, A. Improving water-efficient irrigation: Prospects and difficulties of innovative practices. *Agric. Water Manag.* **2014**, *146*, 84–94. [CrossRef]
5. Bonachela, S.; González, A.M.; Fernández, M.D. Irrigation scheduling of plastic greenhouse vegetable crops based on historical weather data. *Irrig. Sci.* **2006**, *25*, 53–62. [CrossRef]
6. Lea-Cox, J.D.; Ross, D.S.; Teffeau, K.M. A Water and Nutrient Management Planning Process for Container Nursery and Greenhouse Production Systems in Maryland. *J. Environ. Hortic.* **2001**, *19*, 230–236.
7. Cahn, M.D.; Johnson, L.F. New Approaches to Irrigation Scheduling of Vegetables. *Horticulturae* **2017**, *3*, 1–20. [CrossRef]
8. Belayneh, B.E.; Lea-Cox, J.D.; Lichtenberg, E. Costs and benefits of implementing sensor-controlled irrigation in a commercial pot-in-pot container nursery. *Horttechnology* **2013**, *23*, 760–769. [CrossRef]
9. Qiu, R.; Kang, S.; Du, T.; Tong, L.; Hao, X.; Chen, R.; Chen, J.; Li, F. Effect of convection on the Penman–Monteith model estimates of transpiration of hot pepper grown in solar greenhouse. *Sci. Hortic.* **2013**, *160*, 163–171. [CrossRef]

10. Zimmermann, D.; Reus, R.; Westhoff, M.; Gessner, P.; Bauer, W.; Bamberg, E.; Bentrup, F.W.; Zimmermann, U. A novel, non-invasive, online-monitoring, versatile and easy plant-based probe for measuring leaf water status. *J. Exp. Bot.* **2008**, *59*, 3157–3167. [CrossRef]

11. Bacci, L.; Battista, P.; Cardarelli, M.; Carmassi, G.; Rouphael, Y.; Incrocci, L.; Malorgio, F.; Pardossi, A.; Rapi, B.; Colla, G. Modelling Evapotranspiration of Container Crops for Irrigation Scheduling. In *Evapotranspiration—From Measurements to Agricultural and Environmental Applications*; Gerosa, G., Ed.; IntechOpen Limited: London, UK, 2011; pp. 263–282. ISBN 978-953-307-512-9.

12. Van der Linden, A.M.A.; Hoogsteen, M.J.J.; Boesten, J.J.T.I.; Van Os, E.A.; Wipfler, E.L. *Fate of Plant Protection Products in Soilless Cultivations after Drip Irrigation: Measured vs. Modelled Concentrations*; National Institute for Public Health and the Environment: Bilthoven, The Netherlands, 2016; pp. 2–61.

13. Llorach-Massan, P.; Muñoz, P.; Riera, M.R.; Gabarrell, X.; Rieradevall, J.; Montero, J.I.; Villalba, G. N_2O emissions from protected soilless crops for more precise food and urban agriculture life cycle assessments. *J. Clean. Prod.* **2017**, *149*, 1118–1126. [CrossRef]

14. Kläring, H.K. Strategies to control water and nutrient supplies to greenhouse crops. A review. *Agronomie* **2001**, *21*, 311–321. [CrossRef]

15. Schiattone, M.I.; Viggiani, R.; Di Venereb, D.; Sergiob, L.; Cantore, V.; Todorovic, M.; Perniola, M.; Candido, V. Impact of irrigation regime and nitrogen rate on yield, quality and water use efficiency of wild rocket under greenhouse conditions. *Sci. Hortic.* **2018**, *229*, 182–192. [CrossRef]

16. Incrocci, L.; Marzialetti, P.; Incrocci, G.; Di Vita, A.; Balendonck, J.; Bibbiani, C.; Spagnol, S.; Pardossi, A. Substrate water status and evapotranspiration irrigation scheduling in heterogenous container nursery crops. *Agric. Water Manag.* **2014**, *131*, 30–40. [CrossRef]

17. Fulcher, F.A.; Buxton, J.W.; Geneve, R.L. Developing a physiological-based, on-demand irrigation system for container production. *Sci. Hortic.* **2012**, *138*, 221–226. [CrossRef]

18. Daccache, A.; Ciurana, J.S.; Rodriguez Diaz, J.A.; Knox, J.W. Water and energy footprint of irrigated agriculture in the Mediterranean region. *Environ. Res. Lett.* **2014**, *9*, 1–12. [CrossRef]

19. Egea, G.; Fernández, J.E.; Alcon, F. Financial assessment of adopting irrigation technology for plant-based regulated deficit irrigation scheduling in super high-density olive orchards. *Agric. Water Manag.* **2017**, *187*, 47–56. [CrossRef]

20. Montesano, F.F.; Van Iersel, M.W.; Boari, F.; Cantore, V.; D'Amato, G.; Parente, A. Sensor-based irrigation management of soilless basil using a new smart irrigation system: Effects of set-point on plant physiological responses and crop performance. *Agric. Water Manag.* **2018**, *203*, 20–29. [CrossRef]

21. Pawlowski, A.; Sánchez-Molina, J.A.; Guzmán, J.L.; Rodríguez, F.; Dormido, S. Evaluation of event-based irrigation system control scheme for tomato crops in greenhouses. *Agric. Water Manag.* **2017**, *183*, 16–25. [CrossRef]

22. Sezen, S.M.; Celikel, G.; Yazar, A.; Tekin, S.; Kapur, B. Effect of irrigation management on yield and quality of tomatoes grown in different soilless media in a glasshouse. *Sci. Res. Essays* **2010**, *5*, 41–48.

23. Zeng, C.Z.; Bie, Z.L.; Yuan, B.Z. Determination of optimum irrigation water amount for drip-irrigated muskmelon (*Cucumis melo* L.) in plastic greenhouse. *Agric. Water Manag.* **2009**, *96*, 595–602. [CrossRef]

24. Putra, P.A.; Yuliando, H. Soilless Culture System to Support Water Use Efficiency and Product Quality: A Review. *Agric. Agric. Sci. Procedia* **2015**, *3*, 283–288. [CrossRef]

25. Montesano, F.F.; Serio, F.; Mininni, C.; Signore, A.; Parente, A.; Santamaria, P. Tensiometer-Based Irrigation Management of Subirrigated Soilless Tomato: Effects of Substrate Matric Potential Control on Crop Performance. *Front. Plant Sci.* **2015**, *6*, 1–11. [CrossRef] [PubMed]

26. Van Os, E.A.; Gieling, T.H.; Ruijs, M.N.A. Equipment for hydroponic installations. In *Hydroponic Production of Vegetables and Ornamental*; Savvas, D., Passam, H., Eds.; Embryo Publications: Athens, Greece, 2002; pp. 104–140. ISBN 960-8002-12-5.

27. Breś, W.; Kleiber, T.; Trelka, T. Quality of water used for drip irrigation and fertigation of horticultural plants. *Folia Hortic.* **2010**, *22*, 67–74. [CrossRef]

28. Gallardo, M.; Thompson, R.B.; Fernández, M.D. Water requirements and irrigation management in Mediterranean greenhouses: The case of the southeast coast of Spain. In *Good Agricultural Practices for Greenhouse Vegetable Crops*; Plant Production and Protection Paper 217; FAO: Rome, Italy, 2013; pp. 109–136.

29. Chamindu Deepagoda, T.K.K.; Chen Lopez, J.C.; Møldrup, P.; de Jonge, L.W.; Tuller, M. Integral parameters for characterizing water, energy, and aeration properties of soilless plant growth media. *J. Hydrol.* **2013**, *502*, 120–127. [CrossRef]

30. Pardossi, A.; Carmassi, G.; Diara, C.; Incrocci, L.; Maggini, R.; Massa, D. *Fertigation and Substrate Management in Closed Soilless Culture*; University of Pisa, Dipartimento di Bioologia delle Piante Agrarrie (DBPA): Pisa, Italy, 2011; pp. 1–63.

31. Maślanka, M.; Magdziarz, R. The influence of substrate type and chlormequat on the growth and flowering of marigold (*Tagetes* L.). *Folia Hortic.* **2017**, *29*, 189–198. [CrossRef]

32. Raviv, M.; Lieth, J.M. Significance of Soilless Culture in Agriculture. In *Soilless Culture. Theory and Practice*; Raviv, M., Lieth, J.M., Eds.; Elsevier: Amsterdam, The Netherlands, 2008; pp. 1–10. ISBN 978-0-444-52975-6.

33. Raviv, M.; Blom, T.J. The effect of water availability and quality on photosynthesis and productivity of soilless-grown cut roses. *Sci. Hortic.* **2001**, *88*, 257–276. [CrossRef]

34. Adams, P. Nutritional control in hydroponics. In *Hydroponic Production of Vegetables and Ornamental*; Savvas, D., Passam., H., Eds.; Embryo Publications: Athens, Greece, 2002; pp. 211–261. ISBN 960-8002-12-5.

35. Asaduzzaman, Md.; Saifullah, Md.; Mollick, S.R.; Hossain, M.Md.; Halim, G.M.A.; Asao, T. Influence of Soilless Culture Substrate on Improvement of Yield and Produce Quality of Horticultural Crops. In *Soilless Culture-Use of Substrates for the Quality Horticultural Crops*; Asaduzzaman, Md., Ed.; IntechOpen Limited: London, UK, 2015; pp. 1–31.

36. Martínez-Gutiérrez, G.A.; Morales, I.; Aquino-Bolaños, T.; Escamirosa-Tinoco, C.; Hernández-Tolentino, M. Substrate volume and nursery times for earliness and yield of greenhouse tomato. *Emirates J. Food Agric.* **2016**, *28*, 897–902. [CrossRef]

37. Rouphael, Y.; Cardarelli, M.; Rea, E.; Colla, G. The influence of irrigation system and nutrient solution concentration on potted geranium production under various conditions of radiation and temperature. *Sci. Hortic.* **2008**, *118*, 328–337. [CrossRef]

38. Vox, G.; Teitel, M.; Pardossi, A.; Minuto, A.; Tinivella, F.; Schettini, E. Sustainable greenhouse systems. In *Sustainable Agriculture: Technology, Planning and Management*; Salazar, A., Rios, I., Eds.; Nova Science Publishers, Inc.: New York, NY, USA, 2010; pp. 1–78. ISBN 978-1-60876-269-9.

39. Warren, S.L.; Bilderback, T.E. More plant per gallon: Getting more out of your water. *Horttechnology* **2005**, *15*, 14–18. [CrossRef]

40. Leteya, J.; Hoffmanb, G.J.; Hopmansc, J.W.; Grattanc, S.R.; Suarezd, D.; Corwind, D.L.; Ostera, J.D.; Wua, L.; Amrhein, C. Evaluation of soil salinity leaching requirement guidelines Agricultural Water Management Evaluation of soil salinity leaching requirement guidelines. *Agric. Water Manag.* **2011**, *98*, 502–506. [CrossRef]

41. Baille, M.; Baille, A.; Laury, J.C. A simplified model for predicting evapotranspiration rate of nine ornamental species vs. climate factors and leaf area. *Sci. Hortic.* **1994**, *59*, 217–232. [CrossRef]

42. Ayers, R.S.; Westcot, D.W. *Water Quality for Agriculture*; Irrigation and Drainage Paper 29; FAO: Rome, Italy, 1985; pp. 1–131. ISBN 92-5-102263-1.

43. Ben-Gal, A.; Ityel, E.; Dudley, L.; Cohen, S.; Yermiyahu, U.; Presnov, E.; Zigmond, L.; Shani, U. Effect of irrigation water salinity on transpiration and on leaching requirements: A case study for bell peppers. *Agric. Water Manag.* **2008**, *95*, 587–597. [CrossRef]

44. Skaggs, R.K. Predicting drip irrigation use and adoption in a desert region. *Agric. Water Manag.* **2001**, *51*, 125–142. [CrossRef]

45. Rouphael, Y.; Colla, G. Growth, yield, fruit quality and nutrient uptake of hydroponically cultivated zucchini squash as affected by irrigation systems and growing seasons. *Sci. Hortic.* **2005**, *105*, 177–195. [CrossRef]

46. Harmanto; Salokhe, V.M.; Babel, M.S.; Tantau, H.J. Water requirement of drip irrigated tomatoes grown in greenhouse in tropical environment. *Agric. Water Manag.* **2005**, *71*, 225–242. [CrossRef]

47. Bianchi, A.; Masseroni, D.; Thalheimer, M.; Medici, L.O.; Facchi, A. Field irrigation management through soil water potential measurements: A review. *Ital. J. Agrometeorol.* **2017**, *2*, 25–38.

48. Nikolaou, G.; Neocleous, D.; Katsoulas, N.; Kittas, C. Irrigation management techniques used in soilless cultivation. In *Advances in Hydroponic Research*; Webster, D.J., Ed.; Nova Science Publishers, Inc.: New York, NY, USA, 2017; pp. 1–33. ISBN 978-1-53612-131-5.

49. Baille, A. Water management in soilless cultivation in relation to inside and outside climatic conditions and type of substrate. *Italus Hortus.* **2001**, *8*, 16–22.

50. Jones, H.G. Irrigation scheduling: Advantages and pitfalls of plant-based methods. *J. Exp. Bot.* **2004**, *55*, 2427–2436. [CrossRef]

51. Katsoulas, N.; Elvanidi, A.; Ferentinos, K.P.; Kacira, M.; Bartzanas, T.; Kittas, C. Crop reflectance monitoring as a tool for water stress detection in greenhouses: A review. *Biosyst. Eng.* **2016**, *151*, 374–398. [CrossRef]

52. Lizarraga, A.; Boesveld, H.; Huibers, F.; Robles, C. Evaluating irrigation scheduling of hydroponic tomato in Navarra, Spain. *Irrig. Drain.* **2003**, *52*, 177–188. [CrossRef]

53. Silber, A.; Xu, G.; Levkovitch, I.; Soriano, S.; Bilu, A.; Wallach, R. High irrigation frequency: The effect on plant growth and on uptake of water and nutrients. *Plant Soil* **2003**, *253*, 466–477. [CrossRef]

54. Beeson, R.C., Jr. Weighing lysimeter systems for quantifying water use and studies of controlled water stress for crops grown in low bulk density substrates. *Agric. Water Manag.* **2011**, *98*, 967–976. [CrossRef]

55. Libardi, L.G.P.; de Faria, R.T.; Dalri, A.B.; de Souza Rolim, G.; Palaretti, L.F.; Coelho, A.P.; Martins, I.P. Evapotranspiration and cropcoefficient (Kc) of presprouted sugarcane plantlets for greenhouse irrigation management. *Agric. Water Manag.* **2019**, *212*, 306–316. [CrossRef]

56. Vera-Repulloa, J.A.; Ruiz-Peñalverb, L.; Jiménez-Buendíaa, M.; Rosillob, J.J.; Molina-Martínez, J.M. Software for the automatic control of irrigation using weighing-drainage lysimeters. *Agric. Water Manag.* **2015**, *151*, 4–12. [CrossRef]

57. Çakir, R.; Kanburoglu-Çebi, U.; Altintas, S.; Ozdemir, A. Irrigation scheduling and water use efficiency of cucumber grown as a spring-summer cycle crop in solar greenhouse. *Agric. Water Manag.* **2017**, *180*, 78–87. [CrossRef]

58. Abou-Hadid, A.F.; El-Shinawy, M.Z.; El-Oksh, I.; Gomaa, H.; El-Beltagy, A.S. Studies on Water Consumption of Sweet Pepper Plant Under Plastic Houses. *Acta Hortic.* **1994**, *366*, 365–372. [CrossRef]

59. Zhang, C.; Gao, H.; Deng, X.; Lu, Z.; Lei, Y.; Zhou, H. Design method and theoretical analysis for wheel-hub driving solar tractor. *Emirates J. Food Agric.* **2016**, *28*, 903–911. [CrossRef]

60. Nikolaou, G.; Neocleous, D.; Katsoulas, N.; Kittas, C. Effect of irrigation frequency on growth and production of a cucumber crop under soilless culture. *Emirates J. Food Agric.* **2017**, *29*, 863–871. [CrossRef]

61. Jovicich, E.; Cantliffe, D.J.; Stoffella, P.J.; Haman, D.Z. Bell pepper fruit yield and quality as influenced by solar radiation-based irrigation and container media in a passively ventilated greenhouse. *HortScience* **2007**, *42*, 642–652.

62. Pardossi, A.; Incrocci, L.; Incrocci, G.; Fernando, M.; Bacci, L.; Rapi, B.; Marzialetti, P.; Hemming, J.; Balendonck, J. Root Zone Sensors for Irrigation Management in Intensive Agriculture. *Sensors* **2009**, *9*, 2809–2835. [CrossRef] [PubMed]

63. Nemali, K.S.; Montesano, F.; Dove, S.K.; Van Iersel, M.W. Calibration and performance of moisture sensors in soilless substrates: ECH_2O and Theta probes. *Sci. Hortic.* **2007**, *112*, 227–234. [CrossRef]

64. Murray, J.D.; Lea-Cox, J.D.; Ross, D.S. Time domain reflectometry accurately monitors and controls irrigation water applications in soilless substrates. *Acta Hortic.* **2004**, *633*, 75–82. [CrossRef]

65. Mavrogianopoulos, G.N. Irrigation dose according to substrate characteristics, in hydroponic systems. *Open Agric.* **2015**, *1*, 1–6. [CrossRef]

66. Dorai, M.; Papadopoulos, A.P.; Gosselin, A. Influence of electric conductivity management on greenhouse tomato yield and fruit quality. *Agronomie* **2001**, *4*, 367–383. [CrossRef]

67. Liopa-Tsakalidi, A.; Barouchas, P.; Salahas, G. Response of Zucchini to the Electrical Conductivity of the Nutrient Solution in Hydroponic Cultivation. *Agric. Agric. Sci. Procedia* **2015**, *4*, 459–462. [CrossRef]

68. White, S.; Raine, S.R. *A Grower Guide to Plant Based Sensing for Irrigation Scheduling*; National Centre for Engineering in Agriculture Publication 1001574/6; USQ: Toowoomba, Australia, 2008; pp. 1–52.

69. Shin, J.H.; Son, J.E. Development of a real-time irrigation control system considering transpiration, substrate electrical conductivity, and drainage rate of nutrient solutions in soilless culture of paprika (*Capsicum annuum* L.). *Eur. J. Hortic. Sci.* **2015**, *80*, 271–279. [CrossRef]

70. Nikolaou, G.; Neocleous, D.; Katsoulas, N.; Kittas, C. Dynamic assessment of whitewash shading and evaporative cooling on the greenhouse microclimate and cucumber growth in a Mediterranean climate. *Ital. J. Agrometeorol.* **2018**, *2*, 15–26.

71. Prenger, J.J.; Ling, P.P.; Hansen, R.C.; Keener, H.H. Plant response-based irrigation in a greenhouse: System evaluation. *Trans. ASAE Am. Soc. Agric. Eng.* **2005**, *48*, 1175–1183. [CrossRef]

72. Nikolaou, G.; Neocleous, D.; Katsoulas, N.; Kittas, C. Modelling transpiration of soilless greenhouse cucumber and its relationship with leaf temperature in a mediterranean climate. *Emirates J. Food Agric.* **2017**, *29*, 911–920. [CrossRef]

73. De Swaef, T.; Steppe, K. Linking stem diameter variations to sap flow, turgor and water potential in tomato. *Funct. Plant Biol.* **2010**, *37*, 429–438. [CrossRef]

74. Ehret, D.L.; Lau, A.; Bittman, S.; Lin, W.; Shelford, T. Automated monitoring of greenhouse crops. *Agronomie* **2001**, *21*, 403–414. [CrossRef]

75. Kim, M.; Kim, S.; Kim, Y.; Choi, Y.; Seo, M. Infrared Estimation of Canopy Temperature as Crop Water Stress Indicator. *Korean J. Soil Sci. Fertil.* **2015**, *48*, 499–504. [CrossRef]

76. Nuruddin, M.; Madramootoo, C.A.; Dodds, G.T. Effects of Water Stress at Different Growth Stages on Greenhouse Tomato Yield and Quality. *HortScience* **2003**, *38*, 1389–1393.

77. Alomran, A.M.; Louki, I.I.; Aly, A.A.; Nadeem, M.E. Impact of deficit irrigation on soil salinity and cucumber yield under greenhouse condition in an arid environment. *J. Agric. Sci. Technol.* **2013**, *15*, 1247–1259.

78. Saleh, S.; Liu, G.; Liu, M.; Ji, Y.; He, H.; Gruda, N. Effect of Irrigation on Growth, Yield, and Chemical Composition of Two Green Bean Cultivars. *Horticulturae* **2018**, *4*, 1–10. [CrossRef]

79. Saha, U.K.; Papadopoulos, A.P.; Hao, X.; Khosla, S. Irrigation strategies for greenhouse tomato production on rockwool. *HortScience* **2008**, *43*, 484–493.

80. Zegbe, J.A.; Behboudian, M.H.; Clothier, B.E. Yield and fruit quality in processing tomato under partial rootzone drying. *Eur. J. Hortic. Sci.* **2006**, *71*, 252–258.

81. Schröder, F.G.; Lieth, J.H. Irrigation control in hydroponics. In *Hydroponic Production of Vegetables and Ornamentals*; Savvas, D., Passam, H., Eds.; Embryo Publications: Athens, Greece, 2002; pp. 263–297. ISBN 960-8002-12-5.

82. Katsoulas, N.; Kittas, C.; Dimokas, G.; Lykas, C. Effect of irrigation frequency on rose flower production and quality. *Biosyst. Eng.* **2006**, *93*, 237–244. [CrossRef]

83. Pires, R.C.M.; Furlani, P.R.; Ribeiro, R.V.; Bodine, J.; Décio, S.; Emílio, L.; André, L.; Torre Neto, A. Irrigation frequency and substrate volume effects in the growth and yield of tomato plants under greenhouse conditions. *Sci. Agric.* **2011**, *68*, 400–405. [CrossRef]

84. Rodriguez-Ortega, W.M.; Martinez, V.; Rivero, R.M.; Camara-Zapata, J.M.; Mestre, T.; Garcia-Sanchez, F. Use of a smart irrigation system to study the effects of irrigation management on the agronomic and physiological responses of tomato plants grown under different temperatures regimes. *Agric. Water Manag.* **2016**, *183*, 158–168. [CrossRef]

85. Tsirogiannis, I.; Katsoulas, N.; Kittas, C. Effect of irrigation scheduling on gerbera flower yield and quality. *HortScience* **2010**, *45*, 265–270.

86. Romero, R.; Muriel, J.L.; García, I.; Muñoz de la Peña, D. Research on automatic irrigation control: State of the art and recent results. *Agric. Water Manag.* **2012**, *114*, 59–66. [CrossRef]

87. Challa, H.; Bakker, J.C. Crop growth. In *Greenhouse Climate Control: An Integrated Approach*; Bakker, J.C., Bot, G.P.A., Challa, H., Van de Braak, N.J., Eds.; Wageningen Academic Publishers: Wageningen, The Netherlands, 1995; pp. 15–97. ISBN 978-90-74134-17-0.

88. Lea-Cox, J.D.; Bauerle, W.L.; Van Iersel, M.W.; Kantor, G.F.; Bauerle, T.L.; Lichtenberg, E.; King, D.M.; Crawford, L. Advancing wireless sensor networks for irrigation management of ornamental crops: An overview. *Horttechnology* **2013**, *23*, 717–724. [CrossRef]

89. Shelford, T.J.; Lau, A.K.; Ehret, D.L.; Chieng, S.T. Comparison of a new plant-based irrigation control method with light-based irrigation control for greenhouse tomato production. *Can. Biosyst. Eng.* **2004**, *1*, 1–6.

90. Raju, K.S.; Kumar, D.N. Multicriterion decision making in irrigation planning. *Irrig. Drain.* **2005**, *54*, 455–465. [CrossRef]

91. Londra, P.A. Simultaneous determination of water retention curve and unsaturated hydraulic conductivity of substrates using a steady-state laboratory method. *HortScience* **2010**, *45*, 1106–1112.

92. Bougoul, S.; Boulard, T. Water dynamics in two rockwool slab growing substrates of contrasting densities. *Sci. Hortic.* **2006**, *107*, 399–404. [CrossRef]

93. Schindler, U.; Müller, L.; Eulenstein, F. Hydraulic Performance of Horticultural Substrates-1. Method for Measuring the Hydraulic Quality Indicators. *Horticulturae* **2017**, *3*, 1–7. [CrossRef]

94. Altland, J.E.; Owen, J.S.; Fonteno, W.C. Developing Moisture Characteristic Curves and Their Descriptive Functions at Low Tensions for Soilless Substrates. *J. Amer. Soc. Hortic. Sci.* **2010**, *135*, 563–567.

95. Fields, J.S.; Fonteno, W.C.; Jackson, B.E.; Heitman, J.L.; Owen, J.S. Hydrophysical properties, moisture retention, and drainage profiles of wood and traditional components for greenhouse substrates. *HortScience* **2014**, *49*, 827–832.

96. De Pascale, S.; Barbieri, G.; Rouphael, Y.; Gallardo, M.; Orsini, F.; Pardossi, A. Irrigation management: Challenges and opportunities. In *Good Agricultural Practices for Greenhouse VEGETABLE production in the South East European Countries for Greenhouse Vegetable*; Plant Production and Protection Paper 230; FAO: Rome, Italy, 2013; pp. 79–105.

97. Raviv, M.; Wallach, R.; Silber, A.; Medina, S.; Krasnovsky, A. The effect of hydraulic characteristics of volcanic materials on yield of roses grown in soilless culture. *J. Am. Soc. Hortic. Sci.* **1999**, *124*, 205–209.

98. Hosseini, S.M.M.M.; Ganjian, N.; Pisheh, Y.P. Estimation of the water retention curve for unsaturated clay. *Can. J. Soil Sci.* **2011**, *91*, 543–549. [CrossRef]

99. Bilderback, T.E.; Warren, S.L.; Owen, J.S.; Albano, J.P. Healthy Substrates Need Physicals Too! *Horttechnology* **2005**, *15*, 747–751. [CrossRef]

100. Nowak, J.S. Changes of Physical Properties in Rockwool and Glasswool Slabs During Hydroponic Cultivation of Roses. *J. Fruit Ornam. Plant Res.* **2010**, *18*, 349–360.

101. Jones, H.G.; Tardieu, F. Modelling water relations of horticultural crops: A review. *Sci. Hortic.* **1998**, *74*, 21–46. [CrossRef]

102. De Jong van Lier, Q. Field capacity, a valid upper limit of crop available water? *Agric. Water Manag.* **2017**, *193*, 214–220. [CrossRef]

103. Snyder, R.L. *Irrigation Scheduling: Water Balance Method*; University of California, Department of Land, Air and Water Resources Atmospheric Science Davis: Berkeley, CA, USA, 2014; pp. 1–39.

104. Greenwood, D.J.; Zhang, K.; Hilton, H.W.; Thompson, A.J. Opportunities for improving irrigation efficiency with quantitative models, soil water sensors and wireless technology. *J. Agric. Sci.* **2010**, *148*, 1–16. [CrossRef]

105. Carmassi, G.; Bacci, L.; Bronzini, M.; Incrocci, L.; Maggini, R.; Bellocchi, G.; Massa, D.; Pardossi, A. Modelling transpiration of greenhouse gerbera (*Gerbera jamesonii* H. Bolus) grown in substrate with saline water in a Mediterranean climate. *Sci. Hortic.* **2013**, *156*, 9–18. [CrossRef]

106. Medrano, E.; Lorenzo, P.; Sánchez-Guerrero, M.C.; Montero, J.I. Evaluation and modelling of greenhouse cucumber-crop transpiration under high and low radiation conditions. *Sci. Hortic.* **2005**, *105*, 163–175. [CrossRef]

107. Andrew, L.; Enthoven, N.; Kaarsemaker, R. *Best Practice Guidelines for Greenhouse Water Management*; GRODAN & Priva: Roermond, The Netherlands, 2016; pp. 1–38.

108. Kittas, C. Solar radiation of a greenhouse as a tool to its irrigation control. *Int. J. Energy Res.* **1990**, *14*, 881–892. [CrossRef]

109. Zhang, Z.K.; Liu, S.Q.; Liu, S.H.; Huang, Z.J. Estimation of Cucumber Evapotranspiration in Solar Greenhouse in Northeast China. *Agric. Sci. China* **2010**, *9*, 512–518. [CrossRef]

110. Shin, J.H.; Park, J.S.; Son, J.E. Estimating the actual transpiration rate with compensated levels of accumulated radiation for the efficient irrigation of soilless cultures of paprika plants. *Agric. Water Manag.* **2014**, *135*, 9–18. [CrossRef]

111. Lee, A. Reducing or eliminating fruit physiological disorders with correct root zone management. *Practical Hydroponics & Greenhouses*, May/June 2010; 53–59.

112. Katsoulas, N.; Kittas, C. Greenhouse Crop Transpiration Modelling. In *Evapotranspiration—From Measurements to Agricultural and Environmental Applications*; Gerosa, G., Ed.; IntechOpen Limited: London, UK, 2011; pp. 311–328. ISBN 978-953-307-512-9.

113. Lovelli, S.; Perniola, M.; Arcieri, M.; Rivelli, R.; Di Tommaso, T. Water use assessment in muskmelon by the Penman-Monteith 'one-step' approach. *Agric. Water Manag.* **2008**, *95*, 1153–1160. [CrossRef]

114. Morille, B.; Migeon, C.; Bournet, P.E. Is the Penman-Monteith model adapted to predict crop transpiration under greenhouse conditions? Application to a New Guinea Impatiens crop. *Sci. Hortic.* **2013**, *152*, 80–91. [CrossRef]

115. Fazlil Ilahi, W.F. Evapotranspiration Models in Greenhouse. Master's Thesis, Irrigation and Water Engineering Group, Wageningen University, Wageningen, The Netherlands, 2009; pp. 1–52.

116. Simba, F.M. A Flexible Plant Based Irrigation Control for Greenhouse Crops. A Thesis submitted in partial fulfillment for the requirements of the Master of Science, University of Zimbabwe, Harare, Zimbabwe, 2010.

117. Allen, R.; Pereira, L.; Raes, D.; Smith, M. *Crop Evapotranspiration Guidelines for Computing Crop Water Requirements*; FAO Irrigation and Drainage Paper, 56; FAO: Rome, Italy, 1998; pp. 1–289.

118. Abdel-Razzak, H.; Wahb-Allah, M.; Ibrahim, A.; Alenazi, M.; Alsadon, A. Response of cherry tomato to irrigation levels and fruit pruning under greenhouse conditions. *J. Agric. Sci. Technol.* **2016**, *18*, 1091–1103.

119. Fernández, M.D.; Bonachela, S.; Orgaz, F.; Thompson, R.; López, J.C.; Granados, M.R.; Gallardo, M.; Fereres, E. Measurement and estimation of plastic greenhouse reference evapotranspiration in a Mediterranean climate. *Irrig. Sci.* **2010**, *28*, 497–509. [CrossRef]

120. Markou, M.; Papadavid, G. Norm input -output data for the main crop and livestock enterprises of Cyprus. *Agric. Econ.* **2007**, *46*, 0379–0827.

121. Christou, A.; Dalias, P.; Neocleous, D. Spatial and temporal variations in evapotranspiration and net water requirements of typical Mediterranean crops on the island of Cyprus. *J. Agric. Sci.* **2017**, *1*, 1188–1197. [CrossRef]

122. Liu, H.J.; Cohen, S.; Tanny, J.; Lemcoff, J.H.; Huang, G. Estimation of banana (*Musa* sp.) plant transpiration using a standard 20 cm pan in a greenhouse. *Irrig. Drain. Syst.* **2008**, *22*, 311–323. [CrossRef]

123. Blanco, F.F.; Folegatti, M.V. Evaluation of evaporation measuring-equipments for estimating evapotranspiration within a greenhouse evapotranspiranspiration greenhouse. *Revista Brasileira de Engenharia Agrícola e Ambiental.* **2004**, *8*, 184–188. [CrossRef]

124. Sabeh, N.C. Evaluating and minimizing water use by greenhouse evaporative cooling systems in a semi-arid climate. In *Partial Fulfillment of the Requirements for the Degree of Doctor of Philosophy*; University of Arizona: Tucson, AZ, USA, 2007.

125. Seginer, I.; Kantz, D.; Levav, N.; Peiper, U.M. Night-time transpiration in greenhouses. *Sci. Hortic.* **1990**, *41*, 265–276. [CrossRef]

126. Mao, X.; Liu, M.; Wang, X.; Liu, C.; Hou, Z.; Shi, J. Effects of deficit irrigation on yield and water use of greenhouse grown cucumber in the North China Plain. *Agric. Water Manag.* **2003**, *61*, 219–228. [CrossRef]

127. Thompson, R.B.; Gallardo, M.; Valdez, L.C.; Fernández, M.D. Using plant water status to define threshold values for irrigation management of vegetable crops using soil moisture sensors. *Agric. Water Manag.* **2007**, *88*, 147–158. [CrossRef]

128. Buttaroa, D.; Santamariab, P.; Signoreb, A.; Cantorea, V.; Boari, F.; Montesano, F.F.; Parente, A. Irrigation Management of Greenhouse Tomato and Cucumber Using Tensiometer: Effects on Yield, Quality and Water Use. *Agric. Agric. Sci. Procedia* **2015**, *4*, 440–444. [CrossRef]

129. Contreras, J.I.; Alonso, F.; Cánovas, G.; Baeza, R. Irrigation management of greenhouse zucchini with different soil matric potential level. Agronomic and environmental effects. *Agric. Water Manag.* **2017**, *183*, 26–34. [CrossRef]

130. Depardieu, C.; Prémont, V.; Boily, C.; Caron, J. Sawdust and bark-based substrates for soilless strawberry production: Irrigation and electrical conductivity management. *PLoS ONE.* **2016**, *11*, e0154104. [CrossRef] [PubMed]

131. Gurovich, L.A.; Ton, Y.; Vergara, L.M. Irrigation scheduling of avocado using phytomonitoring techniques. *Cienc. E Investig. Agrar.* **2006**, *33*, 117–124.

132. Seelig, H.-D.; Stoner, R.J.; Linden, J.C. Irrigation control of cowpea plants using the measurement of leaf thickness under greenhouse conditions. *Irrig. Sci.* **2012**, *30*, 247–257. [CrossRef]

133. Sarlikioti, V.; Meinen, E.; Marcelis, L.F.M. Crop Reflectance as a tool for the online monitoring of LAI and PAR interception in two different greenhouse Crops. *Biosyst. Eng.* **2011**, *108*, 114–120. [CrossRef]

134. Ozcep, F.; Yıldırım, E.; Tezel, O.; Asci, M.; Karabulut, S. Correlation between electrical resistivity and soil-water content based artificial intelligent techniques. *Int. J. Phys. Sci.* **2010**, *5*, 47–56.

135. Morimoto, T.; Hashimoto, Y. An Intelligent Control Technique Based on Fuzzy Controls, Neural Networks and Genetic Algorithms for Greenhouse Automation. *IFAC Artif. Intell. Agnculture* **1998**, *31*, 61–66. [CrossRef]

136. Mohapatra, A.G.; Lenka, S.K. Neural Network Pattern Classification and Weather Dependent Fuzzy Logic Model for Irrigation Control in WSN Based Precision Agriculture. *Phys. Procedia* **2016**, *78*, 499–506. [CrossRef]

137. Saylan, L.; Kimura, R.; Caldag, B.; Akalas, N. Modeling of Soil Water Content for Vegetated Surface by Artificial Neural Network and Adaptive Neuro-Fuzzy Inference System. *Ital. J. Agrometeorol.* **2017**, *22*, 37–44.

138. Pérez-Castro, A.; Sánchez-Molina, J.A.; Castilla, M.; Sánchez-Moreno, J.; Moreno-Úbeda, J.C.; Magán, J.J. cFertigUAL: A fertigation management app for greenhouse vegetable crops. *Agric. Water Manag.* **2017**, *183*, 186–193. [CrossRef]

139. Sánchez-Molina, J.A.; Rodríguez, F.; Guzmán, J.L.; Ramírez-Arias, J.A. Water content virtual sensor for tomatoes in coconut coir substrate for irrigation control design. *Agric. Water Manag.* **2015**, *151*, 114–125. [CrossRef]
140. Ben Ali, R.; Bouadila, S.; Mami, A. Development of a Fuzzy Logic Controller applied to an agricultural greenhouse experimentally validated. *Appl. Therm. Eng.* **2018**, *141*, 798–810. [CrossRef]
141. Guirado-Clavijo, R.; Sanchez-Molina, J.A.; Wang, H.; Bienvenido, F. Conceptual Data Model for IoT in a Chain-Integrated Greenhouse Production: Case of the Tomato Production in Almeria (Spain). *IFAC-PapersOnLine* **2018**, *51*, 102–107. [CrossRef]
142. Rodríguez, F.; Castilla, M.; Sánchez, J.A.; Pawlowski, A. Semi-virtual Plant for the Modeling Control and Supervision of batch-processes. An example of a greenhouse irrigation system. *IFAC-PapersOnLine* **2015**, *48*, 123–128. [CrossRef]

Review

Effects of Biochar on Container Substrate Properties and Growth of Plants—A Review

Lan Huang [1] and Mengmeng Gu [2,*]

[1]   Department of Horticultural Sciences, Texas A&M University, College Station, TX 77843, USA;
      huanglan92@tamu.edu
[2]   Department of Horticultural Sciences, Texas A&M AgriLife Extension Service,
      College Station, TX 77843, USA
*   Correspondence: mgu@tamu.edu; Tel.: +1-979-845-8567

Received: 15 November 2018; Accepted: 24 January 2019; Published: 1 February 2019

Abstract: Biochar refers to a processed, carbon-rich material made from biomass. This article provides a brief summary on the effects of biochar on container substrate properties and plant growth. Biochar could be produced through pyrolysis, gasification, and hydrothermal carbonization of various feedstocks. Biochar produced through different production conditions and feedstocks affect its properties and how it performs when incorporated in container substrates. Biochar incorporation affects the physical and chemical properties of container substrates, including bulk density, total porosity, container capacity, nutrient availability, pH, electrical conductivity and cation exchange capacity. Biochar could also affect microbial activities. The effects of biochar incorporation on plant growth in container substrates depend on biochar properties, plant type, percentage of biochar applied and other container substrates components mixed with biochar. A review of the literature on the impact of biochar on container-grown plants without other factors (such as irrigation or fertilization rates) indicated that 77.3% of the studies found that certain percentages of biochar addition in container substrates promoted plant growth, and 50% of the studies revealed that plant growth decreased due to certain percentages of biochar incorporation. Most of the plants tested in these studies were herbaceous plants. More plant species should be tested for a broader assessment of the use of biochar. Toxic substances (heavy metals, polycyclic aromatic hydrocarbons and dioxin) in biochars used in container substrates has rarely been studied. Caution is needed when selecting feedstocks and setting up biochar production conditions, which might cause toxic contaminants in the biochar products that could have negative effects on plant growth.

Keywords: container substrates; physical properties; chemical properties; biomass

---

1. Introduction

Biochar refers to processed, carbon-rich material derived from biomass [1–3]. Recent research has shown that biochar can be used as a replacement for commonly-used container substrates [4–8]. Container substrates are often soilless, making it easy to achieve consistency. Primary substrate components include peat moss, vermiculite, perlite, bark, and compost [9]. Peat moss is an excellent substrate component; it has essential characteristics such as low pH, high cation exchange capacity (CEC), and appropriate aeration and good container capacity [10–12], which are ideal for horticultural container application. However, intensive extraction of peat from peatlands can damage natural habitats and release CO_2 into the atmosphere if the disturbed peatland is left unrestored [13]. The United Kingdom government has thus proposed reducing the use of peat [14]. The cost of this commonly-used substrate is also high due to the extreme cost of transportation, fuel for extraction, and processing [9,15]. Therefore, it is beneficial and necessary to search for alternative environmentally-friendly and local substrate components [9,16]. Research has shown that biochar could be a potential alternative to commonly-used substrates. Using biochars (a byproduct of bioenergy production) in agriculture adds value to bioenergy production [17]. Biochar could offer economic advantages over other commonly-used substrates, if produced on site. Extensive research has shown that replacing a certain percentage of commonly-used container substrates with biochar could increase plant growth in certain conditions [18–22].

However, biochars are variable, and their impact on container substrates could vary. It would be of interest to examine the characteristics of biochars, their incorporation in container substrates, and their effects on diverse types of container-grown plants. In this review, we provide a brief summary of the effects of biochar on container substrate properties and plant growth, and discuss the potential mechanism behind their effects. This review examines factors related to the impact of biochar, which include feedstock sources, production conditions, percentage of biochar applied, other substrate components mixed with biochar, and plant species. These factors can help address the general hypothesis that incorporation of biochar may not always have beneficial effects on container substrate properties or plant growth.

2. Biochar Production

There are many variables prior to, during, and after production of biochar. These factors will eventually affect biochar properties and its effect on plant growth and container properties when incorporated in container substrates.

2.1. Biochar Production Methods

There are three main processes to produce biochar: pyrolysis, gasification and hydrothermal carbonization. Pyrolysis is the thermal decomposition of biomass by heating (around 400 °C to 600 °C) without oxygen [23–25]. Compared to pyrolysis, gasification is conducted under small amounts of oxygen at relatively higher temperatures (around 700 °C to 1200 °C) [2]. Gasification produces smaller quantities of biochar with lower carbon (C) content than pyrolysis [2,25,26]. Hydrothermal carbonization uses water and catalysts at lower temperatures (180 to 300 °C) under high pressure to convert biomass to a different type of biochar product, hydrochar [27,28]. Hydrochars are acidic, and have low surface areas, less aromatic compounds, and higher CEC than those produced by pyrolysis and gasification [28,29]. Production temperature significantly influences the characteristics of biochars (Table 1). Biochar made from pruning waste at 500 °C had higher pH and different container capacity, total porosity, electrical conductivity (EC) and CEC, when compared to biochar produced at 300 °C [20]. Biochars made from different production processes can have different physical and chemical properties.

Utilizing biochar in agriculture adds values to biomass pyrolysis and gasification. The main purpose of fast pyrolysis is to produce syngas and bio-oil [17,23] and gasification syngas [30], with biochar being the byproduct. Syngas mainly includes carbon monoxide and hydrogen [31]. It could be used to provide energy for other pyrolysis processes. Bio-oil could be burned to produce heat or further processed to be used as fuel [32]. A specific process and its heating rate could be modified to produce desirable products. For example, gasification has higher yields of syngas and energy than pyrolysis [33]. Liquid bio-oil produced by pyrolysis has higher energy density and is cheaper and easier to transport; however, it is corrosive, which makes it difficult to store for a long time [31]. Slow pyrolysis produces more biochar and syngas, and fast pyrolysis more bio-oil [34]. The residence time (the amount of time taken in the pyrolysis procedure) of slow pyrolysis is from 5 min to 30 min, while that of fast pyrolysis is from seconds to less than a second, and the temperatures are higher [23,35]. Raising pyrolysis temperature can decrease biochar yield [36,37]. Production conditions can be adjusted on the basis of whether the desirable products are biochar or bioenergy products (bio-oil or syngas). Low temperatures and slow pyrolysis could be used to produce more biochar than the other products.

Pre-treatment of feedstocks has been reported to have a significant influence on biochar ash content, yield, and properties. Pre-treatment of biomass, such as washing with water or acid, could help remove some ash culprits in feedstocks to reduce fouling, and improve the quality of biomass feedstock and final biochar products [37,38]. Rahman et al. [37] tested the effectiveness of different pre-treatments by comparing the EC of the initial washing medium and leachate collected after treatments. The result showed that the leachate EC of palm kernel shell increased, when pre-treated with dilute acid, dilute alkali, and distilled water. The highest increase in EC was found using dilute acid pre-treatment as a result of the removal of soil and alkaline metal by the acid solution and degradation of the biomass chemical composition. The ash content of palm kernel shell was reduced when pretreated with distilled water or diluted acid. The ash content increased with alkaline pre-treatment since abundant sodium ions in alkaline medium prevented ions from leaching into the medium and ions were bound and tied up by the biomass particles, which resulted in a high amount of ash content [37]. Torrefaction pre-treatment, which is a low temperature thermal conversion conducted without oxygen aiming to reduce moisture content of the biomass, could increase biochar yield during pyrolysis because the pretreatment predisposed carbon and oxygen content to remain as solids [39]. Another study showed that paper mill sludge as biochar feedstock, pre-treated with phosphoric acid and torrefaction, followed by pyrolysis, resulted in reduced volatile matter content, increased inorganic matter, and increased biochar yield [40]. It was shown that biochar made from feedstock with pretreatments such as light bio-oil or phosphoric acid may have larger surface areas and more porous structure [41,42], which could influence the effects of biochar on air space, nutrient and water-holding ability, and microbial activity. Biochar made from bark pre-treated with tannery slurry as an alkaline treatment could have a higher $NH4^+$ absorption capacity, as well as more surface functional groups (carboxyl and carbonyl groups) formed than untreated ones [43], causing increased CEC of biochar. Silica enrichment was also found in biochar made from rick husk pretreated with bio-oil or HCl [42].

In addition to pre-treatments of feedstocks, post-treatments could also change biochar properties. Some biochars could contain toxic compounds such as polycyclic aromatic hydrocarbons (PAHs) during production. Drying biochars at temperature of 100 °C, 200 °C, and 300 °C significantly decreased the amount of PAHs in biochars, which indicated that the release of PAHs from biochars was due to the increased opening of the pores and diffusion of PAHs from the pores after the thermal treatment [44]. Biochar could be treated and mixed with other substances. Dumroese et al. [7] dry-blended biochar with wood flour, polylactic acid, and starch to form pelleted biochar, which is preferred over the original fine-textured and dusty form for its handling convenience and even incorporation. McCabe et al. [45] evenly blended soybean-based bioplastics with biochar in a pelletized form as a source of nutrients in container substrates.

2.2. Biochar Feedstocks

In addition to production conditions, biochars could be made from varying feedstocks, which would contribute to differences in physical and chemical properties (Table 1). The feedstocks could be waste materials such as green waste [18], forest waste [46,47], wheat straw [5], sugarcane bagasse [48], rice hull [49], crab shell [50] and *Eucalyptus saligna* wood chips (byproduct of construction, fuel-wood and pulp wood) [51]. Biochars could also be made from non-waste materials such as holm oak [52], conifer wood [53], citrus wood [54] and pine wood [6,55–57]. The crab shell biochar and oak chip biochar have different pH ECs, and C, nitrogen (N), phosphorus (P) and potassium (K) content, although they were made by the same production method and temperature [50]. Straw biochar had a higher pH, exchangeable cations, and K content compared to a wood biochar [5]. The biochars made from sewage sludges of two different municipal plants also had slightly different pHs and N content [19]. Biochar could have high P and K content and could be used as P and K fertilizers, when made from rice hulls with high content of the minerals [49]. It was shown that biochar properties were related to the properties of the original feedstock [49]. The biochars made from different feedstocks could have different physical and chemical properties, which should be taken into account when they are incorporated in containers.

Table 1. Summary of the feedstock, production condition and properties of the biochars used in container substrates.

Biochar Feedstock	Production Temp (°C)	CC (%)	AS (%)	TP (%)	BD (g cm^{-3})	pH	EC (dS m^{-1})	CEC (cmol kg^{-1})	N (%)	C (%)	P (%)	K (%)	Na (%)	Ca (%)	Mg (%)	S (%)	Reference
Citrus wood [z]	n	n	n	n	n	7.6	1.6	n	0.6	70.6	0.0008	0.37	0.32	0.02	0.01	0.07	[54]
Coir (coconut husk fiber)	450	64	33	97	0.14	8.2	1.0	153	1.3	n	0.17	1.89	4.83	0.33	0.73	n	[21]
Conifer wood	450	n	n	92	0.64	8.5	0.4	n	n	n	n	n	n	n	n	n	[53]
Crab shell [z]	200–250	n	n	n	n	8.8	0.005	n	3.6	28.7	0.03	0.61	0.04	0.18	0.08	n	[50]
Eucalyptus saligna wood chip	550	n	n	n	n	8.8	0.2	n	0.3	83.6	0.02	0.24	n	2.13	0.11	0.05	[51]
Forest waste	n	n	n	n	n	9.6	0.7	n	0.7	59.5	0.08	0.87	0.04	2.90	0.24	0.07	[46,47]
Green waste	550	n	n	n	n	7.7	n	250	0.3	77.5	n	n	n	n	n	0.00	[58]
Green waste (willow, pagoda tree and poplar)	n	27	22	49	0.44	8.0	0.9	n	1.2	50.4	0.01	0.47	n	n	n	n	[18]
Green waste (tomato crop)	550	n	28	n	0.13	10.4	3.3	524	n	55.0	n	n	n	n	n	n	[59]
Hardwood pellets	n	n	n	n	0.38	8.0	1.1	n	n	n	0.0005	0.04	0.001	0.02	0.002	n	[5]
Holm oak	650	51	29	80	0.32	9.3	0.5	n	0.9	n	0.18	0.77	n	3.76	0.40	n	[52]
Mixed hardwood (oak, elm, and hickory)	450	n	n	n	0.28	n	n	n	n	n	0.29	3.59	0.02	38.28	0.97	n	[60]
Mixed hardwood	n	60	24	85	0.15	11.2	2.0	n	0.2	n	0.05	0.64	0.01	2.75	0.13	0.02	[61]
Mixed softwood [y]	800	n	n	n	n	10.9	0.5	19	n	n	0.02	n	n	n	n	n	[62]
Oak chip [z]	200–250	n	n	n	n	5.1	0.3	n	0.1	52.2	0.09	0.10	0.06	1.03	0.08	n	[50]
Olive mill waste	500	n	n	n	n	9.7	9.2	n	0.6	59.5	0.90	6.42	0.05	3.40	0.61	0.17	[47]
Pine chip [z]	200–250	n	n	n	n	6.4	0.03	n	0.3	53.7	0.05	0.65	0.05	0.23	0.08	n	[50]

Table 1. *Cont.*

Biochar Feedstock	Production Temp (°C)	CC (%)	AS (%)	TP (%)	BD (g cm^{-3})	pH	EC (dS m^{-1})	CEC (cmol kg^{-1})	N (%)	C (%)	P (%)	K (%)	Na (%)	Ca (%)	Mg (%)	S (%)	Reference
Pine cone [z]	200–250	n	n	n	n	5.1	1.2	n	0.6	53.2	0.01	0.16	0.04	0.36	0.05	n	[50]
Pine wood	450	49	34	83	0.17	n	n	n	0.4	48.1	n	0.10	n	0.50	0.30	n	[6,57]
Pine wood	450	47	36	83	0.18	5.4	0.2	n	n	n	n	n	n	n	n	n	[55,56]
Poplar [y]	1100–1200	57	34	91	n	9.7	0.2	n	0.7	n	0.51	0.98	n	4.31	7.64	n	[63]
Pruning wastes	300	17	4	21	0.18	7.5	0.3	26	1.2	66.2	0.004	n	n	n	n	n	[20]
Pruning wastes	500	35	4	39	0.18	10.3	1.0	16	1.2	77.7	0.01	n	n	n	n	n	[20]
Rice husk	500	n	n	n	n	10.2	0.8	50	0.3	20.5	n	n	n	n	n	n	[64]
Rice husk [z]	n	n	n	n	0.30	7.3	n	n	1.1	n	0.10	0.50	n	n	n	n	[65]
Rice husk [z]	200–250	n	n	n	n	6.3	0.4	n	0.6	45.4	1.21	0.27	0.73	15.80	1.04	n	[50]
Rice hull [y]	815–871	n	n	n	0.20	10.5	n	n	0.2	17.7	0.30	0.98	n	0.35	0.15	0.03	[49,66,67]
Sewage sludge [x]	450	n	n	n	n	7.9	1.1	n	1.1	n	n	n	n	n	n	n	[19]
Sewage sludge [x]	450	n	n	n	n	7.5	1.1	n	3.1	n	n	n	n	n	n	n	[19]
Southern yellow pine	400	n	n	n	n	6.0	n	n	n	n	0.03	0.29	n	0.06	0.12	0.08	[68]
Spruce wood [y]	1100–1200	29	63	92	n	11.1	0.3	n	0.2	n	0.05	0.74	n	1.34	0.17	n	[63]
Sugarcane bagasse [z]	343	n	n	n	0.11	5.8	n	n	n	n	n	n	n	n	n	n	[48]
Sugarcane bagasse [z]	343	n	n	n	0.11	6.1	n	n	n	n	n	n	n	n	n	n	[48]
Switchgrass [z]	1000	n	n	n	0.10	10.8	3.5	n	1.3	79.0	1.20	6.60	n	n	n	n	[69]
Wheat straw	600	n	n	n	0.31	10.0	1.0	n	1.0	79.3	n	n	n	n	n	n	[70]
Wheat straw	n	n	n	n	0.24	9.5	2.5	n	n	n	0.003	0.10	0.002	0.004	0.0009	n	[5]

Note: Production temp: production temperature; CC: container capacity; AS: air space; TP: total porosity; BD: bulk density; EC: electrical conductivity; CEC: cation exchange capacity. Pyrolysis was the biochar production method, unless indicated otherwise. "n" means not available. [z]: Biochar production method was not available. [y]: Biochar was produced from gasification. [x]: Two different sewage sludges were selected from two municipal plants.

3. Effects of Biochar on Container Substrates

3.1. Physical Properties

3.1.1. Bulk Density

The addition of biochar affects the physical properties of container substrates. Biochars have a higher bulk density than commonly-used substrate components, such as peat moss, perlite, and vermiculite. Using biochar to replace certain percentages of peat could thus increase the bulk density of the substrates [5,7,18,71,72].

3.1.2. Container Capacity, Air Space and Total Porosity

Biochar incorporation in container substrates may affect container capacity, air space, and total porosity. Particle size distribution of the substrate components is important for determining their physical properties [73]. Due to the differing particle sizes of biochars and substrate components, the effects of biochar incorporation on the physical properties of a container substrate will vary. Container capacity is the maximum percent volume of water a substrate can hold after gravity drainage [74]. Container substrates absorb water in small pores (micropores) between, or inside component particles [10]. Méndez et al. [75] showed that the incorporation of 50% (by vol.) biochar with peat increased container capacity, compared to those with 100% peat substrate due to increased micropores after biochar incorporation. Similar to these results, Zhang et al. [21] also reported that mixing 20% or 35% (w/w) biochar with compost made from green waste increased container capacity. Yet, some research has shown that the incorporation of biochar in container substrates had no effect on container capacity [5,18]. The differing results after biochar incorporation could be due to the different particle sizes of the biochars and the substrate components used. Besides container capacity, biochar incorporation could also affect air space. Air space is the proportion of air-filled large pores (macropores) after drainage [10]. Méndez et al. [75] showed that the incorporation of 50% (by vol.) biochar with peat increased the air space compared to 100% peat substrate. In this study, the percentage of particle size larger than 2 mm was 29% (w/w) for biochar but 8.8% for peat. Thus, the increased air space was caused by an increased number of macropores due to the incorporation of biochar with larger particle size. Zhang et al. [21] confirmed this by showing that mixing biochar with compost increased the percentage of particles larger than 2 mm and thus increased the air space. Total porosity is the sum of air space and container capacity. The effect of biochar on total porosity is related to its effect on air space and container capacity. Substituting peat with 50% biochar (by vol.) made from green waste had no effect on total porosity [18]. Méndez et al. [75] concluded that the addition of biochar produced from deinking sludge increased the total porosity. Zhang et al. [21] also showed that mixing biochar with compost increased the total porosity. Vaughn et al. [5] showed that the effects of biochar on total porosity were mixed and there was no specific trend, when mixing biochar with peat. In summary, biochar incorporation could impact total porosity, air space, and container capacity.

3.2. Chemical Properties

3.2.1. pH

In general, biochar is effective at increasing the pH of container substrates since the pH of biochars used in most research is neutral to basic [21,53,58,59]. Biochar could buffer acidity due to the negative charge on the surface of biochar [76]. However, the pH of biochars could be acidic. The pH of the biochar depends on the nature of the feedstock and the temperatures during biochar production. The lower the temperature of production, the lower the pH of the biochar. The pH of oak wood biochar was 4.8 when produced at 350 °C [24]. Khodadad et al. [77] also showed pH of biochar made from pyrolysis of oak and grass at 250 °C was 3.5. Lima et al. [78] showed that the pH was around 5.9 for biochars made from pecan shell at 350 °C and switchgrass at 250 °C.

3.2.2. Electrical Conductivity

Biochar incorporation could increase container substrate EC due to high EC of the biochar used. The EC of biochar was affected by the biochar functional groups (such as fused-ring aromatic structures and anomeric O-C-O carbons), metal oxide precipitates and binding of metals [24,79]. Hossain et al. [80] also found that as pyrolysis temperature increased, EC of the sludge biochars decreased. When incorporating biochar in container substrate, Vaughn et al. [5] showed that mixing 5%, 10%, and 15% (by vol.) pelletized wheat straw and hardwood biochars with container substrates containing peat moss and vermiculite increased the EC. Tian et al. [18] also found that adding 50% (by vol.) biochar made from green waste to peat moss media significantly increased EC. The increased substrate EC after biochar incorporation could be due to the high pH, large surface area, and charge density of the biochar [70].

3.2.3. Cation Exchange Capacity

Biochar incorporation could affect CEC and nutrient availability, which is related to the original properties of biochar itself. Surface functional groups, such as carboxylate, carbonyl and ether are responsible for the CEC of biochar [81]. Different biochars have different chemical functional groups. Vaughn et al. [5] found that some volatile materials were removed and wood cellulosic polymers were carbonized in wood biochar after pyrolysis, while wheat straw biochar was less carbonized and had more chemical functionality, which serves as exchange sites for nutrient absorption. It was shown that CEC was higher in a 25% biochar and 75% peat moss mix (by vol.) than that in 100% peat moss [22]. Some biochars can even provide nutrients to the plants due to the high concentration of certain nutrients in the original feedstocks. Some forms of biochars can serve as a source of P and K, which leads to increased availability of these minerals in container substrates and improved fertility [49,66,82].

3.3. Effects on the Microbial Activities

Biochar incorporation may affect microbial activity and biomass in containers. Adding biochar can increase pH, available water content, and influx of nutrients as discussed above, thus stimulating microbial communities and increasing microbial biomass. Warnock et al. [83] also indicated that porous biochar with a high surface area could provide shelter for microorganisms. Saito [84] showed that biochar could serve as a microhabitat for arbuscular mycorrhizal fungi. Higher mycorrhizal colonization and plant growth were shown in mixes of biochar and soil in container experiments [85]. However, only a limited amount of research investigated the effects of biochar on microbial activity or inoculation with mycorrhizae in soilless substrates. Increased mycorrhizal colonization was found in containers containing sand and clay in a ratio of 3:1 (by vol.) with activated biochar (2 g per container) [86]. Inoculation with arbuscular mycorrhizas fungus significantly increased *Pelargonium zonale* plant growth in containers with 0%, 30% or 70% (by vol.) biochar with the rest being peat [87]. Biochars produced at different temperatures may have different surface areas and adsorption abilities [88], which could lead to different levels of nutrient retention and effects on microbial activities.

4. Effects of Biochar on Plant Growth in Container Substrates

There is an increasing amount of research on the effects of biochar on container-grown plant growth that shows the potential for biochar to be a replacement for commonly-used soilless container substrate components including peat moss, bark, vermiculite, perlite, coir, etc. Mixing biochar in container substrates may have a positive impact on plant growth due to beneficial effects like improved container physical and chemical properties and enhanced nutrient and water retention, as mentioned above. Tian et al. [18] found that mixing biochar made from green waste with peat (50% each, by vol.) increased total biomass and leaf surface area of *Calathea rotundifolia* cv. Fasciata when compared to that of peat substrates alone, because of improved substrate properties and increased nutrient retention after biochar incorporation. Replacing 10% (by vol.) of peat with sewage sludge biochar enhanced lettuce (*Lactuca sativa*) biomass production by 184%–270% when compared to 100% peat-based substrate, due to increased N, P and K concentrations and microbial activities [19]. Incorporation of biochar produced from pruning waste at 300 °C (pH = 7.53) and 500 °C (pH = 10.3) into peat substrates at the ratio of 50% and 75% (by vol.) increased lettuce biomass when compared to those in peat alone (pH = 6.14), probably because the increased pH after biochar incorporation was more ideal for many crops [20]. Graber et al. [54] tested the effects of mixing three ratios of citrus wood biochar (1%, 3% or 5%, w/w) with commercial container substrates (a mixture of coconut fiber and tuff at a 7:3 ratio by vol.) on the growth of peppers (*Capsicum annuum*) and tomatoes (*Solanum lycopersicum*). The effects included increased leaf area, shoot dry weight (after detaching the fruits), numbers of flowers and fruit of pepper and increased plant height and leaf size of tomato plants compared to those in commercial container substrates. Graber et al. [54] indicated two possible reasons for the responses, increased beneficial microbial populations or low doses of biochar chemicals stimulating plant growth (hormesis). Mixing 20% or 35% (w/w) biochar made from coir in composted green waste medium increased plant height, root and shoot length, and root fresh and dry weight of *Calathea insignis* when compared to one without any biochar incorporation, effects due to increased water retention, optimized total porosity, aeration porosity, water-holding porosity, nutrients, and microbial activities [21]. Overall, increased plant growth after biochar incorporation could be attributed to increased availability of nutrients and improved water retention, both desirable substrate properties.

However, biochar incorporation may not always improve plant growth. Not all biochars are the same (Table 1). The effects of biochars on container-grown plants are variable (Tables 2–5) depending on multiple factors. There are distinct interactions between biochar and different substrate components. Different biochars, biochar incorporation rate, and other components mixed with biochar can contribute to differing results. Furthermore, individual plant responses to biochar also vary. Across studies of the effects of biochar alone on plant growth, without other factors such as irrigation or fertilization rates, (Tables 2–4), 77.3% reported that some biochar addition to container substrates could promote plant growth, and 50% revealed that plant growth or dry weight was suppressed by some biochar in container substrates. Most studies (69.4%) in Tables 2–5 investigated plant growth in container substrates with biochar for 12 weeks or less than 12 weeks. The length of the experiments in these studies varied from 3 weeks to 7 months. Many mechanisms of biochar-plant interactions are not fully understood.

4.1. Different Plant Species

The impact of biochar on plant growth differs by species since different plants have different suitable growth conditions or different tolerance to certain stresses. Mixing potato anaerobic digestate with acidified wood pellet biochar (1:1, by vol.) led to higher fresh and dry weight of tomatoes than a peat: vermiculite control, but led to lower fresh and dry weight of marigold (*Calendula officinalis*) plant [71]. The EC of potato anaerobic digestate is high (7.1 dS m^{-1}). The different fresh and dry weight responses of tomato and marigold could be due to the salt tolerances of these two plants [71]. Choi et al. [57] also showed that mixes with 20% pine bark and 80% biochar (by vol.) led to higher chrysanthemum (*Chrysanthemum nankingense*) fresh and dry weight, but lower tomato plant fresh and dry weight when compared to the control. The reduced tomato plant fresh weight and dry weight was because tomato usually requires more nutrients than other plants and biochar can hold or capture nutrients. Furthermore, 80% biochar mixes had no effect on lettuce (*Lactuca sativa*) and basil (*Ocimum basilicum*) fresh and dry weights. Altland and Locke [67] also showed that mixes of 20% (by vol.) gasified rice hull biochar with Sunshine Mix #2 fertilized with 100 mg L^{-1} N using ammonium nitrate and 0.9 kg m^{-3} Micromax caused a smaller *Pelargonium x hortorum* shoot dry weight but increased shoot dry weight of tomato plants when compared to the control (Sunshine Mix #2) fertilized at the rate of 100 mg L^{-1} N with a commercial complete fertilizer with micronutrients.

Table 2. Summary of the effects of biochar made from different feedstocks mixed with other substrate components on container-grown plants, with percentage of biochar in container substrates less than 50% (by vol.).

Plant Species	Non-Biochar Components	Biochar Feedstock	Percentage (%, by vol.) of Biochar and the Effects on Plants' Dry Weight/Growth Index (DW/GI) [z]								Reference
			1	5	10	15	20	25	30	40	
Buxus sempervirens × *Buxus microphylla*	Pine bark and 24 g osmocote 18N–6P–12K	Switchgrass			=/n			=/n			[69]
Calendula offcinalis	Coir	Forest waste			=/n			=/n			[46]
Chrysanthemum nankingense	Pine bark	Pine wood					=/=			=/=	[57]
Cucumis melo	Sunshine commercial growing medium	Standard sugarcane bagasse						=/= [y]			[48]
		Sugarcane bagasse using a pneumatic transport system						=/= [y]			
Cucurbita pepo	Sunshine commercial growing medium	Standard sugarcane bagasse						=/= [y]			
		Sugarcane bagasse using a pneumatic transport system						=/= [y]			
Euphorbia × *lomi*	Peat	Conifer wood					n/= [y]			n/+ [y]	[53]
Euphorbia pulcherrima	Sunshine Mix #1	Pine wood					+/=			=/=	[55]
Hydrangea paniculata	Pine bark and 24 g osmocote 18N–6P–12K	Switchgrass			=/n			-/n			[69]
Lactuca sativa	Peat	Sewage sludge			+/+ [y]						[19]
Lactuca sativa 'Black Seeded Simpson'	Pine bark	Pine wood					n/=			n/=	[57]
Lilium longiflorum	Sunshine Mix #1	Pine wood					= [x] /= y			= [x] /= y	[56]
Ocimum basilicum 'Genovese'	Pine bark	Pine wood					=/n			=/n	[57]
Ocimum basilicum	Peat	Softwood from spruce wood							=/n		[63]
		Harwood from poplar							=/n		
Ocimum basilicum	5% vermicompost (VC) with the rest being Berger BM7	Mixed hardwood					=/=			=/=	[61]
	10% VC with the rest being Berger BM7						=/=			=/=	
	15% VC with the rest being Berger BM7						=/=			+/=	
	20% VC with the rest being Berger BM7						=/=			+/=	

Table 2. *Cont.*

Plant Species	Non-Biochar Components	Biochar Feedstock	Percentage (%, by vol.) of Biochar and the Effects on Plants' Dry Weight/Growth Index (DW/GI) [z]								Reference
			1	5	10	15	20	25	30	40	
Petunia hybrida	Coir	Forest waste			=/n			=/n			[46]
Solanum lycopersicum. 'Red Robin'	50% vermiculite with the rest being peat and biochar	Pelletized wheat straw		=/+ [y]	=/+ [y]	=/+ [y]					[5]
		Hardwood pellets		=/+ [y]	=/+ [y]	=/+ [y]					
Solanum lycopersicum 'Cuarenteno'	Coir	Forest waste			=/n			=/n			[47]
		Olive mill waste			+/n			=/n			
Solanum lycopersicum 'Gransol Rijk Zwaan'		Forest waste			=/n			-/n			
		Olive mill waste			=/n			=/n			
Solanum lycopersicum 'Hope'	Pine bark	Pine wood					=/=			+/=	[57]
Solanum lycopersicum 'Roma'	5% VC with the rest being Berger BM7	Mixed hardwood					=/=			=/=	[61]
	10% VC with the rest being Berger BM7						+/=			=/=	
	15% VC with the rest being Berger BM7						=/+			+/=	
	20% VC with the rest being Berger BM7						=/=			=/=	
Tagetes erecta 'Inca II Yellow Hybrid'	50% vermiculite with the rest being peat and biochar	Pelletized wheat straw		=/+ [y]	=/+ [y]	=/+ [y]					[5]
		Hardwood pellets		=/= [y]	=/+ [y]	=/+ [y]					

[z]: "+" means increased; "=" means there was no significant difference; "-" means decreased; "n" means not available. [y]: Result for this was for plant height not growth index. [x]: Result for this was for leaf dry height not total dry weight.

Table 3. Summary of the effects of biochar made from different feedstocks mixed with other substrate components on container-grown plants, with percentage of biochar in container substrates ranging from 50% to 100% (by vol.).

Plant Species	Non-Biochar Components	Biochar Feedstock	Percentage (%, by vol.) of Biochar and Its Effect on Plants' Dry Weight/Growth Index (DW/GI) [z]								Reference
			50	60	66	70	75	80	90	100	
Anethum graveolens	Perlite	Rice husk	+/+ [y]								[64] [w]
Brassica rapa ssp. *pekinensis*			+/+ [y]								
Brassica rapa var. *rosularis*			+/+ [y]								

Table 3. *Cont.*

Plant Species	Non-Biochar Components	Biochar Feedstock	Percentage (%, by vol.) of Biochar and Its Effect on Plants' Dry Weight/Growth Index (DW/GI) [z]								Reference
			50	60	66	70	75	80	90	100	
Calathea rotundifolia cv. Fasciata	Peat	Green waste	+/n								[18]
Chrysanthemum nankingense	Pine bark	Pine wood		=/n				+/n		=/n	[57]
Cucumis melo	Sunshine commercial growing medium	Standard sugarcane bagasse	=/= [y]				-/= [y]			-/= [y]	[48]
		Sugarcane bagasse using a pneumatic transport system	=/+ [y]				=/= [y]			-/= [y]	
Cucurbita pepo	Sunshine commercial growing medium	Standard sugarcane bagasse	=/= [y]				=/= [y]			=/= [y]	
		Sugarcane bagasse using a pneumatic transport system	+/+ [y]				=/= [y]			=/= [y]	
Euphorbia × lomi	Peat	Conifer wood		+/+ [y]				+/+ [y]		n/= [y]	[53]
Euphorbia pulcherrima	Sunshine Mix #1	Pine wood		-/=				-/=		-/-	[55]
Lactuca sativa	Perlite	Rice husk	+/+ [y]								[64] [w]
Lactuca sativa	Peat	Deinking sludge	+/n								[75]
	Coir		-/n								
Lactuca sativa	Peat	Pruning wastes	+/n				+/n				[20]
Lactuca sativa 'Black Seeded Simpson'	Pine bark	Pine wood		n/=				n/=		n/=	[57]
Lilium longiflorum	Sunshine Mix #1	Pinewood		= [x] /= y				= [x] /= y			[56]
Malva verticillata	Perlite	Rice husk	+/+ [y]								[64] [w]
Ocimum basilicum	5% VC with the rest being Berger BM7	Mixed hardwood		=/=							[61]
	10% VC with the rest being Berger BM7			=/=							
Ocimum basilicum	15% VC with the rest being Berger BM7	Mixed hardwood		+/=							[61]
	20% VC with the rest being Berger BM7			=/=							
Ocimum basilicum	5% chicken manure compost (CM) with the rest being Berger BM7	Mixed hardwood		=/=			=/=	-/-			[61]
Ocimum basilicum	5% VC with the rest being Berger BM7			=/=			=/=	-/-			

Table 3. *Cont.*

Plant Species	Non-Biochar Components	Biochar Feedstock	Percentage (%, by vol.) of Biochar and Its Effect on Plants' Dry Weight/Growth Index (DW/GI) [z]								Reference
			50	60	66	70	75	80	90	100	
Ocimum basilicum 'Genovese'	Pine bark	Pine wood		=/n				=/n		-/n	[57]
Solanum lycopersicum 'Red Robin'	Potato digestate	Wood pellet	+/= [y]								[71]
		Pelletized wheat straw	+/= [y]								
		Pennycress presscake	=/= [y]								
Solanum lycopersicum 'Gransol Rijk Zwaan'	Coir	Forest waste	-/n				-/n			-/n	[47]
		Olive mill waste	-/n				-/n			-/n	
Solanum lycopersicum 'Cuarenteno'	Coir	Forest waste	-/n				-/n			-/n	
		Olive mill waste	=/n				=/n			=/n	
Solanum lycopersicum	Faecal sludge based compost	Rice husk	=/= [y]							-/- [y]	[65]
Solanum lycopersicum 'Roma'	5% VC with the rest being Berger BM7	Mixed hardwood		=/=							[61]
	10% VC with the rest being Berger BM7			=/=							
	15% VC with the rest being Berger BM7			=/=							
	20% VC with the rest being Berger BM7			=/=							
Solanum lycopersicum 'Tumbling Tom Red"	5% CM with the rest being Berger BM7	Mixed hardwood		+/=		=/=		-/-			[61]
	5% VC with the rest being Berger BM7			=/=		=/=		=/=			
Solanum lycopersicum 'Hope'	Pine bark	Pine wood		=/=				-/=		-/=	[57]
Tagetes erecta	Potato digestate	Wood pellet	-/= [y]								[71]
		Pelletized wheat straw	=/= [y]								
		Pennycress presscake	-/= [y]								

[z]: "+" means increased; "=" means there was no significant difference; "-" means decreased; "n" means not available. [y]: Result for this was for plant height not growth index. [x]: Result for this was for leaf dry height not total dry weight. [w]: Hydroponic experiment.

Table 4. Summary of the effects of biochar made from different feedstocks mixed with other substrate components on container-grown plants, with percentage of biochar in container substrates measured by weight.

Plant Species	Non-Biochar Components	Biochar Feedstock	Percentage (%, by weight) of Biochar and Its Effect on Plants' Dry Weight/Growth Index (DW/GI) [z]										Reference
			1	2.5	3	5	10	20	35	40	60	80	
Acmena smithii	Growing medium (pine bark, coir, clinker ash and coarse sand) with 3 kg m^{-3} controlled-release fertilizer (CRF)	*Eucalyptus saligna* wood chip		=/n		=/n	=/n						[51]
	Growing medium (pine bark, coir, clinker ash and coarse sand) with 6 kg m^{-3} CRF			+/n		=/n	=/n						
	Growing medium (pine bark, coir, clinker ash and coarse sand) with no CRF			+/n		=/n	=/n						
Calathea insignis	Composted green waste medium	Coir (coconut husk fiber)						+/+ y	+/+ y				[21]
	0.5% humic acid (w/w) with the rest being green waste compost							+/+ y	+/+ y				
	0.7% humic acid (w/w) with the rest being green waste compost							+/+ y	+/= y				
Capsicum annuum	A mixture of coconut fiber and tuff at a ratio of 7:3 (by vol.)	Citrus wood	n/= y		n/= y								[54]
Capsicum annuum	Sphagnum peatmoss-based medium in 50-cell transplant trays	Hardwood including oak, elm, and hickory						=/= y		=/= y	-/- y	-/- y	[60]
	Sphagnum peatmoss-based medium in 72-cell transplant trays							=/+ y		=/= y	=/= y	-/- y	
	Sphagnum peatmoss-based medium in 98-cell transplant trays							=/= y		=/= y	=/= y	-/- y	
Solanum lycopersicum	A mixture of coconut fiber and tuff at a ratio of 7:3 (by vol.)	Citrus wood	n/+ y		n/+ y								[54]
Viola × *hybrida*	Growing medium (pine bark, coir, clinker ash and coarse sand) blended with 3 kg m^{-3} CRF	*Eucalyptus saligna* wood chip		=/n		=/n	=/n						[51]
	Growing medium (pine bark, coir, clinker ash and coarse sand) blended with 6 kg m^{-3} CRF			+/n		=/n	+/n						
	Growing medium (pine bark, coir, clinker ash and coarse sand) with no CRF			-/n		=/n	-/n						
Viola × *wittrockiana*	Growing medium (pine bark, coir, clinker ash and coarse sand) blended with 3 kg m^{-3} CRF	*Eucalyptus saligna* wood chip		-/n		=/n	=/n						[51]
	Growing medium (pine bark, coir, clinker ash and coarse sand) blended with 6 kg m^{-3} CRF			+/n		=/n	=/n						
	Growing medium (pine bark, coir, clinker ash and coarse sand) with no CRF			+/n		=/n	=/n						

[z]: "+" means increased; "=" means there was no significant difference; "-" means decreased; "n" means not available. [y]: Result for this was for plant height not growth index.

Table 5. Other studies testing the effects of biochar mixed with other substrate components on container-grown plants.

Plant Species	Non-Biochar Components	Biochar Feedstock	Biochar Percentage (by vol.)	Effects on Plant Growth [z]	Other Information	Reference
Agrostis stolonifera	Sand	Southern yellow pine	15%	Plant height (=)/DW (=)/FW (=)	Control was mixes with 85% sand and 15% peat	[68]
	85% sand and 10% peat, vermicompost, yard-waste compost, Organimix compost, humus or worm castings		5%	Plant height (=)/DW (=)/FW (=)		
	85% sand and 10% anaerobic biosolids		5%	Plant height (+)/DW (+)/FW (+)		
Helianthus annuus	Pig slurry compost	Holm oak	40% or 80%	Shoot DW (+)	Compared to mixes with 40% or 80% coir with the rest being pig slurry compost, respectively	[52]
	Pig slurry compost		60%	Shoot DW (=)	Control was mixes with 60% coir with the rest being pig slurry compost	
	No		100%	Shoot DW (=)	Control was 100% coir	
Ipomoea aquatica	Spent pig litter compost, vermiculite, perlite and peat	Wheat straw	2%, 4% or 8%	Germination rate (=)		[70]
			10%, 12%, 14% or 16%	Germination rate (-)		
Lactuca sativa	Two parts of single wood species sawdust to one-part poultry manure	Rice husk	50% or 66%	Plant height (-)	Half irrigation (0.1125mm)	[92]
			50%	Plant height (=)	Full irrigation (0.225mm)	
			66%	Plant height (+)	Full irrigation (0.225mm)	
Pelargonium × *hortorum* 'Maverick Red'	Sunshine Mix #2	Rice hull	1% or 10%	Shoot DW (-)	Plants in biochar-added substrates were fertilized with 100 mg L^{-1} N. Control was Sunshine Mix #2 with a fertilizer (20N-4.4P-16.6K-0.15Mg-0.02B-0.01Cu-0.1Fe-0.05Mn-0.01Mo-0.05Zn) at the rate of 100 mg L^{-1} N	[66]
	Sunshine Mix #2 with a micronutrient package (Micromax, The Scotts Co., Marysville, OH) at 0.9 kg m^{-3}	Rice hull	5%,10% or 15%	Shoot DW (=)		[67]
			20%	Shoot DW (-)		
Pelargonium zonale	Peat	N/A	30% or 70%	Plant height (=)/DW (=)	140 mg L^{-1} slow released fertilizer applied	[87]
			30%	Plant height (=)/DW (=)	210 mg L^{-1} slow released fertilizer applied	
			70%	Plant height (-)/DW (-)		
Solanum lycopersicum 'Megabite'	Sunshine Mix #2 with a micronutrient package (Micromax, The Scotts Co., Marysville, OH) at 0.9 kg m^{-3}	Rice hull	5%	Shoot DW (=)	Plants in biochar-added substrates were fertilized with 100 mg L^{-1} N. Control was Sunshine Mix #2 with a fertilizer (20N-4.4P- 16.6K-0.15Mg-0.02B-0.01Cu-0.1Fe-0.05Mn-0.01Mo-0.05Zn) at the rate of 100 mg L^{-1} N	[67]
			10%, 15% or 20%	Shoot DW (+)		
Solanum lycopersicum	Pine (*Pinus radiata* D. Don) sawdust	Tomato crop green waste	25%, 50%, 75% or 100%	Shoot fresh weight (FW) (=)/Fruit number (=)/ Yield (=)	Control was 100% pine sawdust.	[59]

Table 5. *Cont.*

Plant Species	Non-Biochar Components	Biochar Feedstock	Biochar Percentage (by vol.)	Effects on Plant Growth [z]	Other Information	Reference
Sylibum marianum	Pig slurry compost	Holm oak	40%, 60% or 80%	Shoot DW (=)	Compared to mixes with 40%, 60%, or 80% coir with the rest being pig slurry compost, respectively	[52]
	No		100%	Shoot DW (-)	Control was 100% coir	
Tagetes erecta	30% perlite with the rest being peat and biochar	Mixed softwood	10%, 20%, 30%, 40%, 50%, 60% or 70%	Plant height (=)/DW (=)	No pH adjustment; control was 70:30 peat: perlite mixture.	[62]
			10% or 70%	Plant height (-)/DW (=)	pH adjusted to 5.8; control was 70:30 peat: perlite mixture.	
			20%, 30%, 40%, 50% or 60%	Plant height (=)/DW (=)		
Zelkova serrata	Growing medium mixture of peat moss, perlite, and vermiculite at a ratio of 1:1:1 by vol.	Pine chip	20%	Plant height (=)/Stem DW (=)	0.5 or 1 g/L fertilization	[50]
		Oak chip	20%	Plant height (=)/Stem DW (=)	0.5 or 1 g/L fertilization	
		Pine cone	20%	Plant height (-)/Stem DW (-)	0.5 g/L fertilization	
				Plant height (=)/Stem DW (=)	1 g/L fertilization	
		Rice husk	20%	Plant height (+)/Stem DW (+)	0.5 g/L fertilization	
				Plant height (=)/Stem DW (=)	1 g/L fertilization	
		Crab shell	20%	Plant height (-)/Stem DW (-)	0.5 or 1 g/L fertilization	

[z]: "+" means increased; "=" means there was no significant difference; "-" means decreased; "n" means not available.

(20N-4.4P-16.6K-0.15Mg-0.02B-0.01Cu-0.1Fe-0.05Mn-0.01Mo-0.05Zn). The effects of biochar on container-grown plants could be different due to different plant materials.

Most of the plant species used in testing biochars in container substrates have been herbaceous. Only six woody plants have been tested, including Japanese zelkova (*Zelkova serrata*), lilly pilly (*Acmena smithii*), 'Green Velvet' boxwood (*Buxus sempervirens* × *Buxus microphylla*), Pinky Winky hardy hydrangea (*Hydrangea paniculata*), myrtle (*Myrtus communis*) and mastic tree (*Pistacia lentiscus*). Across all studies, the most frequently tested species have been tomato and lettuce. About 30.5% of the studies used tomato plants to test biochars in container substrates and 19.4% used lettuce. Research is needed to test more plant species.

4.2. Different Biochar and Biochar Percentage in Container Substrates

The impact of biochar on plant growth depends on the properties of the biochar used and the percentage of biochar in the substrates. Those factors impact the overall physical and chemical properties of the container substrates, such as pH, container capacity and CEC. Belda et al. [89] showed that mixing 10%, 25% or 50% (by vol.) forest waste biochar with coir led to higher *Myrtus communis* and *Pistacia lentiscus* stem length and dry weight than using olive mill waste biochar. It was shown that *Zelkova serrata* plants in mixes that contained 20% rice husk biochar with the rest of the mixture composed of peat moss, perlite, and vermiculite at a ratio of 1:1:1 (by vol.) were 6 times larger than those in mixes with crab shell biochar, which could be due to the high concentration of nutrients, nutrient absorption ability and water retention ability of rice husk biochar [50]. Webber et al. [48] showed that pneumatic sugarcane bagasse biochar and standard sugarcane bagasse biochar led to different effects on plant growth, due to different physical and chemical compositions of the two biochars, produced by different conditions. Pumpkin (*Cucurbita pepo*) and muskmelon (*Cucumis melo*) both had increased plant height in mixes with 50% pneumatic sugarcane bagasse biochar with the rest being Sunshine commercial growing media (by vol.) compared to the control, while both in mixes with 50% standard sugarcane bagasse biochar showed similar plant height to the control. Webber et al. [48] also indicated that different biochar percentages could affect the results and showed that mixes with 75% or 100% biochar decreased muskmelon plant dry weight, but mixes with 25% or 50% biochar had no effect. Similarly, the aboveground dry weight of *Viola* × *hybrida* showed no significant effects after the incorporation of 5% (w/w) *Eucalyptus saligna* wood chip biochar to growing medium containing pine bark, coir, clinker ash and coarse sand, but aboveground dry weight decreased when mixing 10% (w/w) biochar with the growing medium, when compared to the control [51]. The decreased plant dry weight was due to reduced concentrations of S, P, and Ca caused by the binding ability of the biochar [51]. Fan et al. [70] found that the germination rate of water spinach (*Ipomoea aquatica*) decreased when the biochar incorporation rate in mixes containing spent pig litter compost, vermiculite, perlite and peat increased to 10%, 12%, 14% or 16% (by vol.) due to the high and unsuitable pH and EC after biochar incorporation, while there was no effect on the germination rate if the biochar incorporation rate was 2%, 4% or 8% (by vol.). Conversa et al. [87] showed that mixing peat with biochar at the ratio of 70:30 (by vol.) with slow released fertilizer at a rate of 140 and 210 mg L^{-1} led to increased *Pelargonium* leaf number and similar shoot dry weight compared to the control. However, mixing peat with biochar at the ratio of 30:70 (by vol.) with a high rate of slow release fertilizer (210 mg L^{-1}) showed decreased *Pelargonium* plant growth and flowering traits due to osmotic stress caused by high EC and decreased mycorrhizal activity with this high biochar rate [87]. Awad et al. [64] also showed that mixes with 50% (by vol.) biochar with the rest being perlite led to increased dry weight and growth of Chinese cabbage (*Brassica rapa* ssp. *pekinensis*), dill (*Anethum graveolens*), curled mallow (*Malva verticillata*), red lettuce, and tatsoi (*Brassica rapa* var. *rosularis*) while 100% rice husk biochar decreased plant growth due to high pH of the substrate, low air space, and decreased N availability due to biochar's N absorption ability.

Across studies that mixed biochar in container substrates by volume and tested the effects of biochar on plant growth without other factors (Tables 2 and 3), 72.2% incorporated biochar at 50% or more (by vol.) in container substrates. This suggested that the substitution of the commonly-used substrates or substrate components with a large proportion of biochar is highly desired and, based on the results, achievable. About 36.4% of the studies (Tables 2 and 3) showed that mixing high percentages of biochar (at least 50% by vol.) in container media could improve the growth of some species when compared to the control. All container substrates with biochar percentages lower than 25% (by vol.) led to similar or higher plant growth or dry weight when compared to the control. A biochar incorporation rate as high as 100% (by vol.) in container substrate often led to similar plant growth to the control [48,53,57].

The physical and chemical properties of biochar could determine whether a large proportion of biochar could be used in container substrates to grow plants. When the physical and chemical properties of biochar or substrates with high percentages of biochar are similar to the commercial substrates or are in the ideal range for container-grown plant growth, a high percentage of biochar could be incorporated into the container substrate. The recommended ranges for the physical properties of most substrates used in commercial container plant production are 50%–85% for total porosity, 10%–30% for air space, 45%–65% for container capacity and 0.19 to 0.7 g cm^{-3} for bulk density [72]. Choi et al. [57] has achieved using 100% biochar substrates to replace the 100% pine bark substrates to grow chrysanthemum and lettuce. The container capacity and air space of the biochar were similar to the bark [57]. Although the total porosity of the biochar used was different from that of the bark, it was in the recommended range for container plant production [57]. Guo et al. [56] also succeeded using up to 80% biochar in peat-based commercial substrates, and the physical properties of the biochar substrates were in, or close to, the recommended range for container plant production. Among all properties, pH could be a limiting factor determining the potential use of biochar in containers. Webber et al. [48] made two kinds of biochars, pneumatic sugarcane bagasse biochar and standard sugarcane bagasse biochar, and indicated that these two biochars could be used in containers as high as 100% to grow pumpkin seedlings for 20 days. The pH of these two bicohars were 5.8 and 6.05, respectively. If the pH of the biochar is high, other acidic components should be added to reduce the pH or a high percentage of biochar in a container may not be achievable. It was shown that the addition of 80% (by vol.) biochar (pH = 8.5) to peat (pH = 5.7) increased plant growth due to neutral pH and improved water holding and air structure after biochar addition [53].

4.3. Other Substrate Components Mixed with Biochar in Container Substrates

The other substrate components used with biochar could affect plant growth due to their different physical and chemical properties and their effects on the overall container substrate properties. Substrate components mixed with biochar have included peat, vermiculite, perlite, coir, pine bark, pine sawdust, commercial growing media, compost, composted green waste and potato digestate (Tables 2–5). Gu et al. [90] showed that gomphrena (*Gomphrena globosa*) grown in 5%, 10%, 15%, 20%, 25% and 30% (by vol.) pinewood biochar mixed with the peat-based Sunshine Mix #1 had greater width and height, higher fresh weight and dry weight than those grown in biochar mixed with bark substrates at 43 days after transplanting. The reason for this result could be that peat-based substrates have more organic matter and higher water and nutrient holding capacity than bark-based substrates. Ain Najwa et al. [91] also indicated that the fruit number and fresh weight of tomato in mixes with coco peat and 150 g biochar were higher than in mixes with oil palm fruit bunch (a newly developed organic medium) and 150 g biochar due to different physical and chemical properties of these two substrates. Vaughn et al. [68] showed that creeping bentgrass (*Agrostis stolonifera*) had higher fresh and dry weight and shoot height in mixes with 85% sand, 10% anaerobic biosolids and 5% biochar (by vol.) than the one in mixes with 85% sand, 10% peat and 5% biochar (vol.), due to higher nitrate concentration caused by biosolid incorporation. Méndez et al. [75] also demonstrated that the total biomass and shoot and root weight of lettuce were higher in deinking sludge biochar with peat (50:50 by vol.) than those in biochar mixed with coir (50:50 by vol.). The lower plant biomass in coir with biochar incorporation may be due to the lower CEC, N and P in coir when compared to peat. Fan et al. [70]

investigated the effects of mixed wheat straw biochar with or without superabsorbent polymer on the substrates containing spent pig litter compost, vermiculite, perlite and peat. The germination rate of water spinach decreased when the biochar incorporation rate in the medium without superabsorbent polymer was 10%, 12%, 14% or 16% (by vol.) due to the high and unsuitable pH and EC after biochar incorporation. However, there was no difference on germination rate between the mixes with different percentages of biochar (from 0% to 16% by vol.) when biochar was applied together with superabsorbent polymer. The reason was that the incorporation of superabsorbent polymer increased the porosity and water-holding capacity and also effectively prevented an excessive increase of pH and EC at the high biochar rates [70]. Margenot et al. [62] also showed that mixes with 10%, 20%, 30%, 40%, 50%, 60% or 70% softwood biochar and 30% perlite with the rest being peat (by vol.) led to similar seed germination and plant height compared to control (mixes with 30% perlite and 70% peat by vol.). However, if other components such as calcium hydroxide were added to increase the pH of 10% biochar mixes to 5.8 or pyroligneous acid to decrease substrate (mixes with more than 10% biochar) pH, lower seed germination resulted in mixes with 50%, 60% or 70% biochar and lower plant height in mixes with 10% or 70% biochar.

5. Effect of Potentially Toxic Contaminants in Biochar on Plant Growth

Biochar may contain potentially toxic substances, such as heavy metals and organic contaminants (PAH and dioxin), which are affected by the production conditions and feedstocks used. The incorporation of biochar with a high content of these contaminants is a concern. Various studies have shown reduced plant growth caused by the toxicity of PAHs [93,94], dioxins [95] and heavy metals [96,97]. The utilization of biochar that contains toxic substances could be detrimental, and could influence plant growth and development, leach into groundwater, and have noxious effects on soil function and microorganisms. However, toxic substances (heavy metals, PAHs and dioxin) in biochars used in container substrates have rarely been tested. Attention is needed when choosing biochar feedstocks and biochar production conditions to avoid or minimize the production of toxic substances.

Biochar could contain heavy metals from contaminated feedstocks; however, heavy metals could be transformed to more stable forms after pyrolysis, thus having less effect on plant growth. Heavy metals may remain in biochar made from contaminated feedstock such as cadmium (Cd), copper (Cu), lead (Pb), and zinc (Zn) as observed with contaminated willow leaves and branches [98] or sewage [99]. However, the heavy metals in biochar might have low bioavailability after pyrolysis and a lower risk to plant growth. Jin et al. [100] found most of the heavy metals in sludge biochar after pyrolysis at 400 to 600 °C, including Cu, Zn, Pb, chromium (Cr), manganese (Mn) and nickel (Ni), were in their oxidized and residual forms, which had low bioavailability and thus risks. Similarly, Devi and Saroha [101] found that the bioavailability of heavy metals (Cr, Cu, Ni, Zn and Pb) in paper mill sludge biochar derived from pyrolysis at 200 °C to 700 °C was reduced due to transformation into more stable forms. Buss et al. [102] investigated the effects of 19 types of biochar produced from marginal biomass containing contaminants (such as Cu, Cr, Ni and Zn) on plant growth and found that only five types of biochar in the study showed suppressive effects on plant growth after adding 5% (by weight) of biochar in sand due to high K and pH, not heavy metals.

Although PAHs could be formed in biochars due to production conditions, the amount of PAH in biochars used in many studies has been low and may have had low toxicity for plant growth. Large quantities of PAHs are formed in reactions at high temperatures, especially over 750 °C [103], although no research was found using biochar produced over 750 °C in container substrates. There is also evidence that small amounts of PAHs can be formed in pyrolysis reactors operating between 400 °C and 600 °C [103,104], which is the temperature range that most biochars suitable as container substrate component were produced [6,19–21,51,70,75,90]. Research has shown that PAHs in biochar produced from slow pyrolysis between temperature 250 °C and 900 °C had very low bioavailability [105]. Wiedner et al. [29] also found that all biochars made from gasification of poplar, wheat straw, sorghum and olive, and from pyrolysis of draff (the waste product from the production of beer after separating liquid malt) and miscanthus contained very low content of PAH (below 1.7 mg kg^{-1}) and biochar made from woodchip gasification (15 mg kg^{-1} PAH). Although biochars produced at certain conditions, especially over 750 °C, could contain PAHs, no research was found using these biochars in container substrates to test their effects on substrate properties and plant growth.

Dioxins could be formed in biochar if the feedstock contains chlorine in certain conditions, but dioxin concentration in biochars could be very low and have a negligible effect on plant growth. Dioxins refer to compounds such as polychlorinated dibenzo dioxins (PCDDs) and polychlorinated dibenzo furans (PCDFs), which are persistent organic pollutants [106]. Dioxins could be formed only in biochars made from feedstock containing chlorine, such as straws, grasses, halogenated plastics and food waste containing sodium chloride under specific conditions [103,106]. Dioxins could be produced during two pathways: "precursor" pathway, which begins with the synthesis of dioxin precursors from feedstock containing chlorine at temperatures between 300 °C and 600 °C; and the "de novo" pathway, which occurs between 200 °C and 400 °C in a catalytic reaction with oxygen and carbon [106–108]. However, the dioxin in biochar made from feedstock with chlorine could be very low. Hale et al. [105] investigated the biochars produced at 250 °C to 900 °C via slow pyrolysis, fast pyrolysis and gasification and found that total dioxin concentrations in biochars tested were very low (92 pg g^{-1}) and bioavailable concentrations were below detection limit [105]. Wiedner et al. [29] found that the dioxins in four biochars produced from gasification of poplar and olive residues and pyrolysis of draff and wood chips and two other hydrochars made from leftover food and sewage sludge were all under the limit of detection, except the one made from sewage sludge (14.2 ng kg^{-1}). No evidence was found testing the effect of biochars with dioxin in container substrates on plant growth.

6. Discussion

The incorporation of biochar into container substrates could affect physical and chemical properties of the container substrates and thus contribute to the growth of container-grown plants. Most biochars have a higher bulk density than commonly-used substrates, and thus the incorporation of biochar could increase the bulk density of the container substrate. The effect of biochar on container capacity, air space, and total porosity of the container substrates depends on the particle size distribution of the biochar and the other components in the container. The liming effect of alkaline biochars could adjust the container substrate with low pH to an optimal pH. In addition, biochar incorporation could increase EC, nutrient availability, and CEC.

The effects of biochar on plant growth in container substrates varies as not all biochars are the same. The characteristics of biochars differ according to the feedstock used and the pyrolysis process. Many factors, such as plant species and the ratio of biochar to other container substrate components, can contribute to different results on container substrate properties and plant growth. Across studies testing the effects of biochar on plant growth but not other factors (such as irrigation or fertilization rates) (Tables 2–4), 77.3% of the studies found that plant growth could be increased by the incorporation of certain percentages of biochar in container substrates, and 50% revealed that certain percentages of biochar addition could decrease plant growth. Among studies mixing biochar with container substrates by volume and testing the effects of biochar on plant growth without other factors (Tables 2 and 3), 36.4% showed that container substrates with high percentages of biochar (at least 50% by vol.) could improve plant growth under certain conditions compared to the control. All the container substrates with biochar percentages lower than 25% (by vol.) led to similar or higher plant growth or dry weight when compared to the control. A biochar incorporation rate as high as 100% (by vol.) in container substrates could lead to similar plant growth to the control. The physical and chemical properties of the biochar could determine whether a large proportion of biochar could be used in container substrates to grow plants.

There is no universal standard for using biochar in container substrates for all plants. Many mechanisms of biochar are not fully understood. Research on biochar in container substrates is still in an exploratory state. Most research has focused on testing whether biochar could be used to substitute for commonly-used substrates such as peat, perlite and bark in containers to grow plants, and compared plant growth with a control that had no biochar addition. There is very limited research that tests other properties such as the effect of biochar on disease suppression in container substrates. Research has shown that biochar could impact greenhouse gas emissions in soil, but limited research has been conducted on soilless container substrates. A limited number of published studies have investigated the effect of biochar on microbial activity or inoculation with mycorrhizae in containers. Most of the species used in reported studies testing biochar in container substrates have been herbaceous plants. More plant species should be used to test the effects of biochar to broaden its use. Future studies could be focused on biochars with promising results, to fine-tune the pyrolysis process and incorporate formulae for diverse container substrates.

Author Contributions: This review is a product of the combined effort of both authors. L.H. wrote the original draft and improved it based on M.G.'s advice and assistance. M.G. reviewed, edited and revised the manuscript.

Conflicts of Interest: The authors declare no conflict of interest.

References

1. Lehmann, J. A handful of carbon. *Nature* **2007**, *447*, 143–144. [CrossRef] [PubMed]
2. Hansen, V.; Hauggaard-Nielsen, H.; Petersen, C.T.; Mikkelsen, T.N.; Müller-Stöver, D. Effects of gasification biochar on plant-available water capacity and plant growth in two contrasting soil types. *Soil Tillage Res.* **2016**, *161*, 1–9. [CrossRef]
3. Johannes, L.; Stephen, J. Biochar for environmental management: An introduction. In *Biochar for Environmental Management-Science and Technology*; Earthscan: London, UK, 2009; pp. 1–10.
4. Vaughn, S.F.; Kenar, J.A.; Eller, F.J.; Moser, B.R.; Jackson, M.A.; Peterson, S.C. Physical and chemical characterization of biochars produced from coppiced wood of thirteen tree species for use in horticultural substrates. *Ind. Crop. Prod.* **2015**, *66*, 44–51. [CrossRef]
5. Vaughn, S.F.; Kenar, J.A.; Thompson, A.R.; Peterson, S.C. Comparison of biochars derived from wood pellets and pelletized wheat straw as replacements for peat in potting substrates. *Ind. Crop. Prod.* **2013**, *51*, 437–443. [CrossRef]
6. Yu, F.; Steele, P.H.; Gu, M.; Zhao, Y. Using Biochar as Container Substrate for Plant Growth. U.S. Patent 9,359,267, 7 June 2016.
7. Dumroese, R.K.; Heiskanen, J.; Englund, K.; Tervahauta, A. Pelleted biochar: Chemical and physical properties show potential use as a substrate in container nurseries. *Biomass Bioenergy* **2011**, *35*, 2018–2027. [CrossRef]
8. Northup, J. Biochar as a Replacement for Perlite in Greenhouse Soilless Substrates. Master's Thesis, Iowa State Univversity, Ames, IA, USA, 2013.
9. Landis, T.D.; Morgan, N. Growing Media Alternatives for Forest and Native Plant Nurseries. In *National Proceedings: Forest and Conservation Nursery Associations-2008*. *Proceedings RMRS-P-58*; Dumroese, R.K., Riley, L.E., Eds.; US Department of Agriculture, Forest Service, Rocky Mountain Research Station: Fort Collins, CO, USA, 2009; pp. 26–31.
10. Landis, T.D. Growing media. *Contain. Grow. Media* **1990**, *2*, 41–85.
11. Bohlin, C.; Holmberg, P. Peat: Dominating growing medium in Swedish horticulture. *Acta Hortic.* **2004**, *644*, 177–181. [CrossRef]
12. Fascella, G. Growing substrates alternative to peat for ornamental plants. In *Soilless Culture-Use of Substrates for the Production of Quality Horticultural Crops*; InTech: London, UK, 2015.
13. Rankin, T.; Strachan, I.; Strack, M. Carbon dioxide and methane exchange at a post-extraction, unrestored peatland. *Ecol. Eng.* **2018**, *122*, 241–251. [CrossRef]
14. Carlile, W.; Waller, P. Peat, politics and pressure groups. *Chron. Hortic.* **2013**, *53*, 10–16.
15. Carlile, W.; Cattivello, C.; Zaccheo, P. Organic growing media: Constituents and properties. *Vadose Zone J.* **2015**, *14*. [CrossRef]
16. Wright, R.D.; Browder, J.F. Chipped pine logs: A potential substrate for greenhouse and nursery crops. *HortScience* **2005**, *40*, 1513–1515. [CrossRef]

17. Laird, D.A. The charcoal vision: A win–win–win scenario for simultaneously producing bioenergy, permanently sequestering carbon, while improving soil and water quality. *Agron. J.* **2008**, *100*, 178–181. [CrossRef]

18. Tian, Y.; Sun, X.; Li, S.; Wang, H.; Wang, L.; Cao, J.; Zhang, L. Biochar made from green waste as peat substitute in growth media for *Calathea rotundifola* cv. Fasciata. *Sci. Hortic.* **2012**, *143*, 15–18. [CrossRef]

19. Méndez, A.; Cárdenas-Aguiar, E.; Paz-Ferreiro, J.; Plaza, C.; Gascó, G. The effect of sewage sludge biochar on peat-based growing media. *Biol. Agric. Hortic.* **2017**, *33*, 40–51. [CrossRef]

20. Nieto, A.; Gascó, G.; Paz-Ferreiro, J.; Fernández, J.; Plaza, C.; Méndez, A. The effect of pruning waste and biochar addition on brown peat based growing media properties. *Sci. Hortic.* **2016**, *199*, 142–148. [CrossRef]

21. Zhang, L.; Sun, X.; Tian, Y.; Gong, X. Biochar and humic acid amendments improve the quality of composted green waste as a growth medium for the ornamental plant *Calathea insignis*. *Sci. Hortic.* **2014**, *176*, 70–78. [CrossRef]

22. Headlee, W.L.; Brewer, C.E.; Hall, R.B. Biochar as a substitute for vermiculite in potting mix for hybrid poplar. *Bioenergy Res.* **2014**, *7*, 120–131. [CrossRef]

23. Gvero, P.M.; Papuga, S.; Mujanic, I.; Vaskovic, S. Pyrolysis as a key process in biomass combustion and thermochemical conversion. *Therm. Sci.* **2016**, *20*, 1209–1222. [CrossRef]

24. Lehmann, J.; Rillig, M.C.; Thies, J.; Masiello, C.A.; Hockaday, W.C.; Crowley, D. Biochar effects on soil biota—A review. *Soil Biol. Biochem.* **2011**, *43*, 1812–1836. [CrossRef]

25. Hansen, V.; Müller-Stöver, D.; Ahrenfeldt, J.; Holm, J.K.; Henriksen, U.B.; Hauggaard-Nielsen, H. Gasification biochar as a valuable by-product for carbon sequestration and soil amendment. *Biomass Bioenergy* **2015**, *72*, 300–308. [CrossRef]

26. Bruun, E.W.; Hauggaard-Nielsen, H.; Ibrahim, N.; Egsgaard, H.; Ambus, P.; Jensen, P.A.; Dam-Johansen, K. Influence of fast pyrolysis temperature on biochar labile fraction and short-term carbon loss in a loamy soil. *Biomass Bioenergy* **2011**, *35*, 1182–1189. [CrossRef]

27. Libra, J.A.; Ro, K.S.; Kammann, C.; Funke, A.; Berge, N.D.; Neubauer, Y.; Titirici, M.-M.; Fühner, C.; Bens, O.; Kern, J. Hydrothermal carbonization of biomass residuals: A comparative review of the chemistry, processes and applications of wet and dry pyrolysis. *Biofuels* **2011**, *2*, 71–106. [CrossRef]

28. Kalderis, D.; Kotti, M.; Méndez, A.; Gascó, G. Characterization of hydrochars produced by hydrothermal carbonization of rice husk. *Solid Earth* **2014**, *5*, 477–483. [CrossRef]

29. Wiedner, K.; Rumpel, C.; Steiner, C.; Pozzi, A.; Maas, R.; Glaser, B. Chemical evaluation of chars produced by thermochemical conversion (gasification, pyrolysis and hydrothermal carbonization) of agro-industrial biomass on a commercial scale. *Biomass Bioenergy* **2013**, *59*, 264–278. [CrossRef]

30. Zheng, W.; Sharma, B.; Rajagopalan, N. Using biochar as a soil amendment for sustainable agriculture. 2010. Available online: http://hdl.handle.net/2142/25503 (accessed on 31 January 2019).

31. Roos, C.J. *Clean Heat and Power Using Biomass Gasification for Industrial and Agricultural Projects*; Northwest CHP Application Center: Olympia, WA, USA, 2010.

32. Bridgwater, A.; Meier, D.; Radlein, D. An overview of fast pyrolysis of biomass. *Org. Geochem.* **1999**, *30*, 1479–1493. [CrossRef]

33. Ahmed, I.; Gupta, A. Syngas yield during pyrolysis and steam gasification of paper. *Appl. Energy* **2009**, *86*, 1813–1821. [CrossRef]

34. Fuchs, M.R.; Garcia-Perez, M.; Small, P.; Flora, G. Campfire Lessons: Breaking down the Combustion Process to Understand Biochar Production and Characterization. Available online: https://www.biochar-journal.org/en/ct/47 (accessed on 31 January 2019).

35. Panwar, N.; Kothari, R.; Tyagi, V. Thermo chemical conversion of biomass–eco friendly energy routes. *Renew. Sustain. Energy Rev.* **2012**, *16*, 1801–1816. [CrossRef]

36. Li, J.; Li, Y.; Wu, Y.; Zheng, M. A comparison of biochars from lignin, cellulose and wood as the sorbent to an aromatic pollutant. *J. Hazard. Mater.* **2014**, *280*, 450–457. [CrossRef] [PubMed]

37. Rahman, A.A.; Sulaiman, F.; Abdullah, N. Influence of washing medium pre-treatment on pyrolysis yields and product characteristics of palm kernel shell. *J. Phys. Sci.* **2016**, *27*, 53–75.

38. Jenkins, B.; Bakker, R.; Wei, J. On the properties of washed straw. *Biomass Bioenergy* **1996**, *10*, 177–200. [CrossRef]

39. Boateng, A.; Mullen, C. Fast pyrolysis of biomass thermally pretreated by torrefaction. *J. Anal. Appl. Pyrolysis* **2013**, *100*, 95–102. [CrossRef]

40. Reckamp, J.M.; Garrido, R.A.; Satrio, J.A. Selective pyrolysis of paper mill sludge by using pretreatment processes to enhance the quality of bio-oil and biochar products. *Biomass Bioenergy* **2014**, *71*, 235–244. [CrossRef]

41. Chu, G.; Zhao, J.; Huang, Y.; Zhou, D.; Liu, Y.; Wu, M.; Peng, H.; Zhao, Q.; Pan, B.; Steinberg, C.E. Phosphoric acid pretreatment enhances the specific surface areas of biochars by generation of micropores. *Environ. Pollut.* **2018**, *240*, 1–9. [CrossRef] [PubMed]

42. Zhang, S.; Xiong, Y. Washing pretreatment with light bio-oil and its effect on pyrolysis products of bio-oil and biochar. *RSC Adv.* **2016**, *6*, 5270–5277. [CrossRef]

43. Hina, K.; Bishop, P.; Arbestain, M.C.; Calvelo-Pereira, R.; Maciá-Agulló, J.A.; Hindmarsh, J.; Hanly, J.; Macìas, F.; Hedley, M. Producing biochars with enhanced surface activity through alkaline pretreatment of feedstocks. *Soil Res.* **2010**, *48*, 606–617. [CrossRef]

44. Kołtowski, M.; Oleszczuk, P. Toxicity of biochars after polycyclic aromatic hydrocarbons removal by thermal treatment. *Ecol. Eng.* **2015**, *75*, 79–85. [CrossRef]

45. McCabe, K.G.; Currey, C.J.; Schrader, J.A.; Grewell, D.; Behrens, J.; Graves, W.R. Pelletized soy-based bioplastic fertilizers for container-crop production. *HortScience* **2016**, *51*, 1417–1426. [CrossRef]

46. Fornes, F.; Belda, R.M. Biochar versus hydrochar as growth media constituents for ornamental plant cultivation. *Sci. Agric.* **2018**, *75*, 304–312. [CrossRef]

47. Fornes, F.; Belda, R.M.; Fernández de Córdova, P.; Cebolla-Cornejo, J. Assessment of biochar and hydrochar as minor to major constituents of growing media for containerized tomato production. *J. Sci. Food Agric.* **2017**, *97*, 3675–3684. [CrossRef] [PubMed]

48. Webber, C.L., III; White, P.M., Jr.; Spaunhorst, D.J.; Lima, I.M.; Petrie, E.C. Sugarcane biochar as an amendment for greenhouse growing media for the production of cucurbit seedlings. *J. Agric. Sci.* **2018**, *10*, 104. [CrossRef]

49. Locke, J.C.; Altland, J.E.; Ford, C.W. Gasified rice hull biochar affects nutrition and growth of horticultural crops in container substrates. *J. Environ. Hortic.* **2013**, *31*, 195–202.

50. Cho, M.S.; Meng, L.; Song, J.-H.; Han, S.H.; Bae, K.; Park, B.B. The effects of biochars on the growth of *Zelkova serrata* seedlings in a containerized seedling production system. *For. Sci. Technol.* **2017**, *13*, 25–30. [CrossRef]

51. Housley, C.; Kachenko, A.; Singh, B. Effects of *Eucalyptus saligna* biochar-amended media on the growth of *Acmena smithii*, *Viola* var. *Hybrida*, and *Viola*× *wittrockiana*. *J. Hortic. Sci. Biotechnol.* **2015**, *90*, 187–194. [CrossRef]

52. Sáez, J.; Belda, R.; Bernal, M.; Fornes, F. Biochar improves agro-environmental aspects of pig slurry compost as a substrate for crops with energy and remediation uses. *Ind. Crop. Prod.* **2016**, *94*, 97–106. [CrossRef]

53. Dispenza, V.; De Pasquale, C.; Fascella, G.; Mammano, M.M.; Alonzo, G. Use of biochar as peat substitute for growing substrates of *Euphorbia* × *lomi* potted plants. *Span. J. Agric. Res.* **2017**, *14*, 0908. [CrossRef]

54. Graber, E.R.; Harel, Y.M.; Kolton, M.; Cytryn, E.; Silber, A.; David, D.R.; Tsechansky, L.; Borenshtein, M.; Elad, Y. Biochar impact on development and productivity of pepper and tomato grown in fertigated soilless media. *Plant Soil* **2010**, *337*, 481–496. [CrossRef]

55. Guo, Y.; Niu, G.; Starman, T.; Volder, A.; Gu, M. Poinsettia growth and development response to container root substrate with biochar. *Horticulturae* **2018**, *4*, 1. [CrossRef]

56. Guo, Y.; Niu, G.; Starman, T.; Gu, M. Growth and development of easter lily in response to container substrate with biochar. *J. Hortic. Sci. Biotechnol.* **2018**, *94*, 80–86. [CrossRef]

57. Choi, H.-S.; Zhao, Y.; Dou, H.; Cai, X.; Gu, M.; Yu, F. Effects of biochar mixtures with pine-bark based substrates on growth and development of horticultural crops. *Hortic. Environ. Biotechnol.* **2018**, *59*, 345–354. [CrossRef]

58. Park, J.H.; Choppala, G.K.; Bolan, N.S.; Chung, J.W.; Chuasavathi, T. Biochar reduces the bioavailability and phytotoxicity of heavy metals. *Plant Soil* **2011**, *348*, 439–451. [CrossRef]

59. Dunlop, S.J.; Arbestain, M.C.; Bishop, P.A.; Wargent, J.J. Closing the loop: Use of biochar produced from tomato crop green waste as a substrate for soilless, hydroponic tomato production. *HortScience* **2015**, *50*, 1572–1581. [CrossRef]

60. Nair, A.; Carpenter, B. Biochar rate and transplant tray cell number have implications on pepper growth during transplant production. *HortTechnology* **2016**, *26*, 713–719. [CrossRef]

61.	Huang, L. Effects of Biochar and Composts on Substrates Properties and Container-Grown Basil (*Ocimum basilicum*) and Tomato (*Solanum lycopersicum*). Master's Thesis, Texas A&M University, College Station, TX, USA, 2018.

62.	Margenot, A.J.; Griffin, D.E.; Alves, B.S.; Rippner, D.A.; Li, C.; Parikh, S.J. Substitution of peat moss with softwood biochar for soil-free marigold growth. *Ind. Crop. Prod.* **2018**, *112*, 160–169. [CrossRef]

63.	Bedussi, F.; Zaccheo, P.; Crippa, L. Pattern of pore water nutrients in planted and non-planted soilless substrates as affected by the addition of biochars from wood gasification. *Biol. Fertil. Soils* **2015**, *51*, 625–635. [CrossRef]

64.	Awad, Y.M.; Lee, S.-E.; Ahmed, M.B.M.; Vu, N.T.; Farooq, M.; Kim, I.S.; Kim, H.S.; Vithanage, M.; Usman, A.R.A.; Al-Wabel, M. Biochar, a potential hydroponic growth substrate, enhances the nutritional status and growth of leafy vegetables. *J. Clean. Prod.* **2017**, *156*, 581–588. [CrossRef]

65.	Nartey, E.G.; Amoah, P.; Ofosu-Budu, G.K.; Muspratt, A.; Kumar Pradhan, S. Effects of co-composting of faecal sludge and agricultural wastes on tomato transplant and growth. *Int. J. Recycl. Org. Waste Agric.* **2017**, *6*, 23–36. [CrossRef]

66.	Altland, J.E.; Locke, J.C. Gasified rice hull biochar is a source of phosphorus and potassium for container-grown plants. *J. Environ. Hortic.* **2013**, *31*, 138–144.

67.	Altland, J.E.; Locke, J.C. High rates of gasified rice hull biochar affect geranium and tomato growth in a soilless substrate. *J. Plant Nutr.* **2017**, *40*, 1816–1828. [CrossRef]

68.	Vaughn, S.F.; Dinelli, F.D.; Jackson, M.A.; Vaughan, M.M.; Peterson, S.C. Biochar-organic amendment mixtures added to simulated golf greens under reduced chemical fertilization increase creeping bentgrass growth. *Ind. Crop. Prod.* **2018**, *111*, 667–672. [CrossRef]

69.	Jahromi, N.B.; Walker, F.; Fulcher, A.; Altland, J.; Wright, W.C. Growth response, mineral nutrition, and water utilization of container-grown woody ornamentals grown in biochar-amended pine bark. *HortScience* **2018**, *53*, 347–353. [CrossRef]

70.	Fan, R.; Luo, J.; Yan, S.; Zhou, Y.; Zhang, Z. Effects of biochar and super absorbent polymer on substrate properties and water spinach growth. *Pedosphere* **2015**, *25*, 737–748. [CrossRef]

71.	Vaughn, S.F.; Eller, F.J.; Evangelista, R.L.; Moser, B.R.; Lee, E.; Wagner, R.E.; Peterson, S.C. Evaluation of biochar-anaerobic potato digestate mixtures as renewable components of horticultural potting media. *Ind. Crop. Prod.* **2015**, *65*, 467–471. [CrossRef]

72.	Bilderback, T.E.; Warren, S.L.; Owen, J.S.; Albano, J.P. Healthy substrates need physicals too! *HortTechnology* **2005**, *15*, 747–751. [CrossRef]

73.	Noguera, P.; Abad, M.; Puchades, R.; Maquieira, A.; Noguera, V. Influence of particle size on physical and chemical properties of coconut coir dust as container medium. *Commun. Soil Sci. Plant Anal.* **2003**, *34*, 593–605. [CrossRef]

74.	Fonteno, W.; Hardin, C.; Brewster, J. *Procedures for Determining Physical Properties of Horticultural Substrates Using the NCSU Porometer*; Horticultural Substrates Laboratory, North Carolina State University: Raleigh, NC, USA, 1995.

75.	Méndez, A.; Paz-Ferreiro, J.; Gil, E.; Gascó, G. The effect of paper sludge and biochar addition on brown peat and coir based growing media properties. *Sci. Hortic.* **2015**, *193*, 225–230. [CrossRef]

76.	Initiative, I.B. Frequently Asked Questions about Biochar. Available online: https://biochar-international.org/faqs/#1521824701674-1093ca88-fd85 (accessed on 31 January 2019).

77.	Khodadad, C.L.; Zimmerman, A.R.; Green, S.J.; Uthandi, S.; Foster, J.S. Taxa-specific changes in soil microbial community composition induced by pyrogenic carbon amendments. *Soil Biol. Biochem* **2011**, *43*, 385–392. [CrossRef]

78.	Lima, I.; Steiner, C.; Das, K. Characterization of designer biochar produced at different temperatures and their effects on a loamy sand. *Ann. Environ. Sci.* **2009**, *3*, 195–206.

79.	Li, X.; Shen, Q.; Zhang, D.; Mei, X.; Ran, W.; Xu, Y.; Yu, G. Functional groups determine biochar properties (pH and EC) as studied by two-dimensional 13c nmr correlation spectroscopy. *PLoS ONE* **2013**, *8*, e65949. [CrossRef] [PubMed]

80.	Hossain, M.K.; Strezov, V.; Chan, K.Y.; Ziolkowski, A.; Nelson, P.F. Influence of pyrolysis temperature on production and nutrient properties of wastewater sludge biochar. *J. Environ. Manag.* **2011**, *92*, 223–228. [CrossRef] [PubMed]

81.	Lawrinenko, M.; Laird, D.A. Anion exchange capacity of biochar. *Green Chem.* **2015**, *17*, 4628–4636. [CrossRef]

82. Altland, J.; Locke, J. Biochar affects macronutrient leaching from a soilless substrate. *HortScience* **2012**, *47*, 1136–1140. [CrossRef]

83. Warnock, D.D.; Lehmann, J.; Kuyper, T.W.; Rillig, M.C. Mycorrhizal responses to biochar in soil–concepts and mechanisms. *Plant Soil* **2007**, *300*, 9–20. [CrossRef]

84. Saito, M. Charcoal as a micro-habitat for VA mycorrhizal fungi, and its practical implication. *Agric. Ecosyst. Environ.* **1990**, *29*, 341–344. [CrossRef]

85. Ezawa, T.; Yamamoto, K.; Yoshida, S. Enhancement of the effectiveness of indigenous arbuscular mycorrhizal fungi by inorganic soil amendments. *Soil Sci. Plant Nutr.* **2002**, *48*, 897–900. [CrossRef]

86. Dubchak, S.; Ogar, A.; Mietelski, J.; Turnau, K. Influence of silver and titanium nanoparticles on arbuscular mycorrhiza colonization and accumulation of radiocaesium in *Helianthus annuus*. *Span. J. Agric. Res.* **2010**, *8*, 103–108. [CrossRef]

87. Conversa, G.; Bonasia, A.; Lazzizera, C.; Elia, A. Influence of biochar, mycorrhizal inoculation, and fertilizer rate on growth and flowering of pelargonium (*Pelargonium zonale* L.) plants. *Front. Plant Sci.* **2015**, *6*, 429. [CrossRef] [PubMed]

88. Masiello, C.A.; Chen, Y.; Gao, X.; Liu, S.; Cheng, H.-Y.; Bennett, M.R.; Rudgers, J.A.; Wagner, D.S.; Zygourakis, K.; Silberg, J.J. Biochar and microbial signaling: Production conditions determine effects on microbial communication. *Environ. Sci. Technol.* **2013**, *47*, 11496–11503. [CrossRef] [PubMed]

89. Belda, R.M.; Lidón, A.; Fornes, F. Biochars and hydrochars as substrate constituents for soilless growth of myrtle and mastic. *Ind. Crop. Prod.* **2016**, *94*, 132–142. [CrossRef]

90. Gu, M.; Li, Q.; Steele, P.H.; Niu, G.; Yu, F. Growth of 'fireworks' gomphrena grown in substrates amended with biochar. *J. Food Agric. Environ.* **2013**, *11*, 819–821.

91. Ain Najwa, K.; Wan Zaliha, W.; Yusnita, H.; Zuraida, A. Effect of different soilless growing media and biochar on growth, yield and postharvest quality of lowland cherry tomato (*Solanum lycopersicum* var. Cerasiforme). *Trans. Malaysian Soc. Plant Physiol.* **2014**, *22*, 53–57.

92. Abubakari, A.-H.; Atuah, L.; Banful, B.K. Growth and yield response of lettuce to irrigation and growth media from composted sawdust and rice husk. *J. Plant Nutr.* **2018**, *41*, 221–232. [CrossRef]

93. Somtrakoon, K.; Chouychai, W. Phytotoxicity of single and combined polycyclic aromatic hydrocarbons toward economic crops. *Russ. J. Plant Physiol.* **2013**, *60*, 139–148. [CrossRef]

94. Huang, X.-D.; Zeiler, L.F.; Dixon, D.G.; Greenberg, B.M. Photoinduced toxicity of PAHs to the foliar regions of *Brassica napus* (canola) and *Cucumbis sativus* (cucumber) in simulated solar radiation. *Ecotoxicol. Environ. Saf.* **1996**, *35*, 190–197. [CrossRef] [PubMed]

95. Hanano, A.; Almousally, I.; Shaban, M.; Moursel, N.; Shahadeh, A.; Alhajji, E. Differential tissue accumulation of 2,3,7,8-tetrachlorinated dibenzo-p-dioxin in *Arabidopsis thaliana* affects plant chronology, lipid metabolism and seed yield. *Bmc Plant Biol.* **2015**, *15*, 193. [CrossRef] [PubMed]

96. John, R.; Ahmad, P.; Gadgil, K.; Sharma, S. Heavy metal toxicity: Effect on plant growth, biochemical parameters and metal accumulation by *Brassica juncea* L. *Int. J. Plant Prod.* **2012**, *3*, 65–76.

97. Peralta, J.; Gardea-Torresdey, J.; Tiemann, K.; Gomez, E.; Arteaga, S.; Rascon, E.; Parsons, J. Uptake and effects of five heavy metals on seed germination and plant growth in alfalfa (*Medicago sativa* L.). *Bull. Environ. Contam. Toxicol.* **2001**, *66*, 727–734. [CrossRef] [PubMed]

98. Lievens, C.; Carleer, R.; Cornelissen, T.; Yperman, J. Fast pyrolysis of heavy metal contaminated willow: Influence of the plant part. *Fuel* **2009**, *88*, 1417–1425. [CrossRef]

99. Bridle, T.; Pritchard, D. Energy and nutrient recovery from sewage sludge via pyrolysis. *Water Sci. Technol.* **2004**, *50*, 169–175. [CrossRef] [PubMed]

100. Jin, J.; Li, Y.; Zhang, J.; Wu, S.; Cao, Y.; Liang, P.; Zhang, J.; Wong, M.H.; Wang, M.; Shan, S.; et al. Influence of pyrolysis temperature on properties and environmental safety of heavy metals in biochars derived from municipal sewage sludge. *J. Hazard. Mater.* **2016**, *320*, 417–426. [CrossRef] [PubMed]

101. Devi, P.; Saroha, A.K. Risk analysis of pyrolyzed biochar made from paper mill effluent treatment plant sludge for bioavailability and eco-toxicity of heavy metals. *Bioresour. Technol.* **2014**, *162*, 308–315. [CrossRef] [PubMed]

102. Buss, W.; Graham, M.C.; Shepherd, J.G.; Mašek, O. Risks and benefits of marginal biomass-derived biochars for plant growth. *Sci. Total Environ.* **2016**, *569*, 496–506. [CrossRef] [PubMed]

103. Shackley, S.; Sohi, S.; Brownsort, P.; Carter, S.; Cook, J.; Cunningham, C.; Gaunt, J.; Hammond, J.; Ibarrola, R.; Mašek, O. *An Assessment of the Benefits and Issues Associated with the Application of Biochar to Soil*; Department for Environment, Food and Rural Affairs, UK Government: London, UK, 2010.

104. McGrath, T.; Sharma, R.; Hajaligol, M. An experimental investigation into the formation of polycyclic-aromatic hydrocarbons (PAH) from pyrolysis of biomass materials. *Fuel* **2001**, *80*, 1787–1797. [CrossRef]

105. Hale, S.E.; Lehmann, J.; Rutherford, D.; Zimmerman, A.R.; Bachmann, R.T.; Shitumbanuma, V.; O'Toole, A.; Sundqvist, K.L.; Arp, H.P.H.; Cornelissen, G. Quantifying the total and bioavailable polycyclic aromatic hydrocarbons and dioxins in biochars. *Environ. Sci. Technol.* **2012**, *46*, 2830–2838. [CrossRef] [PubMed]

106. Wilson, K.; Reed, D. *IBI White Paper—Implications and Risks of Potential Dioxin Presence in Biochar*; International Biochar Initiative: Canandaigua, NY, USA, 2012. Available online: https://www.biochar-international.org/wp-content/uploads/2018/04/IBI_White_Paper-Implications_of_Potential_%20Dioxin_in_Biochar.pdf (accessed on 31 January 2019).

107. Everaert, K.; Baeyens, J. The formation and emission of dioxins in large scale thermal processes. *Chemosphere* **2002**, *46*, 439–448. [CrossRef]

108. Garcia-Perez, M.; Metcalf, J. The Formation of Polyaromatic Hydrocarbons and Dioxins During Pyrolysis: A Review of the Literature with Descriptions of Biomass Composition, Fast Pyrolysis Technologies and Thermochemical Reactions. 2008. Available online: https://research.wsulibs.wsu.edu:8443/xmlui/bitstream/handle/2376/5966/TheFormationOfPolyaromaticHydrocarbonsAndDioxinsDuringPyrolysis.pdf?sequence=1&isAllowed=y (accessed on 31 January 2019).

 *horticulturae*

Review

Light and Microbial Lifestyle: The Impact of Light Quality on Plant–Microbe Interactions in Horticultural Production Systems—A Review

Beatrix W. Alsanius [1,*], Maria Karlsson [1], Anna Karin Rosberg [1], Martine Dorais [2], Most Tahera Naznin [1], Sammar Khalil [1] and Karl-Johan Bergstrand [3]

[1] Department of Biosystems and Technology, Microbial Horticulture Unit, Swedish University of Agricultural Sciences, SLU, P.O. Box 103, SE-230 53 Alnarp, Sweden; maria.e.karlsson@slu.se (M.K.); anna.karin.rosberg@slu.se (A.K.R.); naznin.most.tahera@slu.se (M.T.N.); sammar.khalil@slu.se (S.K.)

[2] Département de Phytologie, Centre de Recherche et d'Innovation sur les Végétaux (CRIV), Université Laval, Pavillon Envirotron, Local 1216, Québec City, QC G1V 0A6, Canada; martine.Dorais@fsaa.ulaval.ca

[3] Department of Biosystems and Technology, Horticultural Crop Physiology Unit, Swedish University of Agricultural Sciences, SLU, P.O. Box 103, SE-230 53 Alnarp, Sweden; karl-johan.bergstrand@slu.se

* Correspondence: beatrix.alsanius@slu.se; Tel.: +46-40-415336

Received: 15 January 2019; Accepted: 23 May 2019; Published: 28 May 2019

Abstract: Horticultural greenhouse production in circumpolar regions (>60° N latitude), but also at lower latitudes, is dependent on artificial assimilation lighting to improve plant performance and the profitability of ornamental crops, and to secure production of greenhouse vegetables and berries all year round. In order to reduce energy consumption and energy costs, alternative technologies for lighting have been introduced, including light-emitting diodes (LED). This technology is also well-established within urban farming, especially plant factories. Different light technologies influence biotic and abiotic conditions in the plant environment. This review focuses on the impact of light quality on plant–microbe interactions, especially non-phototrophic organisms. Bacterial and fungal pathogens, biocontrol agents, and the phyllobiome are considered. Relevant molecular mechanisms regulating light-quality-related processes in bacteria are described and knowledge gaps are discussed with reference to ecological theories.

Keywords: abiotic factors; biocontrol agent (BCA); controlled environment; ecological theory; greenhouse; molecular mechanisms; non-phototrophic bacteria; pathogens; phyllosphere; plant metabolism; plant morphology

1. Introduction

Plants are meta-organisms colonized with microorganisms, including bacteria, fungi, algae, archaea, protozoa, viruses, and, on rare occasions, nematodes. Depending on the environmental and plant-related conditions prevailing in the various habitats surrounding different plant organs (e.g., soil/growing medium, atmosphere), different compartments (so-called spheres) differing in microbial colonization patterns and community structure have been identified. The very well-researched zone affected by the root (*rhizosphere*) consists of an outer layer (*ectorhizosphere*), the root surface (*rhizoplane*), and the interior of the root (*endorhizosphere*). Likewise, aboveground plant parts constitute three spheres, the *phyllosphere, caulosphere,* and *carposphere,* which denote zones affected by the leaf, stem, and fruit, respectively. The phyllosphere is divided into the epiphytically colonized leaf *surface* and the leaf *endosphere.*

The phyllosphere and its microbiota have received increasing attention during recent years [1–33], because this can be a powerful tool to improve plant health, growth, development, and human health metabolites. Studies have been conducted on a wide range of scales, from parts of a leaf to intact leaves, entire canopies of individual plants, and crop stands. Studies on plant stands tend to use the term phyllosphere in a wider sense, including also the caulosphere and carposphere. The phyllosphere can be divided into the epiphytically colonized leaf surface and the leaf endosphere. The leaf surface is a hostile environment for microbes due to exposure to diurnally and seasonally fluctuating environmental and plant physiological conditions and their interactions (e.g., ambient temperature, irradiation, and water and nutrient availability) [30,34,35]. In contrast, the leaf endosphere offers a nutritionally rich and shielded environment [35]. Plant leaves host 10^6–10^7 bacteria/cm^2 leaf surface [30], with microbially available nutrients (organic carbon sources) being the driving force. However, nutrients are not evenly distributed on the leaf surface, so leaves are not covered with an even biofilm, but rather with patches containing assemblages of microorganisms [17,30,35–37]. While the leaf microbiota is affected by external conditions in the habitat, it is also able to respond proactively to suboptimal conditions through the use of light receptor proteins and to modify its habitat to shield itself from harmful environmental effects and to optimize nutrient acquisition and chances of survival [30,35].

Controlled environments, such as greenhouses, polytunnels, and plant factories, reduce the amplitude of fluctuations in the crop environment, which in turn affects plant performance and the structure and function of the associated microbiome. Greenhouse-covering materials and shade netting alter prevailing environmental conditions (e.g., temperature, relative humidity, and carbon dioxide (CO_2) concentration), but also conditions at the crop level and in the crop phyllosphere, as they influence greenhouse light transmission, reflection, absorption, and diffusion within the canopy [38–40]. Figure 1 summarizes the most important growth parameters affecting the phyllosphere of greenhouse crops.

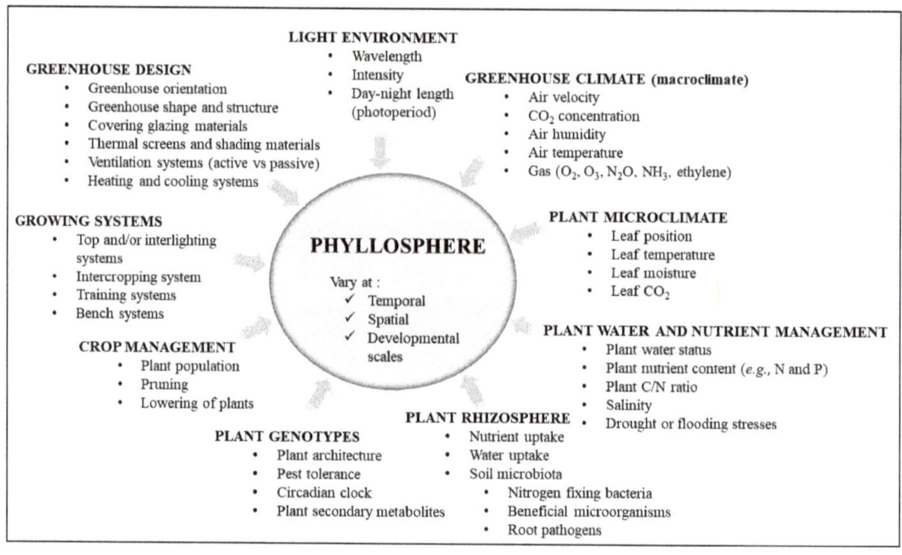

Figure 1. Most important growth parameters affecting the phyllosphere of greenhouse crops. (*Illustration: M. Dorais*).

To compensate for light deprivation under naturally low light conditions and to optimize plant development and quality with respect to crop and market demands, additional artificial assimilation lighting is necessary. Different types of lamps are available (Table 1). Alternative technologies, among these light-emitting diodes (LED), have been introduced during recent years as a measure to reduce energy consumption and costs.

Table 1. Commonly used sources for artificial assimilation lighting (high-pressure sodium, HPS; metal halide; light tube; continuous spectrum polychromatic light-emitting diode (LED). Parameters of importance for plant–microbe interactions are displayed (ultraviolet light, UV; photosynthetically active radiation, PAR). Heat emissions directed towards (↓) or away from (↑) the crop are indicated by arrows.

Lamp Type	Effect (W)	Infra-Red	UV	PAR [1]	Direction of Heat Emissions	References
HPS [2]	400	High	Low	1.6	↓	[41]
Metal halide [3]	400	High	Low	n/a	↓	[42]
Light tube [4]	58	Medium	Low	n/a	↓	
LED 1 [5]	630	None	None	1.85	↑	[43]
LED 2 [6]	550	None	None	2.5	↑	[44]
LED 3 [7]	400	None	None	2.3	↑	[45]

[1] Spectral distribution for different lamp types shown in Figure S1. [2] Philips Master, Philips, Eidhoven, the Netherlands; [3] Philips Master HPI-T plus; Philips, Eindhoven, the Netherlands. [4] Osram G13 T8 58W 840, Osram, Munich, Germany; [5] Heliospectra EOS, Heliospectra AB, Gothenburg, Sweden; [6] Senmatic FL300 Grow, Senmatic A/S, Soendersoe, Denmark; [7] Valoya RX400, Valoya Oy, Helsinki, Finland.

In the horticultural and controlled environment context, light and plant interactions, including light intensity, light quality (*light spectrum*), and day length, have been well-researched (Figure S2), but studies in these disciplines only rarely consider the fact that plants are meta-organisms. In plant microbiology studies, on the other hand, there has been an increasing focus on the phyllosphere in recent years. Prompted by advances in culture-independent techniques, many of these studies focus on the community structure and microbial biodiversity on a descriptive level, but rarely include ecological theories or concepts [19]. Although such studies are often carried out under controlled climate conditions, description and monitoring of environmental factors receive little attention (Figure S3). In fact, the leaf surface and the phyllosphere are often considered a matrix with limited interactions, rather than part of a living and aging system, and very few studies explicitly consider the impact of light quality on the phyllosphere microbiota under greenhouse conditions. With respect to artificial assimilation lighting, the architecture of the plant and crop stand and the position of the light source (top and/or intracanopy lighting) are important. With a rosette-like leaf organization, all leaves are fully exposed to the administered light, whereas only the most outer leaf layer of cushion-forming plants and plants within dense crop stands is exposed, irrespective of top or intercrop irradiation. Leaves inside the canopy are shaded and, thus, dominated by green light (wavelength: 500–565 nm).

The bacterial community structure in the phyllosphere has received more attention than the fungal community structure. Examples of the bacterial community structure of various greenhouse-grown crops are shown in Figure 2. With respect to foliar pathogens, the focus in previous research has been on alternative control of fungi using different wavelengths of light (*light qualities*), rather than on bacteria [46–50]. However, different light technologies influence biotic and abiotic conditions in the plant environment [51]. Modifications in the cropping environment, induced by light intensity and quality and by daylength, influence the structure but also the function of the leaf-associated microbiome [51–53]. Microbes switch lifestyle to adapt to light qualities, as a matter of life and death (i.e., to enable metabolism, function, survival, growth, and nutrient acquisition) [50,52,53].

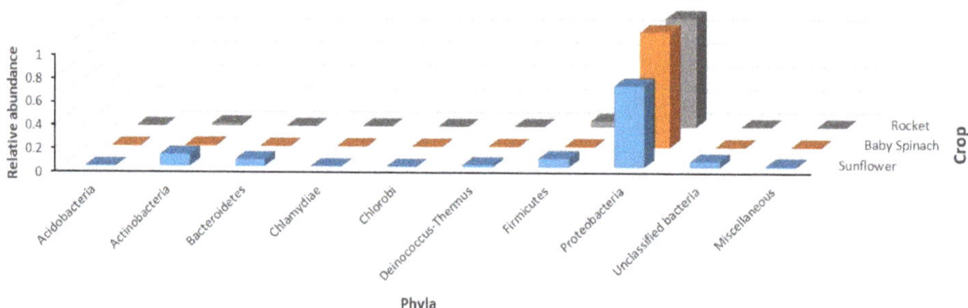

Figure 2. Bacterial phyllosphere community structure of some greenhouse crops artificially illuminated with high-pressure sodium lamps (HPS). Blue: sunflower, *Helianthus annuus* L. [51]. Orange: baby leaf spinach (*Spinacia oleacea*); grey: rocket (*Diplotaxis tenuifolia*) [Alsanius, unpublished data]. (*Illustration: B. Alsanius*).

In this review, we consider the impact of light quality on plant–microbe interactions in light of current ecological theories and concepts. In particular, we focus on the following research questions:

(i)   Which light-dependent plant processes and mechanisms are decisive for phyllosphere colonizers?

(ii)   Which morphological plant characteristics are modified by light quality and consequently influence the structure and/or function of the phyllosphere microbiome?

(iii) Which light-quality-dependent microbial processes and mechanisms affect plant traits?

(iv) Which ecological principles and theories apply to microbiome effects in the phyllosphere with regard to artificial illumination?

2. Materials and Methods

In this literature review, we followed recommendations developed for systematic reviews and meta-analyses [54] and covered the literature in a 30-year period (1988–2018). All keywords and keyword combinations are listed in Table S1. Searches were performed in Web of Knowledge (WoK) using all WoK databases (Web of Science Core Collection, Biosis Citation Index, CABI, Current Contents Connect, Data Citation Index, Derwent Innovation Index, KCI-Korean Journal Database, MEDLINE, Russian Science Citation Index, SciELO Citation Index, Zoological Record).

3. Abiotic Effects of Light on the Leaf Microbiota

3.1. Impact of Lighting Technology on Leaf Temperature, Leaf Moisture, and Humidity

The light environment affects the environment of the leaf surface in several ways. In greenhouse conditions, the most profound effect on the leaf microbiota caused by artificial lighting is due to changes in leaf microclimate [55]. Differences in the amount of infrared (IR) light emitted by different types of light sources is the major cause of these light-source-dependent changes in the leaf microclimate [55]. Conventional high-intensity discharge (HID) lamps, including metal halide (MH) and high-pressure sodium (HPS) lamps, emit most of their waste heat as IR radiation, radiated in the same direction as visible light [55]. In contrast, LED-based light sources mainly produce sensible heat, which has to be cooled away from the fixture using fans, heat sinks, or water cooling [55]. It is well-documented that lighting using HID lamps results in higher leaf temperatures than lighting using LED lamps, e.g., one study [56] reported 0.5–0.7 °C higher air temperature in the canopy in plants illuminated with HPS lights compared with plants illuminated with LED lights [56]. Another study found that air temperatures were around 1 °C higher within the crop stand of potted ornamentals when HPS lighting was applied, compared with LED lighting [57]. Moreover, the relative humidity (RH) in the canopy has been found to be around 5%-units lower when HPS lights are applied compared with LED lights [57]. Interactions between UV radiation and relative humidity have also been observed [58]. It was observed that attacks of powdery mildew in roses can be reduced to practically zero when applying 24-h lighting, which has been explained by constant moisture conditions on the leaves preventing conidia from germinating [59]. On the other hand, higher relative humidity (lower vapor pressure deficit) is generally known to increase the incidence of infection and sporulation of *Botrytis cinerea* [60]. The ambient air temperature also affects the incidence of Botrytis infection in tomato, with an optimum at 15 °C [60].

A number of studies have also demonstrated lower leaf temperatures when using LEDs instead of HPS lamps [58]. Leaf temperatures exceeding the ambient air temperature create air movement within the canopy, thus removing humidity from the boundary layer of the leaf and supplying CO_2 to the boundary layer. Leaf temperatures higher than the ambient air temperature also eliminate the risk of condensation on the leaf surfaces at dew point temperatures close to the ambient air temperature.

When producing plants in closed environments (i.e., plant factories), high leaf temperatures can be a problem [61]. However, in greenhouse production, leaf temperatures during winter are often sub-optimal due to losses of radiant heat through the greenhouse roof. The need for supplementary lighting typically arises during periods of the year where the greenhouse also needs supplementary heating due to low outdoor temperatures. In addition, increased light intensities should typically be accompanied by higher ambient temperatures [62].

3.2. Effects of Ultraviolet (UV) Light

The amount of ultraviolet (UV) light emitted by light fixtures affects the conditions for microbial growth on leaf surfaces. Greenhouse-covering materials normally filter out a large proportion of the UV light, making the greenhouse a UV-deficient environment. In particular, conventional glass panes filter out most UV light, whereas some plastic films have good transmittance of both UV-A and, in some cases, UV-B light [63–65]. Conventional greenhouse HID fixtures normally emit negligible amounts of UV light, but it is possible to supply UV light by using UV lamps [48].

3.3. Effects of Far Red (FR) Light

At the other end of the light spectrum, far-red (FR) light (710–850 nm) and particularly the red-to-far-red ratio (R:FR photoequilibrium) of light perceived by phytochromes can strongly affect the conditions for the leaf microbiota via physiological processes affecting plant architectural development, flowering, photosynthesis, plant nutrition, and plant tolerance to biotic and abiotic stresses [58,66]. The R:FR ratio varies within the day (e.g., from 0.6 at the beginning and end of the day to 1.0–1.3 at noon) and it is strongly reduced within the canopy (e.g., to 0.03). Greenhouse-covering materials, such as FR-absorbing plastic film, also impact R:FR ratio (e.g., increasing it from 1.0 under natural light up to 5.7) and plant development [67,68].

The spectral distribution of the light also has direct effects on photosynthesis and, thereby, the availability to microbes of carbon sources within the leaf and on the leaf surface. The blue and red parts of the spectrum are generally considered more efficient for photosynthesis than the yellow and green parts [69]. However, more recent research suggests that green light with its better penetration contributes significantly to photosynthesis in the deeper layers of the canopy [70]. Using light sources emitting just red and blue light is, therefore, not recommended [71,72].

4. Plant-Mediated Effects of Light on the Leaf Microbiota

Light is one of the most important environmental factors affecting plant growth, development, and metabolite content. Light within a broad spectrum range (400–700 nm) is essential for plant photosynthesis, plant growth, and crop productivity, while specific light spectra trigger different intracellular processes via diverse photoreceptors that modify gene expression, metabolism, plant morphology, and functions [58,73–75]. Figure 3 summarizes plant processes affected by light that can be targeted to promote beneficial phyllosphere components, contributing to greenhouse crop productivity and plant resilience to abiotic and biotic stresses. Modifications in plant architecture, plant morphology, and plant physiology processes will then directly or indirectly impact the leaf microclimate, such as leaf moisture and temperature, as well as habitat resource availability (e.g., carbon, nitrogen, and phosphorus compounds) for the phyllosphere microbiota. However, plant–light interactions are often plant-species-dependent.

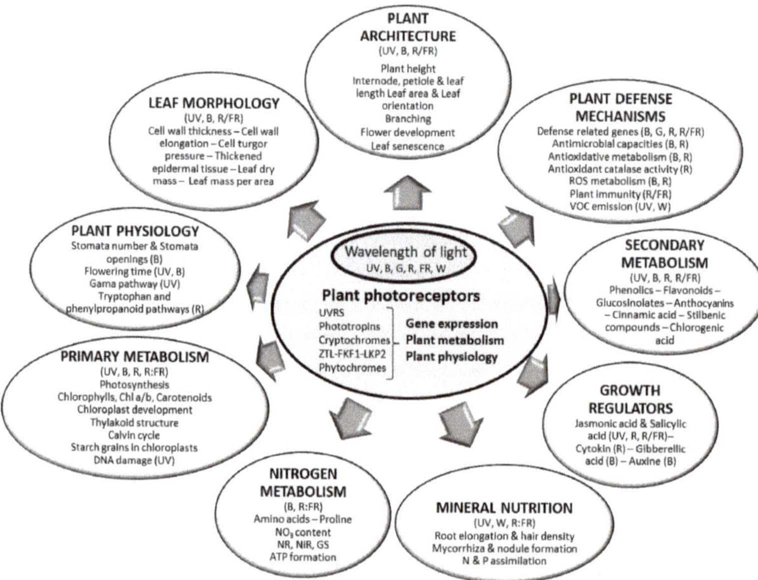

Figure 3. Plant processes affected by light that can be targeted to promote beneficial phyllosphere components, contributing to greenhouse crop productivity. (*Illustration: M. Dorais*).

4.1. Plant–Light Interactions

4.1.1. Plant Architecture and Leaf Morphology

Modulation of light spectra (e.g., R:FR ratio) to control plant architecture and leaf morphology is a well-known technique used by producers of ornamental plants to improve the shape and appearance of their plants, while assimilation lighting (400–700 nm) of vegetables improves crop productivity. However, both these artificial lighting regimens modify the canopy microclimate and plant structure. Blue light controls cell elongation and is an essential signal for the plant to adjust its growth to the surrounding light conditions [75]. Although light responses in horticultural crops differ between genotypes, enhanced amounts of UV light generally increase the thickness of the leaf cuticle [64], which in turn strengthens plant resistance to attacks from fungal pathogens [65,66]. However, for other species and under other exposure conditions, UV radiation can increase plant susceptibility to fungal pathogens [67,68]. A reduction in stem elongation as a result of UV exposure is often observed in greenhouse crops [76], which in turn modifies the plant microclimate. For example, UV-A and blue light increase shoot length and internode length in cucumber, but have the opposite effect on tomato and no effect on rose, while UV-A reduces stem length in rose and internode length in poinsettia [77–81]. Similarly, enrichment of natural and/or HPS light with blue and red light limits stem length in ornamental crops [82]. In combination, UV-B and blue light reduce leaf area in cucumber, leafy vegetables, and rose, while UV-B in combination with red light increases leaf area of pepper compared with monochromatic red light [56,80,83–85]. Blue light also increases leaf mass area, leaf and stem thickness, and shoot dry mass in cucumber [79,86,87], but reduces shoot dry mass in leafy vegetables [56,84], whereas UV-B increases leaf thickness in lettuce [85]. Plant responses to a light spectrum may be within a small wavelength interval. For example, it has been shown that 430–450 nm gives a greater leaf and stem growth increase in green perilla than 455–470 nm [88]. A combination of blue, red, and far-red light increases dry matter in cucumber and tomato compared with HPS lamps, particularly at a low blue:red ratio [89,90]. Exposure to UV light also alters epicuticular wax, which protects the plant against pathogen invasion [91]. Additionally, it induces morphological changes in trichomes of leaves [92]. Branching is often promoted by UV-B and blue light, while flowering induction, precocity, and duration are species-dependent. Furthermore, elevated parts of far-red light (lowered R:FR ratio) increase elongation of plants due to greater internode elongation [93], while a high R:FR ratio results in plants with a compact growth habit [58], creating a more shaded and moist leaf surface due to slower air movements within the canopy. However, plant ability to respond to R:FR ratio (e.g., via phytochromes PHYA, PHYB) is variety- and species-dependent. Leaf expansion can be promoted or inhibited by FR light [93,94], which may be related to competition for resources between the leaf and stem growth or auxin-induced cytokinin breakdown in leaf primordia. A low R:FR ratio may also decrease leaf mass per area and leaf duration, and cause leaf hyponasty and solar leaf tracking. Reduced branching in many species due to inhibition of bud outgrowth via phytohormones (i.e., auxin, strigolactones, cytokinins, ABA) has been observed under a low R:FR ratio. Flowering of many crops is accelerated under a low R:FR [58], but this varies according to the species [95].

4.1.2. Photosynthesis

Plant growth and metabolite accumulation depend on photosynthesis, which is suboptimal at very weak [96] or excessive light intensity [97]. Light intensity and light quality both have a very strong impact on plant photosynthesis, while daylength may affect the plant circadian clock and primary metabolism via cumulative carbon biomass. Plants modulate their photosynthesis pigments to the prevailing light spectrum and intensity. Although species-dependent, the chlorophyll (chl) content and the chl a/b ratio usually increase with blue light [77,86,98]. For different species, blue and red light increase the plant content of carotenoids, such as lutein and β-carotene, while UV may reduce it [84,89,99–101]. Blue light increases photosynthetic activity when used together with other wavelengths, but reduces it when used alone [77,84,86], whereas UV-B decreases plant photosynthesis efficiency [102]. Studies of several specific wavelengths (from 405 to 700 nm) on photosynthesis of tomato, lettuce, and petunia plants have revealed higher photosynthesis with the blue region (range 417–450 nm) and red region (range 630–680 nm) than the green region (501, 520, 575, 595 nm) [103]. A blue and red light combination allows for higher photosynthetic activity than monochromatic light of either, which can be harmful for plants [104,105]. Opening of the stomata, which are a natural entry point for leaf microorganisms, is driven by blue light, although red light also promotes stomatal opening. Blue light is also involved in chloroplast movement within the cell to increase photosynthetic ability under different light conditions. An increased number of stomata and length of palisade tissue cells have been observed under blue light compared with red or green light [87]. Higher numbers of grana lamellae and more stacked thylakoid membranes have been observed in cucumber grown under low blue light radiation [98]. Blue light also prevents accumulation in the chloroplasts of starch grains, which block the incoming light. In addition to its effect on leaf area, leaf orientation, and leaf branching, thereby modifying crop photosynthesis, a low R:FR ratio may reduce stomatal conductance, stomatal density, chlorophyll content, chloroplast development, thylakoid structure and protein composition, and the activity of some enzymes of the Calvin cycle [58]. As FR light negatively affects root hair density, mycorrhizal colonization, and ATP formation, the R:FR ratio influences plant mineral nutrition. Moreover, a low R:FR ratio promotes nutrient allocation to the shoot at the expense of roots [58].

4.1.3. Primary and Secondary Metabolism

The light spectrum also influences the accumulation of plant primary and secondary metabolites [106]. For example, accumulation of soluble sugars, starch, soluble protein, and polyphenols is higher when crops are grown under monochromatic red or blue than white light [107–113], which may impact the phyllobiome. Indeed, the phyllosphere microbiota uses leaf surface resources, such as amino acids, carbohydrates, and organic acids, passively leaked by plants [114]. A combination of red, blue, and white light enhances soluble sugar and nitrate concentrations in basil plants [115]. However, a low R:FR ratio downregulates the activity of the key enzymes involved in nitrogen assimilation (nitrate reductase, nitrite reductase, and glutamine synthase), which may impact cell metabolism [58]. Blue light and mixtures of red, blue, and green light increase ascorbic acid accumulation in leafy vegetables [84,116,117], while UV-A may reduce ascorbic acid content [118]. Anthocyanin leaf content is usually promoted by UV-A, UV-B, and blue light [75]. In particular, UV-A, blue, and red light increase the anthocyanin level in leafy vegetables, while green light reverses blue-light-induced anthocyanin accumulation [94,118–120]. Sulfur-containing secondary metabolites, such as glucosinolate, which can protect the plant against predation and pathogens, may also be promoted by blue light or a mixture of red, blue, and green light [121–123]. The R:FR ratio also impacts accumulation of phenolics in different species [115,124]. In addition, light affects the synthesis, profile, and emission of volatile organic compounds (VOCs) by plants, which increases plant attractiveness to plant parasitoids and orientation of predators [124,125]. Release of VOCs from leaves increases when they are exposed to UV, white light, or a low R:FR ratio, although this effect may be species-specific.

4.1.4. Plant Defense Mechanisms

Light, such as UV-B, affects several plant hormones, notably jasmonate (involved in response to attack by necrotrophic pathogens) and salicyclic acid (involved in response to attack by biotrophic microbial pathogens), that coordinate the plant immune response to environmental stresses [66]. Blue light also induces pathogenesis-related gene expression [126], while red light induces salicylic acid content and expression of salicylic-acid-regulating *PR-1* and *WRKY* genes in pathogen-inoculated cucumber plants [127,128]. On the other hand, a low R:FR ratio resulting in high plant population or plant shading may affect plant immunity, which has been linked in some species to reduced transcription of salicylic-acid-responsive genes or to decreased jasmonate sensitivity and reduced biosynthesis of tryptophan-derived secondary metabolites [129,130]. In particular, a low R:FR ratio inhibits salicylic acid and jasmonic-acid-mediated disease resistance in *Arabidopsis* plants [130,131]. Furthermore, a low R:FR ratio may induce high levels of gibberellin and auxin, which are involved in internode elongation, while gibberellin and ethylene are implicated in petiole extension [58,132].

Light spectrum has a strong effect on the antioxidant properties of horticultural plants [109,110, 133–136]. For example, UV-B light can cause plant DNA damage, resulting in a cascade of protective events, such as flavonoid synthesis and expression of chalcone synthase genes and photolyase genes [75]. However, UV-B may have no effect or may reduce flavonoid accumulation in some species [100]. Compared with white light, blue and red light increase the activity of various reactive oxygen species (ROS)-scavenging enzymes and reducing substances (reduced glutathione (GSH) and ascorbic acid (ASA)) [137], which play an important role in plant defense mechanisms against plant pathogens in species such as tomato [138]. In tomato, blue light promotes leaf accumulation of proline, polyphenolic compounds, and antioxidants, and ROS scavenger activities, which might be partly related to inhibition of gray mold disease. On the other hand, red- and green-light-treated tomato plants have been shown to exhibit lower proline content [110]. Light of specific wavelengths (UV, blue, red) also promotes synthesis of stilbenic compounds compared with white light [46,139,140]. Stilbenes, which are low-molecular-weight phenolics, play an important role in plant defense responses by overcoming fungal pathogen attacks [110,141]. Similarly, red light induces cinnamic acid synthesis and increases plant resistance via the tryptophan and phenylpropanoid pathways [142]. Moreover, high gamma-aminobutyric acid levels are promoted by plant UV-B exposure, resulting in higher bacterial diversity in the phyllosphere and lower plant resistance to fungal disease [143]. On the other hand, plants exposed to a low R:FR are more sensitive to pathogens due to changes in leaf morphology, chlorophyll content, and downregulation of jasmonate and salicylic acid [58,66].

4.2. Direct Plant–Microbe Interactions Induced by Light

Physiological changes in the plant caused by different light qualities have a major impact on the phyllosphere microbiota. In sunflower plants grown under HPS lamps, white LEDs, or red:blue (80:20 ratio) LEDs, it has been shown that the fungal communities are more affected than the bacterial communities by different light qualities [51]. Although that study did not investigate whether the effects on the microbiota are direct or indirect, other research (presented below) suggests that the effect is often caused by physiological alterations in the plant.

4.2.1. Leaf Leachate

The availability of organic carbon as a prerequisite for microbial colonization in the phyllosphere has been surveyed in several reviews [28,30,36,37]. Microbial phyllosphere communities are limited first by availability of organic carbon sources and only second by availability of organic nitrogen sources [37]. Although the phyllosphere is often characterized as a habitat lacking in nutrients, leaves exude a wide range of carbon compounds, such as carbohydrates, amino acids, organic acids, and sugar alcohols [30]. The availability of these nutrients is highly dependent on photosynthesis, which in turn is highly dependent on light quality and intensity. Leaching of nutrients across the leaf surface occurs in the presence of liquid water, but can also be increased by the phyllosphere microbiota through microbially produced biosurfactants [28]. The most abundant compounds in leachate are photosynthetic compounds, such as glucose, fructose, and sucrose [144,145]. However, the glandular trichomes, which are important sites of leaching, also secrete proteins, oils, secondary metabolites, and mucilage [36,146–148]. Use of red, or red plus blue, LED light has been shown to increase the amount of soluble sugars and proteins in a wide range of plants, as mentioned earlier. This physiological change in the plant, caused by choice of artificial light source, changes the carrying capacity of the leaf and governs which microorganisms are favored by the increase or decrease in compounds specific for their survival. While the microbial community as a whole can utilize a wide range of compounds for colonization and growth, a single microbial species can be quite specific in its metabolism. For example, substrate profiling of *Pseudomonas syringae* has shown that this bacterium uses a restricted number of sugars, organic acids, and amino acids [149]. This implies that, with increased knowledge of the metabolic patterns of specific microorganisms, light could be used as a management tool, not only for plant growth, but also in order to favor microbial species of importance.

4.2.2. Light-Triggered Pathways

While light within the spectral wavelength from 300 to 800 nm can have an effect on plant growth and development, red light seems to have the largest impact relating to defense against microbial pathogens by triggering both plant defense genes and hormonal pathways. The composition of the phyllosphere microbial community is driven by a wide range of factors. However, the plant immune system is thought to play a major role in shaping the community composition. It has been shown that triggering of the salicylic acid pathway leads to reductions in both diversity and population sizes of endophytic bacteria, while epiphytic bacteria are not measurably affected, and that *Arabidopsis thaliana* plants deficient in the jasmonic acid pathway host a greater epiphytic bacterial community diversity [150]. For horticultural species, red and green light have a positive effect on tomato seedlings, with less infection by *Pseudomonas cichorii* JBC1 compared with white light or dark treatment [151]. A similar result was observed for cucumber plants infected with powdery mildew (*Sphaerotheca fuliginea*) and exposed to red light, while no effect was found under green light [128]. This decrease in infection level was related to the upregulation of the defense gene phenylalanine ammonia lyase (*PAL*) and pathogenesis-related protein 1a (PR1a) under red or green light treatments [151]. This leads to the conclusion that light significantly alters the activation of defense-related genes. Differences in results between different studies, however, imply that use of light treatment for control of pathogens has to be customized to the plant–pathogen system.

Downregulation of the salicylic acid and jasmonic acid pathways when plants compete for light against other plants (i.e., a low R:FR ratio) means that the plants become more sensitive to pathogen attack, as shown with *Botrytis cinerea* in *Arabidopsis* [129]. This plant response to a low R:FR ratio has also been reported elsewhere [130]. A low R:FR ratio can be avoided in the greenhouse by spacing out the plants, allowing for more light to enter the lower parts of the canopy. While a high R:FR ratio leads to pathogen susceptibility, use of red light leads to activation of the salicylic acid pathway-mediated systemic acquired resistance (SAR) in *Arabidopsis*, making the plant more resistant to *Pseudomonas syringae* pv. *tomato* [152]. A study on rice has also shown increased resistance to disease, specifically *Bipolaris oryzae*, when rice plants are subjected to red light, with an increasing level of resistance being demonstrated with an increasing dose of red light [142]. However, in rice plants, disease resistance is mediated through the tryptophan and phenylpropanoid pathways, and not by the salicylic acid pathway as suggested in *Arabidopsis*.

4.2.3. Changes in Leaf Physiological Characteristics

Leaf surface properties have a large impact on the establishment and survival of phyllosphere microorganisms [36]. The thickness of the adaxial epidermis layer has been found to be one of the three most important leaf attributes governing the plant–microbe system, where an epidermal layer thicker than 20.77 μm results in lower microbial colonization rates [153].

Changes in epicuticular wax layers and epidermal tissues, in particular, can emerge as consequences of subjecting plants to different light qualities [127]. A difference in effect on leaf morphological characteristics depending on light quality has also been seen between sun-exposed and shaded leaves, with sun-exposed leaves having a thicker cuticle than shaded leaves [154]. It has been suggested that UV-B radiation is the factor responsible for a thicker cuticular wax layer on the leaf surface, with e.g., increased irradiation with UV-B, increasing the wax layer in cucumber, pea, and barley by 25% [155]. A thicker wax layer prevents, or at least delays, pathogen infection, especially for fungal pathogens that use direct penetration as a means of infection. In a detached leaf assay using soybean, it has been shown that disease severity of soybean rust (*Phakopsora pachyrizi*) is negatively correlated with amount of epicuticular wax and that the leaves at the top of the canopy have a higher amount of wax than leaves in middle and lower levels [91].

5. Light-Quality-Mediated Effects on the Leaf Microbiota

Biofilm is a natural way for microorganisms to co-exist on a surface or interface, enclosed in a exopolysaccharide matrix produced by the microbes themselves [156]. Regardless of whether the microorganisms concerned are human or plant pathogens or microbes used as biocontrol agents (BCA), their efficiency depends on how well they establish and survive on a surface. Environmental factors, such as temperature, humidity, and light, are important factors that shape microbial communities. Biofilms protect microbes against antibiotics and harsh environmental factors and maintain nutrient availability [156]. To date, bacteria have been regarded as non-phototrophic organisms and insensitive to light. Only phototrophic bacteria were known to react and respond to light. However, light has been shown to affect bacterial decisions to change from a planktonic single cell motile lifestyle to a surface-attached lifestyle in a multicellular community as biofilm [157]. This is supported by the fact that some of the photo receptor proteins also control mechanisms involved in biofilm formation and these receptors are linked to the GGDEF and EAL protein domains, which are involved in the transition from a planktonic to a sessile life style [158].

5.1. Leaf Pathogens

Irrespective of their growing site (nature, field stand, or controlled environment), plants can be attacked by plant diseases. Amongst the fungal pathogens, grey mold (*Botrytis cinerea*), powdery mildew (*Podosphaera* spp.), and downy mildew (*Peronospora* spp.) are often reported in major greenhouse crops, such as tomatoes, cucumber, strawberries, and ornamental plants (e.g., grey mold in tomato [110];

powdery mildew in strawberries (Fragaria X ananassa), [159,160] and roses (*Rosa* spp.) (*P. pannosa*) [49]; downy mildew (*Pseudoperonospora cubensis*) in cucumber [47]). Different bacterial species, for example *Xanthomonas* spp. and *Pseudomonas* spp., can also cause severe damage to plants [50,161,162]. The idea of using light as a strategy to control leaf pathogens is not new, e.g., 20 years ago, greenhouse experiments with blue-pigmented photoselective sheets showed that these inhibited sporulation and colonization of downy mildew on cucumber [47]. Light regulates biofilm formation, attachment, motility, and virulence of both fungal and bacterial plant pathogens (Table 2), factors which are crucial for establishment on the leaf surface.

Table 2. Summary of different microorganisms, the photoreceptor/s they contain, and the physiological response to different light spectra.

Organism	Light Quality	Wave Length (nm)	Photoreceptor	Photoreceptor Architecture	Effect	Ref.
Acinetobacter baumannii	Blue	415	BLUF, LOV	EAL-GAF-GGDEF-LOV-GGDEF	Biofilm formation, metabolism, virulence	[163]
Bacillus amylolique-faciens	Red Blue	645 458	LOV	LOV-STAS	Swarming motility, biofilm formation, antifungal activity	[164]
Botrytis cinerea	Blue	405	PHY, LOV	PAS-GAF-PHY-HK LOV-PAS, short LOV	Inhibited mycelial growth, virulence	[165]
Pseudomonas aeringiunosa	Blue	405	PHY, LOV	PAS-GAF-PHY-kinase Short LOV	Survival, virulence factors	[166]
P. cichorii	Green	NI	LOV	HATP-HisKA-LOV-RR	Siderophore and phytotoxic lipopeptide production	[167]
P. syringae	Red/Far-red Blue White	680/750 470	PHY, LOV	PAS-GAF-PHY-kinase HATP-HisKA-LOV-RR Short LOV	Decreased swarming motility	[50]
Podosphaera pannosa	Blue	420–520			Reduced germination and conidia formation	[49]
Serratia marcescens	Blue White	470			Antibiotic production	[168]
Sphaerotheca fuliginea	Red	NI [1]			Disease suppression	[127]
Staphylococcus aureus	Blue	405, 470			Growth	[169]
Trichoderma harzianum	Blue	NI			Induced gene expression of *phr1*	[170]
Xanthomonas axonopodis	Light/dark		PHY, LOV, BLUF	PAS-GAF-PHY-PAS LOV-HK	Motility, adhesion, biofilm formation	[161]
Xanthomonas campestris	Red/Far-red Blue White	NI	PHY, LOV	PAS-GAF-PHY-PAS HATP-HisKA-LOV-RR	Growth, motility	[162]

[1] NI = not indicated.

Implementation of LED light as an environmentally friendly tool in indoor production has increased in recent years. In this context, it has been shown that light quality has an impact on growth and development of the conidia of *P. pannosa*, which causes powdery mildew disease on roses [49]. Blue light (420–520 nm) was observed to decrease conidial growth in that study, while far-red light (575–675 nm) had the opposite effect, i.e., it increased pathogen growth. However, the same study could not demonstrate a reduction in conidia development when roses were grown with 18 h daylight complemented with 6 h of blue or red light [49]. Exposure to blue light has been demonstrated to increase the antioxidant and polyphenolic content in tomato plants and thereby control the attachment of *Botrytis cinerea* [110]. Moreover, a study on cucumber plants indicated that light quality affects incidence of powdery mildew and expression of defense-related genes [127]. Bacterial infection in plants can also be suppressed by light of different quality, e.g., green light reduces phytotoxic lipopeptide and siderophore production in *Pseudomonas cichorii*, which might affect survival of this plant pathogen [167]. In *Xanthomonas* spp., light quality has a negative impact on motility, and, thus, host colonization [161,162]. Swarming motility in *Pseudomonas syringae* is suppressed under light conditions (white, blue, red plus far-red) compared with dark treatment, but different light qualities (white, blue, red plus far-red) also have differing effects, with blue light promoting swarming motility [50].

5.2. Microbial Biocontrol Agents

As is the case for deleterious microorganisms, such as pathogens, plant-health-promoting biocontrol agents are also affected by the light regime provided under controlled conditions. A successful BCA should exhibit (i) several antifungal or antibacterial (antagonistic) properties, (ii) ability to spread on the plant surface after application, and (iii) capacity to establish in existing biofilms. However, very little is known about how different light regimes affect BCA when it comes to establishment in the plant canopy. One study demonstrated that, in *Serratia marcescens*, antibiotic pigment prodigiosin concentration in bacterial cells decreases under white and blue light (470 nm) conditions, but growth is not affected [168]. The same study showed that red and far-red light have no effect on the concentration of prodigiosin [168]. Another study [170] isolated the photolyase gene (*phr1*) from *Trichoderma harzianum*, a common soil fungus used as a BCA against phytopathogenic fungi [171] and investigated expression of *phr1* when exposed to blue light. Their results showed that gene expression of photolyase (phr1) is induced very rapidly in both mycelia and conidiphores, and that light induces development of pigmented resistance spores as well as expression of *phr1* [159]. *Bacillus amyloliquefaciens* is another BCA often used in horticulture against soilborne and post-harvest pathogens. In this species, all light quality except blue light affects growth, swarming motility, biofilm formation, and antifungal activity positively [164]. Red light (645 nm) increases biocontrol efficacy and colonization of BCA on fruit surfaces, while blue light (458 nm) has a negative impact on growth, motility, and biofilm formation [164].

5.3. Molecular Interactions

For many years, only phototrophs were considered to respond to light in order to find an optimally illuminated environment for harvesting solar energy [172]. However, the increasing number of papers on bacterial whole genome sequencing has now revealed a large number of putative genes coding for photoreceptor proteins distributed among several taxa. During bacterial evolution, bacteria have evolved photoreceptor proteins that can detect visible light in the environment, in order to protect themselves from damaging UV radiation [172]. Bacteria can also respond to light by switching between the single cell planktonoic lifestyle and the multicellular life style of bacterial communities known as biofilms [173]. Six classes of photoreceptors have been identified in the bacteria photosensory system, based on their structure of their chromophore. These are: cryptochrome, rhodopsin, phytochrome, photoactive yellow protein (PYP), light oxygen voltage receptor protein (LOV), and blue light sensing protein using FAD (BLUF) [172].

Cyclic di-guanosine monophosphate (c-di-GMP), a key role player in the bacterial signal transduction system, regulates bacterial behaviors, such as biofilm formation, virulence, and production of adhesion proteins [174]. It is produced from diguanylate cyclases (DGCs) and is then broken down to 5′-phosphoguanylyl-(3′-5′)-guanosine (pGpG) through hydrolysis by phosphodiesterases (PDEs). Of these, DGCs are associated with the GGDEF photoreceptor domain and PDEs with the EAL domain [175]. Both are involved in light-sensing processes, together with the LOV and BLUF domains [157,158].

Two photoreceptors are involved in blue light sensing in plants and in microbes, namely LOV and BLUF. Both these protein domains have been shown to control attachment, multicellularity, production of adhesion proteins, and virulence (Table 2). Therefore, blue light has been shown to be a promising candidate to combat bacterial and fungal infections in medical science. For example, several studies have reported bactericidal effects on *Pseudomonas aeruginosa* and *Staphylococcus aureus* when exposed to 405 nm light [169]. Furthermore, exposure of methicillin-resistant *Staphylococcus aureus* to 405 and 470 nm light has been shown to bring about a significant reduction in growth [176]. Similar findings have been made in a study on bacteria involved in clinical infections, where both planktonic and bacterial biofilm proved susceptible to blue light, with significant reductions in viability for all tested strains [177]. Blue light exposure in horticulture has been shown to have negative effects on both bacteria and fungi (Table 2). However, blue light conditions have been reported to enhance disease attack caused by the fungus *Sphaerotheca fuliginea* [127]. In *Pseudomonas syringae*, it has been demonstrated that light decreases the swarming motility and that this is regulated by bacteriophytochrome and LOV-HK (Light Oxygen Voltage-Histidine Kinase) [50]. In the plant pathogen *Xanthomonas axonopodis*, LOV is activated by blue light and may be involved in the control of bacterial virulence [161].

All these studies show that light quality has an impact, one way or another, on the behavior of microorganisms. For this to occur, the organism needs to perceive and transmit the signal, which is done by photosensory proteins. As mentioned above, the BLUF and LOV photoreceptor domains are involved in blue light sensing. The LOV domain belongs to the PAS (Per-ARNT-Sim) superfamily connected in a network of conserved domains (GGDEF, EAL, PAS, GAF (cGMP-specific phosphodiesterases, adenylyl cyclases and FhlA), HK, HisKA (histidine protein kinases), and STAS (sulfate transporter and anti-sigma factor antagonist)) [178]. A very extensive bioinformatics study of photoreceptors across kingdoms has been conducted [178]. According to their data for bacteria, different groups of protein architecture dominate across different phyla. Within the Proteobacteria, the combinations EAL-GAF-GGDEF-LOV-PAS and EAL-GGDEF-HAMP-LOV-PAS are the most abundant architectures of photoreceptor proteins, together with HATP-HisKA-LOV. Within the Firmicutes, the combination STAS-LOV is the most common architecture. Across all investigated phyla [169], there are short (150 aa) LOV proteins that can stand alone with a highly conserved motif of five amino acids with a cysteine at position 54 that forms the cysteine–flavin assembly during the LOV photocycle, which is also involved in sensing blue light [179]. The BLUF domain control functions such as photosystem synthesis, biofilm formation, and both swarming and twitching motility [180,181]. It is widespread across the bacterial kingdom, but in anoxygenic and plant-associated species BLUF proteins are not as abundant as LOV proteins. For example, BLUF proteins have not been recovered from Firmicutes, Chloroflexi, or EuArchaea, which instead only carry genes encoding LOV proteins [158]. In many species, BLUF seems to act alone. In *Escherichia coli*, *Klebsiella pneumoniae*, and *Magnetococcus* sp., BLUF is combined with EAL [158,173]. The BLUF-EAL protein YcgF in *E. coli* acts as a direct anti repressor in a blue light response, which in turn activates other proteins important for biofilm formation [182].

Photoactive yellow protein (PYP) is a blue light sensor protein first discovered in halophilic purple phototrophic bacteria [183]. With the increasing number of whole genome sequencing studies, there have been reports of PYP proteins in bacteria other than phototrophs, mostly within in Proteobacteria [184]. Photoactive yellow protein is small, only 125 amino acids long, and is often present as part of the PAS domain [185]. Studies have shown that PYP serves as a photosensor for negative phototaxis [186].

6. Discussion

Day length, light intensity, and light quality affect plant architecture and morphology, plant growth, and plant development. Lighting is a crucial tool for greenhouse horticulture and plant production in controlled environments. It plays a central role for the microclimate in the crop stand, e.g., temperature and relative humidity. Light-related effects on the crop can be direct or indirect. The potential of the plant microbiome to influence crop growth and development and the ability to withstand abiotic and biotic stress has been repeatedly highlighted [28,187,188]. Microbiome-based tools and prediction models have been suggested [187]. Despite the importance of light and illumination in greenhouse horticulture (Figures 2 and 3) and the increasing interest in the phyllosphere microbiome, light-associated factors are only occasionally receiving the attention they deserve in experimental settings or in applied contexts (Figure S3).

At present, knowledge on the influence of light, especially light quality, on phyllosphere–microbe interactions resembles a random mosaic in an otherwise vast field, as previous studies have tended to consider either the behavior of a specific target organism or expression of a light-related gene or receptor, or big metagenomics datasets with limited functional information. To bring phyllosphere studies within the scope of ecological principles and theories, the presence and also the function of microorganisms need to be highlighted. However, it is of utmost importance to discriminate between mechanisms and processes that can be abstracted and those that cannot. In this context, the pathosystem deserves particular attention. In the present literature review, we focused on interactions between selected plant and human pathogens and light quality and considered some of the molecular mechanisms involved. However, in planta studies are rare and often lack sufficient characterization of the growing environment. To understand the interactions between light, especially light quality, plant/crop stand, and microbiome, critical experimental conditions and physiological processes need to be continuously monitored. This requires such disciplines as crop physiology and microbiology/plant pathology to engage with common phenotypic platforms.

The composition and amount of microbially available organic nutrients, a suitable microclimate with respect to temperature and humidity/moisture, and niches providing shelter from deleterious irradiation and unintentional predation are key properties of a suitable microbial phyllosphere environment. Thus, mechanisms affecting these key properties will decide colonization density and composition. Light quality directly or indirectly influences many of these processes (see Figure 2). In this regard, nutrient sources available in the phyllosphere can serve as an example. A few studies have examined the composition and quantity of leaf lysates [144,189,190] and interactions between organic nutrients and microbial proliferation [145]. Two recent studies considering almost 400 different nutrient sources have indicated that the nutritional preferences of some phyllosphere colonizers change in the presence of different light qualities [52,53]. The utilization of compounds themselves, but also their use within carbon, nitrogen, or sulfur metabolism, is moderated in the presence of different light qualities, and, consequently, the secondary metabolites are also moderated. At present, the focus has been on light qualities of major importance for plant photosynthetic activity (blue, red) in such studies. To the best of our knowledge, the impact of other light spectra on microbial utilization of nutrients has not been investigated. Such information, as well as transcriptomics data on the phyllosphere microbial community, needs to be provided, along with plant and environmental monitoring data, in order to reveal the impact of light quality and to assess the potential of light as a tool for habitat management.

The deleterious impact of certain light qualities on plant pathogenic fungi has been investigated since the late 1990s [47], with particular focus on commercially important pathogens (e.g., grey mold and powdery and downy mildew). In contrast to studies on bacteria, these fungi studies primarily concentrate on the disease incidence, while rarely analyzing underlying molecular mechanisms. Such information is needed for non-pathogenic and pathogenic fungi (for endophytic and ectophytic phyllosphere bacteria) in order to implement light quality strategies into greenhouse horticulture and controlled environment production systems. Overall, studies on new, non-chemical control strategies for leaf pathogens are of substantial interest in the development of sustainable horticultural indoor production systems.

It has been suggested that the phyllosphere be used as a platform for the testing of ecological principles [77]. Different literature reviews [187,188] have proposed ecological theories and principles relating to the phytobiome. Given a multidisciplinary and systematic approach, the theories and principles depicted in Table 3 could contribute to a better understanding of light–phyllosphere interactions in greenhouse horticulture and to the development of sustainable growing practices.

Table 3. Ecological theories and principles of interest to use in light-assisted phyllosphere studies.

Theory/Principles	Modes of Action	Potential Research Questions	Light Spectra of Interest
Niche theory			
Priority effects	Pre-emptying of space and resources by the first arriving species	Heterotrophic utilization of leaf lysates/organic compounds and their impact on secondary metabolites	B [1], G [2], Y [3], R [4], R:FR [5]
Competitive dominance	Dominance due to efficient resource use under prevailing stable conditions		
Niche partitioning	Coexistence	Light-quality-associated impact on biofilm community structure Bacterial–fungi symbionts/Suitable microbe combination Plant–microbe and microbe–microbe compatibility	B, G, Y, R, R:FR B, R:FR
Storage effect	Coexistence of microbes within the same ecological community	Storage effects in non-phototrophic non-spore-forming bacterial leaf colonizers	B, G, Y, R, R:FR
Niche modification	Invasion of leaf interior	Light quality as a driver towards an endophytic lifestyle	B, G, Y, R, R:FR
	Biofilm formation	Light quality as a driver for switch from planktonic to biofilm lifestyle	
Complementarity	Diversification of resource requirements leading to less competition between interspecific than conspecific neighbors	Mechanisms of coexistence under various light qualities	B, G, Y, R, R:FR
Resource-based interactions			
Resource competition		Heterotrophic utilization of leaf lysates/organic compounds and their impact on secondary metabolites in microbial aggregate communities	B, G, Y, R, R:FR
Phenotypic plasticity	Formation of different phenotypes under various conditions	Complementary microbe pair for stimulating plant growth and pathogen control	B, R:FR

[1] B = blue, [2] G = green, [3] Y = yellow; [4] R = red; [5] R:FR = red:far red.

Supplementary Materials: The following are available online at http://www.mdpi.com/2311-7524/5/2/41/s1: Figure S1: Relative spectral output from three different light sources: HPS lamps (Philips Master 400 W), fluorescent tubes (Sylvania TLD840 58 W), and LED lights (Valoya B150, spectrum AP673L 144 W). Figure S2: Number of publications considering the topic 'supplementary lighting in greenhouse horticulture'. The literature search considered three keyword combinations, namely artificial lighting*greenhouse*horticulture (102 publications), supplementary lighting*greenhouse*horticulture (201 publications), and artificial illumination*greenhouse* horticulture (21 publications) and was performed in Web of Knowledge (WoK) using all WoK databases (Web of Science Core Collection, Biosis Citation Index, CABI, Current Contents Connect, Data Citation Index, Derwent Innovation Index, KCI-Korean Journal Database, MEDLINE, Russian Science Citation Index, SciELO Citation Index, Zoological Record). The literature search was restricted to 30 years (1988–2018) (dates of performance: November 19 and 20, 2018). Figure S3: Description of light conditions in study output considering the keyword combination "phyllosphere*greenhouse*horticulture". The search was restricted to 30 years (1988–2018) and entailed 27 publications conducted under greenhouse, climate chamber, or polytunnel. The survey was performed in Web of Knowledge (WoK) using all WoK databases (Web of Science Core Collection, Biosis Citation Index, CABI, Current Contents Connect, Data Citation Index, Derwent Innovation Index, KCI-Korean Journal Database, MEDLINE, Russian Science Citation Index, SciELO Citation Index, Zoological Record). Values indicate percentage of studies stating or avoiding information on environmental conditions (relative humidity, temperature), light conditions (day length, light intensity) and use of supplementary lighting, as well as control of description of light spectrum in the plant stand. The proportion of publications discussing the impact of light on the results obtained was also determined (3.9% corresponds to one publication) (dates of performance: November 19 and 20, 2018).

Author Contributions: All authors were involved in writing the original draft, draft preparation, and reviewing and editing this article. In detail, the authors took on the following responsibilities: conceptualization: B.A. and M.D.; methodology: B.A.; formal analysis: all authors; writing (in alphabetical order of surname): original draft preparation, B.A., K.J.B., M.D., M.K., S.K., A.K.R., T.N., writing, review and editing, B.A., K.J.B., M.D., M.K., S.K., A.K.R., T.N.; visualization: B.A., K.J.B., M.D.; project administration: B.A.; funding acquisition: B.A. (PI) in collaboration with M.K. and K.J.B.

Funding: This research was funded by Stiftelsen Lantbruksforskning and the Swedish Research Council for Environment, Agricultural Sciences and Spatial Planning, both Stockholm, Sweden, grant number R-18-25-006 ("Optimized integrated control in greenhouse systems sees the light"; PI: Beatrix Alsanius).

Conflicts of Interest: The authors declare no conflict of interest. The funders had no role in the design of the study; in the collection, analysis, or interpretation of data; in the writing of the manuscript; or in the decision to publish the results.

References

1. Aleklett, K.; Hart, M.; Shade, A. The microbial ecology of flowers: An emerging frontier in phyllosphereresearch. *Botany* **2014**, *92*. [CrossRef]

2. Andreote, F.D.; Gumiere, T.; Durrer, A. Exploring interactions of plant microbiomes. *Sci. Agric.* **2014**, *71*, 528–539. [CrossRef]

3. Beilsmith, K.; Thoen, M.P.M.; Brachi, B.; Gloss, A.D.; Khan, M.H.; Bergelson, J. Genome-wide association studies on the phyllosphere microbiome: Embracing complexity in host-microbe interactions. *Plant J.* **2018**, *97*, 164–181. [CrossRef] [PubMed]

4. Berg, G.; Grube, M.; Schloter, M.; Smalla, K. Unraveling the plant microbiome: Looking back and future perspectives. *Front. Microbiol.* **2014**, *5*, 148. [CrossRef]

5. Brader, G.; Compant, S.; Vescio, K.; Mitter, B.; Trognitz, F.; Ma, L.J.; Sessitsch, A. Ecology and genomic insights into plant-pathogenic and plant-nonpathogenic endophytes. *Ann. Rev. Phytopathol.* **2017**, *55*, 61–83. [CrossRef] [PubMed]

6. Bringel, F.; Couee, I. Pivotal roles of phyllosphere microorganisms at the interface between plant functioning and atmospheric trace gas dynamics. *Front. Microbiol.* **2015**, *6*, 486. [CrossRef] [PubMed]

7. Bulgarelli, D.; Schlaeppi, K.; Spaepen, S.; van Themaat, E.V.L.; Schulze-Lefert, P. Structure and functions of the bacterial microbiota of plants. *Ann. Rev. Plant Biol.* **2013**, *64*, 807–838. [CrossRef] [PubMed]

8. Carvalho, S.D.; Castillo, J.A. Influence of light on plant-phyllosphere interaction. *Front. Plant Sci.* **2018**, *9*, 1482. [CrossRef] [PubMed]

9. Chaudhary, D.; Kumar, R.; Sihag, K.; Rashmi; Kumari, A. Phyllospheric microflora and its impact on plant growth: A review. *Agric. Rev.* **2017**, *38*, 51–59. [CrossRef]

10. Farre-Armengol, G.; Filella, I.; Llusia, J.; Penuelas, J. Bidirectional interaction between phyllospheric Microbiotas and plant volatile emissions. *Trends Plant Sci.* **2016**, *21*, 854–860. [CrossRef]

11. Finkel, O.M.; Castrillo, G.; Paredes, S.H.; Gonzalez, I.S.; Dangl, J.L. Understanding and exploiting plant beneficial microbes. *Curr. Opin. Plant Biol.* **2017**, *38*, 155–163. [CrossRef] [PubMed]

12. Gao, S.; Liu, X.; Dong, Z.; Liu, M.; Dai, L. Advance of phyllosphere microorganisms and their interaction with the outside environment. *Plant Sci. J.* **2016**, *34*, 654–661.

13. Iguchi, H.; Yurimoto, H.; Sakai, Y. Interactions of methylotrophs with plants and other heterotrophic bacteria. *Microorganisms* **2015**, *3*, 137–151. [CrossRef] [PubMed]

14. Jackson, C.R.; Stone, B.W.G.; Tyler, H.L. Emerging perspectives on the natural microbiome of fresh produce vegetables. *Agriculture* **2015**, *5*, 170–187. [CrossRef]

15. Kowalchuk, G.A.; Yergeau, E.; Leveau, J.H.J.; Sessitsch, A.; Bailey, M. Plant-Associated Microbial Communities. In *Environmental Molecular Microbiology*; Lui, W.-T., Jansson, J., Eds.; Caister Academic Press: Norfolk, UK, 2010; pp. 131–148.

16. Lemanceau, P.; Barret, M.; Mazurier, S.; Mondy, S.; Pivato, B.; Fort, T.; Vacher, C. Plant communication with associated microbiota in the spermosphere, rhizosphere and phyllosphere. *Adv. Bot. Res.* **2017**, *82*, 101–133. [CrossRef]

17. Leveau, J.H.J. Microbiology Life on leaves. *Nature* **2009**, *461*, 741. [CrossRef] [PubMed]

18. Markland, S.M.; Kniel, K.E. Human pathogen-plant interactions: Concerns for food safety. *Adv. Bot. Res.* **2015**, *75*, 115–135. [CrossRef]

19. Meyer, K.M.; Leveau, J.H.J. Microbiology of the phyllosphere: A playground for testing ecological concepts. *Oecologia* **2012**, *168*, 621–629. [CrossRef]

20. Mueller, D.B.; Schubert, O.T.; Roest, H.; Aebersold, R.; Vorholt, J.A. Systems-level Proteomics of Two Ubiquitous Leaf Commensals Reveals Complementary Adaptive Traits for Phyllosphere Colonization. *Mol. Cell. Proteom.* **2016**, *15*, 3256–3269. [CrossRef]

21. Mueller, T.; Ruppel, S. Progress in cultivation-independent phyllosphere microbiology. *Fems Microbiol. Ecol.* **2014**, *87*, 2–17. [CrossRef]

22. Muller, D.B.; Vogel, C.; Bai, Y.; Vorholt, J.A. The plant microbiota: Systems-level insights and perspectives. *Ann. Rev. Genet.* **2016**, *50*, 211–234. [CrossRef] [PubMed]

23. Rastogi, G.; Sbodio, A.; Tech, J.J.; Suslow, T.V.; Coaker, G.L.; Leveau, J.H.J. Leaf microbiota in an agroecosystem: Spatiotemporal variation in bacterial community composition on field-grown lettuce. *ISME J.* **2012**, *6*, 1812–1822. [CrossRef] [PubMed]

24. Remus-Emsermann, M.N.P.; Tecon, R.; Kowalchuk, G.A.; Leveau, J.H.J. Variation in local carrying capacity and the individual fate of bacterial colonizers in the phyllosphere. *ISME J.* **2012**, *6*, 756–765. [CrossRef] [PubMed]

25. Saikkonen, K.; Mikola, J.; Helander, M. Endophytic phyllosphere fungi and nutrient cycling in terrestrial ecosystems. *Curr. Sci.* **2015**, *109*, 121–126.

26. Schlaeppi, K.; Bulgarelli, D. The plant microbiome at work. *Mol. Plant Microbe Interact.* **2015**, *28*, 212–217. [CrossRef] [PubMed]

27. Thapa, S.; Prasanna, R. Prospecting the characteristics and significance of the phyllosphere microbiome. *Ann. Microbiol.* **2018**, *68*, 229–245. [CrossRef]

28. Vacher, C.; Hampe, A.; Porte, A.J.; Sauer, U.; Compant, S.; Morris, C.E. The phyllosphere: Microbial jungle at the plant-climate interface. *Ann. Rev. Ecol. Evol. Syst.* **2016**, *47*, 1–24. [CrossRef]

29. Vogel, C.; Bodenhausen, N.; Gruissem, W.; Vorholt, J.A. The Arabidopsis leaf transcriptome reveals distinct but also overlapping responses to colonization by phyllosphere commensals and pathogen infection with impact on plant health. *New Phytol.* **2016**, *212*, 192–207. [CrossRef] [PubMed]

30. Vorholt, J.A. Microbial life in the phyllosphere. *Nat. Rev. Microbiol.* **2012**, *10*, 828–840. [CrossRef]

31. Vorholt, J.A. The phyllosphere microbiome: Responses to and impacts on plants. *Phytopathology* **2014**, *104*, 155.

32. Whipps, J.M.; Hand, P.; Pink, D.; Bending, G.D. Phyllosphere microbiology with special reference to diversity and plant genotype. *J. Appl. Microbiol.* **2008**, *105*, 1744–1755. [CrossRef] [PubMed]

33. Yang, T.; Chen, Y.; Wang, X.X.; Dai, C.C. Plant symbionts: Keys to the phytosphere. *Symbiosis* **2013**, *59*, 1–14. [CrossRef]

34. Beattie, G.A.; Lindow, S.E. The secret life of bacterial pathogens on plants. *Ann. Rev. Phytopathol.* **1998**, *33*, 145–172. [CrossRef] [PubMed]

35. Beattie, G.A.; Lindow, S.E. Bacterial colonization of leaves: A spectrum of strategies. *Phytopathology* **1999**, *89*, 353–359. [CrossRef] [PubMed]

36. Leveau, J.H.J. Microbial communities in the phyllosphere. In *Biology of the Plant Cuticle*; Riederer, M., Müller, C., Eds.; Blackwell Publishing: Oxford, UK, 2006; pp. 334–367.

37. Lindow, S.E.; Brandl, M.T. Microbiology of the phyllosphere. *Appl. Environ. Microbiol.* **2003**, *69*, 1875–1883. [CrossRef] [PubMed]

38. Diaz, B.M.; Biurrun, R.; Moreno, A.; Nebreda, M.; Fereres, A. Impact of ultraviolet-blicking plastic films on insect vectors of virus diseases infesting crisp lettuce. *HortScience* **2006**, *41*, 711–716. [CrossRef]

39. Hemming, S. Use of natural and artificial light in horticulture—Interaction of plant and technology. *Acta Hortic.* **2011**, *907*, 25–36. [CrossRef]

40. Stamps, R.H. Use of colored shade netting in horticulture. *HortScience* **2009**, *44*, 239–241. [CrossRef]

41. Philips. Master Agro 400W E40 1SL/12. 2018. Available online: http://www.lighting.philips.se/prof/konventionella-lampor-och-lysroer/urladdningslampor/hid-horticulture/horti/928144609201_EU/product (accessed on 7 January 2019).

42. Philips. Master HPI-T Plus 400 W/643. Philips Lighting Holding B.V. 2018. Available online: http://www.lighting.philips.com/main/prof/conventional-lamps-and-tubes/high-intensity-discharge-lamps/quartz-metal-halide/master-hpi-t-plus/928483500191_EU/product (accessed on 7 January 2019).

43. Heliospectra. Heliospectra EOS. Heliospectra AB: 2018. Available online: https://www.heliospectra.com/led-grow-lights/eos/ (accessed on 7 January 2019).

44. Senmatic. Senmatic FL300. Senmatic A/S: 2018. Available online: https://www.senmatic.com/horticulture/products/led-fixtures/fl300-grow-white (accessed on 7 January 2019).

45. Valoya. Valoya Product Brochure. Valoya Oy: 2018. Available online: http://www.valoya.com/brochures/ (accessed on 7 January 2019).

46. Ahn, S.Y.; Kim, S.A.; Yun, H.K. Inhibition of *Botrytis cinerea* and accumulation of stilbene compounds by light-emitting diodes of grapevine leaves and differential expression of defense-related genes. *Eur. J. Plant Pathol.* **2015**, *143*, 753–765. [CrossRef]

47. Reuveni, R.; Raviv, M. Control of downy mildew in greenhouse-grown cucumbers using blue photoselective polyethylene sheets. *Plant Dis.* **1997**, *81*, 999–1004. [CrossRef]

48. Suthaparan, A.; Stensvand, A.; Solhaug, K.A.; Torre, S.; Telfer, K.H.; Ruud, A.K.; Mortensen, L.M.; Gadoury, D.M.; Seem, R.C.; Gislerød, H.R. Suppression of cucumber powdery mildew by supplemental UV-B radiation in greenhouses can be augmented or reduced by background radiation quality. *Plant Dis.* **2014**, *98*, 1349–1357. [CrossRef] [PubMed]

49. Suthaparan, A.; Stensvand, A.; Torre, S.; Herrero, M.L.; Pettersen, R.; Gadoury, D.M.; Gislerød, H.R. Continuous lighting reduces conidial production and germinability in the rose powdery mildew pathosystem. *Plant Dis.* **2010**, *94*, 339–344. [CrossRef] [PubMed]

50. Wu, L.; McGrane, R.S.; Beattie, G.A. Light regulation of swarming motility in Pseudomonas syringae integrates signaling pathways mediated by a bacteriophytochrome and a LOV protein. *mBio* **2013**, *4*, e00334-13. [CrossRef] [PubMed]

51. Alsanius, B.W.; Bergstrand, K.J.; Hartmann, R.; Gharaie, S.; Wohanka, W.; Dorais, M.; Rosberg, A.K. Ornamental flowers in new light: Artificial lighting shpares the microbial phyllosphere community structure of greenhouse grown sunflowers (*Helianthus annuus* L.). *Sci. Hortic.* **2017**, *216*, 234–247. [CrossRef]

52. Alsanius, B.W.; Vaas, L.A.I.; Gharaie, S.; Karlsson, M.E.; Rosberg, A.K.; Grudén, M.; Wohanka, W.; Khalil, S.; Windstam, S. Dining in blue light impairs the appetite of some leaf epiphytes. *Manuscript* **2019**.

53. Gharaie, S.; Vaas, L.A.I.; Rosberg, A.K.; Windstam, S.T.; Karlsson, M.E.; Bergstrand, K.J.; Khalil, S.; Wohanka, W.; Alsanius, B.W. Light spectrum modifies the utilization pattern of energy sources in Pseudomonas sp. *PLoS ONE* **2017**, *12*, e0189862. [CrossRef] [PubMed]

54. Moher, D.; Liberati, A.; Tetzlaff, J.; Altman, D.G. The PRISMA Group, Preferred reporting items for systematic reviews and meta-analyses: The PRISMA statement. *PLoS Med.* **2009**, *6*, e1000097. [CrossRef]

55. Nelson, J.A.; Bugbee, B. Analysis of environmental effects on leaf temperature under sunlight, high pressure sodium and light emitting diodes. *PLoS ONE* **2015**, *10*, e0138930. [CrossRef] [PubMed]

56. Hernández, R.; Kubota, C. Physiological responses of cucumber seedlings under different blue and red photon flux ratios using LEDs. *Environ. Exp. Bot.* **2014**, *121*, 66–74. [CrossRef]

57. Bergstrand, K.J.; Schüssler, H.K. Growth, development and photosynthesis of some horticultural plants as affected by different supplementary lighting technologies. *Eur. J. Hortic. Sci.* **2013**, *78*, 119–125.

58. Demotes-Mainard, S.; Péron, T.; Corot, A.; Bertheloot, J.; Le Gourriere, J.; Pelleschi-Travier, S.; Crespel, L.; Morel, P.; Huché-Thélier, L.; Boumaz, R.; et al. Plant responses to red and far-red lights, applications in horticulture. *Environ. Exp. Bot.* **2016**, *121*, 4–21. [CrossRef]

59. Mortensen, L.M. The effect of photon flux density and lighting period on growth, flowering, powdery mildew and water relations of miniature roses. *Am. J. Plant Sci.* **2014**, *5*, 1813–1818. [CrossRef]

60. O'Neill, T.M.; Shtienberg, D.; Elad, Y. Effect of some host and microclimate factors on infection of tomato stems by Botrytis cinerea. *Plant Dis.* **1997**, *81*, 36–40. [CrossRef] [PubMed]

61. Kozai, T.; Niu, G.; Takagaki, M. (Eds.) Physical environmental factors and their properties. In *Plant Factory: An Indoor Vertical Farming System for Efficient Quality Food Production*; Elsevier Academic Press: Amsterdam, The Netherlands, 2016; pp. 129–140.

62. Rünger, W. *Licht und Temperatur im Zierpflanzenbau*; Paul Parey: Berlin, Germany, 1964.

63. Raviv, M.; Antignus, Y. UV radiation effects on pathogens and insect pests of greenhouse-grown crops. *Photochem. Photobiol.* **2004**, *79*, 219–226. [CrossRef] [PubMed]

64. Von Zabeltitz, C. Cladding material. In *Integrated Greenhouse Systems for Mild Climates*; Springer: Heidelberg, Germany, 2011; pp. 145–167.

65. Waaijenberg, D. Design, construction and maintenance of greenhouse structures. *Acta Hortic.* **2004**, *710*, 31–42. [CrossRef]
66. Ballaré, C.L. Light regulation of plant defense. *Ann. Rev. Plant Biol.* **2014**, *65*, 335–363. [CrossRef] [PubMed]
67. Clifford, S.C.; Runkle, E.S.; Langton, F.A.; Mead, A.; Foster, S.A.; Pearson, S.; Heins, R.D. Height control of poinsettia using photoselective filters. *HortScience* **2004**, *39*, 383–387. [CrossRef]
68. Mata, D.A.; Botto, J.F. Manipulation of light environment to produce high-quality poinsettia plants. *HortScience* **2009**, *44*, 702–706. [CrossRef]
69. McCree, K.J. The action spectrum, absorptance and quantum yield of photosynthesis in crop plants. *Agric. Meteorol.* **1972**, *9*, 191–216. [CrossRef]
70. Massa, G.; Graham, T.; Haire, T.; Flemming, C.; Newsham, G.; Wheeler, R. Light-emitting diode light transmission through leaf tissue of seven different crops. *HortScience* **2015**, *50*, 501–506. [CrossRef]
71. Bergstrand, K.J.; Mortensen, L.M.; Suthaparan, A.; Gislerød, H.R. Acclimatisation of greenhouse crops to differing light quality. *Sci. Hortic.* **2016**, *204*, 1–7. [CrossRef]
72. Lin, K.H.; Huang, M.Y.; Huang, W.D.; Hsu, M.H.; Yang, Z.W.; Yang, C.M. The effects of red, blue, and white light-emitting diodes on the growth, development, and edible quality of hydroponically grown lettuce (Lactuca sativa L. var. capitata). *Sci. Hortic.* **2013**, *150*, 86–91. [CrossRef]
73. Fankhauser, C.; Ulm, R. A photoreceptor's on-off switch. *Science* **2016**, *354*, 282–283. [CrossRef] [PubMed]
74. Folta, K.M.; Carvalho, S.D. Photoreceptors and control of horticultural plant traits. *HortScience* **2015**, *50*, 1274–1280. [CrossRef]
75. Huché-Thélier, L.; Crespel, L.; Le Gourrierec, J.; Morel, P.; Sakr, S.; Leduc, N. Light signaling and plant responses to blue light and UV radiation—Perspectives for applications in horticulture. *Environ. Exp. Bot.* **2016**, *121*, 22–38. [CrossRef]
76. Lercari, B.; Bretzel, F.; Piazza, S. Effects of UV Treatments on Stem Growth of Some Greenhouse Crops. *Act Hortic.* **1992**, *327*, 99–104. [CrossRef]
77. Abidi, F.; Girault, T.; Douillet, O.; Guillemain, G.; Sintes, G.; Laffaire, M.; Ahmed, H.B.; Smiti, S.; Huché-Thélier, L.; Leduc, N. Blue light effects on rose photosynthesis and photomorphogenesis. *Plant Biol.* **2013**, *15*, 67–74. [CrossRef]
78. Glowacka, B. The effect of blue light on the height and habit of the tomato (*Lycopersicon esculentum* Mill.) transplant. *Folia Hortic.* **2004**, *16*, 3–10.
79. Piszczek, P.; Głowacka, B. Effect of the colour of light on cucumber (*Cucumis sativus* L.) seedlings. *Veg. Crop. Res. Bull.* **2008**, *68*, 71–80. [CrossRef]
80. Terfa, M.T.; Roro, A.G.; Olsen, J.E.; Torre, S. Effects of UV radiation on growth and postharvest characteristics of three pot rose cultivars grown at different altitudes. *Sci. Hortic.* **2014**, *178*, 184–191. [CrossRef]
81. Torre, S.; Roro, A.G.; Bengtsson, S.; Mortensen, L.; Solhaug, K.A.; Gislerød, H.R.; Olsen, J.E. Control of plant morphology by UV-B and UV-B-temperature interactions. *Acta Hortic.* **2012**, *956*, 207–214. [CrossRef]
82. Islam, M.A.; Kuwar, G.; Clarke, J.L.; Blystad, D.R.; Gislerød, H.R.; Olsen, J.E.; Torre, S. Artificial light from light emitting diodes (LEDs) with a high portion of blue light results in shorter poinsettias compared to high pressure sodium (HPS) lamps. *Sci. Hortic.* **2012**, *147*, 136–143. [CrossRef]
83. Brown, C.S.; Schuerger, A.C.; Sager, J.C. Growth and photomorphogenesis of pepper plants under red light-emitting diodes with supplemental blue or far-red lighting. *J. Am. Soc. Hortic. Sci.* **1995**, *120*, 808–813. [CrossRef] [PubMed]
84. Ohashi-Kaneko, K.; Takase, M.; Kon, N.; Fujiwara, K.; Kurata, K. Effect of light quality on growth and vegetable quality in leaf lettuce, spinach and komatsuna. *Environ. Control Biol.* **2007**, *45*, 189–198. [CrossRef]
85. Wargent, J.J.; Taylor, A.; Paul, N.D. UV supplementation for growth and disease control. *Acta Hortic.* **2006**, *711*, 333–338. [CrossRef]
86. Hogewoning, S.W.; Trouwborst, G.; Maljaars, H.; Poorter, H.; van Ieperen, W.; Harbinson, J. Blue light dose–responses of leaf photosynthesis, morphology, and chemical composition of *Cucumis sativus* grown under different combinations of red and blue light. *J. Exp. Bot.* **2010**, *61*, 3107–3117. [CrossRef]
87. O'Carrigan, A.; Babla, M.; Wang, F.; Liu, X.; Mak, M.; Thomas, R.; Bellotti, B.; Chen, Z.H. Analysis of gas exchange, stomatal behaviour and micronutrients uncovers dynamic response and adaptation of tomato plants to monochromatic light treatments. *Plant Physiol. Biochem.* **2014**, *82*, 105–115. [CrossRef]
88. Lee, J.S.; Lee, C.A.; Kim, Y.H.; Yun, S.J. Shorter wavelength blue light promotes growth of green perilla (*Perilla frutescens*). *Int. J. Agric. Biol.* **2014**, *16*, 1177–1182.

89. Nanya, K.; Ishigami, Y.; Hikosaka, S.; Goto, E. Effects of blue and red light on stem elongation and flowering of tomato seedlings. *Acta Hortic.* **2012**, *956*, 261–266. [CrossRef]

90. Hogewoning, S.W.; Trouwborst, G.; Meinen, E.; van Ieperen, W. Finding the optimal growth-light spectrum for greenhouse crops. *Acta Hortic.* **2012**, *956*, 357–363. [CrossRef]

91. Young, H.M.; George, S.; Narváez, D.F.; Srivastava, P.; Schuerger, A.C.; Wright, D.L.; Marois, J.J. Effect of solar radiation on severity of soybean rust. *Phytopathology* **2012**, *102*, 794–803. [CrossRef] [PubMed]

92. Yamasaki, S.; Noguchi, N.; Mimaki, K. Continuous UV-B irradiation induces morphological changes and the accumulation of polyphenolic compounds on the surface of cucumber cotyledons. *J. Radiat. Res.* **2007**, *48*, 443–454. [CrossRef] [PubMed]

93. Casal, J.J.; Smith, H. The function, action and adaptive significance of phytochrome in light-grown plants. *Plant Cell Environ.* **1989**, *12*, 855–862. [CrossRef]

94. Li, Q.; Kubota, C. Effects of supplemental light quality on growth and phytochemicals of baby leaf lettuce. *Environ. Exp. Bot.* **2009**, *67*, 59–64. [CrossRef]

95. Craig, D.; Runkle, E.S. Using leds to quantify the effect of the red to far-red ratio of night-interruption lighting on flowering of photoperiodic crops. *Acta Hortic.* **2012**, *956*, 179–185. [CrossRef]

96. Solymosi, K.; Schoefs, B. Etioplast and etio-chloroplast formation under natural conditions: The dark side of chlorophyll biosynthesis in angiosperms. *Photosynth. Res.* **2010**, *105*, 143–166. [CrossRef]

97. Niyogi, K.K. Photoprotection revisited: Genetic and molecular approaches. *Ann. Rev. Plant Physiol. Plant Mol. Biol.* **1999**, *50*, 33. [CrossRef]

98. Wang, X.Y.; Xu, X.M.; Cui, J. The importance of blue light for leaf area expansion, development of photosynthetic apparatus, and chloroplast ultrastructure of Cucumis sativus grown under weak light. *Photosynthetica* **2015**, *53*, 1–10. [CrossRef]

99. Lefsrud, M.G.; Kopsell, D.A.; Sams, C.E. Irradiance from distinct wave length light-emitting diodes affect secondary metabolites in kale. *HortScience* **2008**, *43*, 2243–2244. [CrossRef]

100. Hoffmann, A.M.; Noga, G.; Hunsche, M. High blue light improves acclimation and photosynthetic recovery of pepper plants exposed to UV stress. *Environ. Exp. Bot.* **2015**, *109*, 254–263. [CrossRef]

101. Li, J.; Hikosaka, S.; Goto, E. Effects of light quality and photosynthetic photon flux on growth and carotenoid pigments in spinach (*Spinacia oleracea* L.). *Acta Hortic.* **2009**, *907*, 105–110. [CrossRef]

102. Lidon, F.J.C.; Reboredo, F.H.; Leitão, A.E.; Silva, M.M.A.; Duarte, M.P.; Ramalho, J.C. Impact of UV-B radiation on photosynthesis—An overview. *Emir. J. Food Agric.* **2012**, *24*, 546–556. [CrossRef]

103. Naznin, M.T.; Lefsrud, M.; Gagne, J.D.; Schwalb, M.; Bissonnette, B.H. Different wavelengths of LED light affect on plant photosynthesis. *HortScience* **2012**, *47*, S191.

104. Opdam, J.G.; Schoonderbeek, G.G.; Heller, E.B.; Gelder, A. Closed green-house: A starting point for sustainable entrepreneurship in horticulture. *Acta Hortic.* **2005**, *691*, 517–524. [CrossRef]

105. Trouwborst, G.; Hogewoning, S.W.; van Kooten, O.; Harbinson, J.; van Ieperen, W. Plasticity of photosynthesis after the 'red light syndrome' in cucumber. *Environ. Exp. Bot.* **2016**, *121*, 75–82. [CrossRef]

106. Bian, Z.H.; Yang, Q.C.; Liu, W.K. Effects of light quality on the accumulation of phytochemicals in vegetables produced in controlled environments: A review. *J. Sci. Food Agric.* **2015**, *95*, 869–877. [CrossRef]

107. Samuolienè, G.; Viršilè, A.; Brazaitytè, A.; Jankauskienè, J.; Duchovskis, P.; Novickovas, A. Effect of supplementary pre-harvest LED lighting on the antioxidant and nutritional properties of green vegetables. *Acta Hortic.* **2010**, *939*, 85–91. [CrossRef]

108. Britz, S.J.; Sager, J.C. Photomorphogenesis and photoassimilation in soybean and sorghum grown under broad spectrum or blue-deficient light sources. *Plant Physiol.* **1990**, *94*, 448–454. [CrossRef]

109. Johkan, M.; Shoji, K.; Goto, F.; Hashida, S.; Yoshihara, T. Blue light-emitting diode light irradiation of seedlings improves seedling quality and growth after transplanting in red leaf lettuce. *HortScience* **2010**, *45*, 1809–1814. [CrossRef]

110. Kim, K.; Kook, H.S.; Jang, Y.J.; Lee, W.H.; Kamala-Kannan, S.; Chae, J.C.; Lee, K.J. The effect of blue-light emitting diodes on antioxidant properties and resistance to *Botrytis cinerea* in tomato. *J. Plant Pathol. Microbiol.* **2013**, *4*, 203. [CrossRef]

111. Li, H.M.; Tang, C.M.; Xu, Z.G.; Liu, X.Y.; Han, X.L. Effects of different light sources on the growth of non-heading Chinese cabbage (*Brassica campestris* L.). *J. Agric. Sci.* **2012**, *4*, 262–273. [CrossRef]

112. Li, H.M.; Xu, Z.G.; Tang, C.M. Effect of light-emitting diodes on growth and morphogenesis of upland cotton (*Gossypium hirsutum* L.) plantlets in vitro. *Plant Cell Tissue Organ Cult.* **2010**, *103*, 155–163. [CrossRef]

113. Soebo, A.; Krekling, T.; Applegren, M. Light quality affects photosynthesis and leaf anatomy of birch plantlets in vitro. *Plant Cell Tissue Organ Cult.* **1995**, *41*, 177–185. [CrossRef]

114. Knief, C.; Delmotte, N.; Chaffron, S.; Stark, M.; Innerebner, G.; Wassmann, R.; von Mering, C.; Vorholt, J.A. Metaproteogenomic analysis of microbial communities in the phyllosphere and rhizosphere of rice. *ISME J.* **2012**, *6*, 1378–1390. [CrossRef] [PubMed]

115. Bantis, F.; Ouzounis, T.; Radoglou, K. Artificial LED lighting enhances growth characteristics and total phenolic content of Ocimum basilicum, but variably affects transplant success. *Sci. Hortic.* **2016**, *198*, 277–283. [CrossRef]

116. Samuolienè, G.; Sirtautas, R.; Brazaitytè, A.; Duchovskis, P. LED lighting and seasonality effects antioxidant properties of baby leaf lettuce. *Food Chem.* **2012**, *134*, 1494–1499. [CrossRef]

117. Chen, W.H.; Xu, Z.G.; Liu, X.Y.; Yang, Y.; Wang, Z.H.M.; Song, F.F. Effect of LED light source on the growth and quality of different lettuce varieties. *Acta Bot. Boreali Occident. Sin.* **2011**, *31*, 1434–1440.

118. Liu, W.K.; Yang, Q.C. Effects of supplemental UV-A and UV-C irradiation on growth, photosynthetic pigments and nutritional quality of pea seedlings. *Acta Hortic.* **2012**, *956*, 657–663. [CrossRef]

119. Mizuno, T.; Amaki, W.; Watanabe, H. Effects of monochromatic light irradiation by LED on the growth and anthocyanin contents in leaves of cabbage seedlings. *Acta Hortic.* **2009**, *907*, 179–184. [CrossRef]

120. Zhang, T.; Folta, K.M. Green light signaling and adaptive response. *Plant Signal Behav.* **2012**, *7*, 75–78. [CrossRef]

121. Kopsell, D.A.; Sams, C.E. Increases in shoot tissue pigments, glucosinolates, and mineral elements in sprouting broccoli after exposure to short-duration blue light from light-emitting diodes. *J. Am. Soc. Hortic. Sci.* **2013**, *138*, 31–37. [CrossRef]

122. Kopsell, D.A.; Sams, C.E.; Barickman, T.C.; Morrow, R.C. Sprouting broccoli accumulate higher concentrations of nutritionally important metabolites under narrow-band light-emitting diode lighting. *J. Am. Soc. Hortic. Sci.* **2014**, *139*, 469–477. [CrossRef]

123. Zukalova, H.; Vasak, J.; Nerad, D.; Stranc, P. The role of glucosinolates of Brassica genus in the crop system. *Rostlinna Výroba* **2002**, *48*, 181–189. [CrossRef]

124. Colquhoun, T.A.; Schwieterman, M.L.; Gilbert, J.L.; Jaworski, E.A.; Langer, K.M.; Jones, C.R.; Rushing, G.V.; Hunter, T.M.; Olmstead, J.; Clark, D.G.; et al. Light modulation of volatile organic compounds from petunia flowers and select fruits. *Postharvest Biol. Technol.* **2013**, *86*, 37–44. [CrossRef]

125. Vänninen, I.; Pinto, D.M.; Nissinen, A.I.; Johansen, N.S.; Shipp, L. In the light of new greenhouse technologies: 1. Plant-mediated effects of artificial lighting on arthropods and tritrophic interactions. *Ann. Appl. Biol.* **2010**, *157*, 393–414. [CrossRef]

126. Wu, L.; Yang, H.Q. Cryptochrome 1 is implicated in promoting R protein- mediated plant resistance to Pseudomonas syringae in Arabidopsis. *Mol. Plant* **2010**, *3*, 539–548. [CrossRef] [PubMed]

127. Wang, H.; Jiang, Y.P.; Yu, H.J.; Xia, X.J.; Shi, K.; Zhou, Y.H.; Yu, J.Q. Light quality affects incidence of powdery mildew, expression of defence-related genes and associated metabolism in cucumber plants. *Eur. J. Plant Pathol.* **2010**, *127*, 125–135. [CrossRef]

128. Shibuya, T.; Itagaki, K.; Tojo, M.; Endo, R.; Kitaya, Y. Fluorescent illumination with high red-to-far-red ratio improves resistance of cucumber seedlings to powdery mildew. *HortScience* **2011**, *46*, 429–431. [CrossRef]

129. Cargnel, M.D.; Demkura, P.V.; Ballare, C.L. Linking phytochrome to plant immunity: Low red: Far-red ratios increase Arabidopsis susceptibility to Botrytis cinerea by reducing the biosynthesis of indolic glucosinolates and camalexin. *New Phytol.* **2014**, *204*, 342–354. [CrossRef] [PubMed]

130. De Wit, M.; Spoel, S.H.; Sanchez-Perez, G.F.; Gommers, C.M.M.; Pieterse, C.M.J.; Voesenek, L.; Pierik, R. Perception of low red:far-red ratio compromises both salicylic acid- and jasmonic acid-dependent pathogen defences in Arabidopsis. *Plant J.* **2013**, *75*, 90–103. [CrossRef] [PubMed]

131. Moreno, J.E.; Tao, Y.; Chory, J.; Ballare, C.L. Ecological modulation of plant defense via phytochrome control of jasmonate sensitivity. *Proc. Natl. Acad. Sci. USA* **2009**, *106*, 4935–4940. [CrossRef] [PubMed]

132. Kurepin, L.V.; Emery, R.J.N.; Pharis, R.P.; Reid, D.M. Uncoupling light quality from light irradiance effects in Helianthus annuus shoots: Putative roles for plant hormones in leaf and internode growth. *J. Exp. Bot.* **2007**, *58*, 2145–2157. [CrossRef] [PubMed]

133. Lee, M.K.; Arasu, M.V.; Park, S.; Byeon, D.H.; Chung, S.O.; Park, S.U.; Lim, Y.P.; Kim, S.J. LED lights enhance metabolites and antioxidants in chinese cabbage and kale. *Braz. Arch. Biol. Technol.* **2016**, *59*, e16150546. [CrossRef]

134. Muneer, S.; Kim, E.J.; Park, J.S.; Lee, J.H. Influence of green, red and blue light emitting diodes on multiprotein complex proteins and photosynthetic activity under different light intensities in lettuce leaves (*Lactuca sativa* L.). *Int. J. Mol. Sci.* **2014**, *15*, 4657–4670. [CrossRef] [PubMed]

135. Naznin, M.T.; Lefsrud, M.; Gravel, V.; Hao, X. Different ratios of red and blue LEDs light affect on coriander productivity and antioxidant properties. *Acta Hortic.* **2016**, *1134*, 223–229. [CrossRef]

136. Wu, M.C.; Hou, C.Y.; Jiang, C.M.; Wang, Y.T.; Wang, C.Y.; Chen, H.H.; Chang, H.M. A novel approach of LED light radiation improves the antioxidant activity of pea seedlings. *Food Chem.* **2007**, *101*, 1753–1758. [CrossRef]

137. Chen, L.; Zhao, F.; Zhang, M.; Hong-Hui, L.; Xi, D.H. Effects of light quality on the interaction between cucumber mosaic virus and Nicotiana tabacum. *J. Phytopathol.* **2015**, *163*, 1002–1013. [CrossRef]

138. Xu, H.; Fu, Y.; Li, T.; Wang, R. Effects of different LED light wavelengths on the resistance of tomato against Botrytis cinerea and the corresponding physiological mechanisms. *J. Integr. Agric.* **2017**, *16*, 106–114. [CrossRef]

139. Bavaresco, L.; Fregoni, C.; van Zeller de Macedo Basto Gonçalves, M.I.; Vezzulli, S. Physiology and molecular biology of grapevine stilbenes—An update. In *Grapevine Molecular Physiology & Biotechnology*, 2nd ed.; Roubelakis-Angelakis, K.A., Ed.; Springer Science and Business Media B.V.: Amsterdam, The Netherlands; New York, NY, USA, 2009; pp. 341–364.

140. Ahn, S.Y.; Kim, S.A.; Choi, S.J.; Yun, H.K. Comparison of accumulation of stilbene compounds and stilbene related gene expression in two grape berries irradiated with different light sources. *Hortic. Environ. Biotechnol.* **2015**, *56*, 36–43. [CrossRef]

141. Jeandet, P.; Douillt-Breuil, A.C.; Bessis, R.; Debord, S.; Sbaghi, M.; Adrian, M. Phytoalexins from the vitaceae: Biosynthesis, phytoalexin gene expression in transgenic plants, antifungal activity, and metabolism. *J. Agric. Food Chem.* **2002**, *50*, 2731–2741. [CrossRef]

142. Parada, R.Y.; Mon-nai, W.; Ueno, M.; Kihara, J.; Arase, S. Red-light-induced resistance to brown spot disease caused by *Bipolaris oryzae* in rice. *J. Phytopathol.* **2014**, *163*, 116–123. [CrossRef]

143. Balint-Kurti, P.; Simmons, S.J.; Blum, J.E.; Ballaré, C.L.; Stapleton, A.E. Maize leaf epiphytic bacteria diversity patterns are genetically correlated with resistance to fungal pathogen infection. *Mol. Plant Microbe Interact.* **2010**, *23*, 473–484. [CrossRef] [PubMed]

144. Aruscavage, D.; Phelan, P.L.; Lee, K.; LeJeune, J.T. Impact of changes in sugar exudate created by biological damage to tomato plants on the persistence of *Escherichia coli* O157:H7. *J. Food Sci.* **2010**, *75*, M187–M192. [CrossRef] [PubMed]

145. Leveau, J.H.J.; Lindow, S.E. Appetite of an epiphyte: Quantitative monitoring of bacterial sugar consumption in the phyllosphere. *Proc. Natl. Acad. Sci. USA* **2001**, *98*, 3446–3453. [CrossRef] [PubMed]

146. Aschenbrenner, A.K.; Amrehn, E.; Bechtel, L.; Spring, O. Trichome differentiation on leaf primordia of Helianthus annuus (Asteraceae): Morphology, gene expression and metabolite profile. *Planta* **2015**, *41*, 837–846. [CrossRef]

147. Aschenbrenner, A.K.; Horakh, S.; Spring, O. Linear glandular trichomes of *Helianthus* (Asteraceae): Morphology, localization, metabolite activity and occurrence. *AoB Plants* **2013**, *5*. [CrossRef]

148. Huang, S.S.; Kirchoff, B.K.; Liao, J.P. The capitate and peltate glandular trichomes of Lavandula pinnata L. (Lamiaceae): Histochemistry, ultrastructure, and secretion. *J. Torrey Bot. Soc.* **2008**, *135*, 155–167. [CrossRef]

149. Rico, A.; Preston, G.M. *Pseudomonas syringae* pv. *tomato* DC3000 uses constitutive and apoplast-induced nutrient assimilation pathways to catabolize nutrients that are abundant in the tomato apoplast. *Mol. Plant Microbe Interact.* **2008**, *21*, 269–282. [CrossRef]

150. Kniskern, J.M.; Traw, M.B.; Bergelson, J. Salicylic acid and jasmonic acid signaling defense pathways reduce natural bacterial diversity on Arabidopsis thaliana. *Mol. Plant Microbe Interact.* **2007**, *20*, 1512–1522. [CrossRef]

151. Nagendran, R.; Lee, Y.H. Green and red light reduces the disease severity by *Pseudomonas cichorii* JBC1 in tomato plants via upregulation of defense-related gene expression. *Phytopathology* **2015**, *105*, 412–418. [CrossRef]

152. Islam, S.Z.; Babadoost, M.; Bekal, S.; Lambert, K. Red light-induced systemic disease resistance against root-knot nematode *Meloidogyne javanica* and *Pseudomonas syringae* pv. *tomato* DC 3000. *J. Phytopathol.* **2008**, *156*, 708–714. [CrossRef]

153. Yadav, R.K.P.; Karamanoli, K.; Vokou, D. Bacterial colonization of the phyllosphere of Mediterranean perennial species as influenced by leaf structural and chemical features. *Microb. Ecol.* **2005**, *50*, 185–196. [CrossRef] [PubMed]

154. Gratani, L.; Covone, F.; Larcher, W. Leaf plasticity in response to light of three evergreen species of the Mediterranean maquis. *Trees* **2006**, *20*, 549–558. [CrossRef]

155. Steinmüller, D.; Tevini, M. Action of ultraviolet radiation (UV-B) upon cuticular waxes in some crop plants. *Planta* **1985**, *164*, 557–564. [CrossRef] [PubMed]

156. Garrett, T.R.; Bhakoo, M.; Zhang, Z. Bacterial adhesion and biofilms on surfaces. *Prog. Nat. Sci.* **2008**, *18*, 1049–1056. [CrossRef]

157. Gomelsky, M.; Hoff, W.D. Light helps bacteria make important lifestyle decisions. *Trends Microbiol.* **2011**, *19*, 441–448. [CrossRef] [PubMed]

158. Losi, A.; Gärtner, W. Bacterial bilin- and flavin-binding photoreceptors. *Photochem. Photobiol. Sci.* **2008**, *7*, 1168–1178. [CrossRef]

159. Pertot, I.; Fiamingo, F.; Amsalem, L.; Maymon, M.; Freeman, S.; Gobbin, D.; Elad, Y. Sensitivity of two Podosphaera aphanis populations to disease control agents. *J. Plant Pathol.* **2007**, *89*, 85–96.

160. Xiao, C.; Chandler, C.; Price, J.; Duval, J.; Mertely, J.; Legard, D. Comparison of epidemics of Botrytis fruit rot and powdery mildew of strawberry in large plastic tunnel and field production systems. *Plant Dis.* **2001**, *85*, 901–909. [CrossRef]

161. Kraiselburd, I.; Alet, A.I.; Tondo, M.L.; Petrocelli, S.; Daurelio, L.D.; Monzón, J.; Ruiz, O.A.; Losi, A.; Orellano, E.G. A LOV protein modulates the physiological attributes of *Xanthomonas axonopodis* pv. *citri* relevant for host plant colonization. *PLoS ONE* **2012**, *7*, e38226. [CrossRef]

162. Mao, D.; Tao, J.; Li, C.; Luo, C.; Zheng, L.; He, C. Light signalling mediated by Per-ARNT-Sim domain-containing proteins in *Xanthomonas campestris* pv. *campestris*. *FEMS Microbiol. Lett.* **2012**, *326*, 31–39. [CrossRef] [PubMed]

163. Müller, G.L.; Tuttobene, M.; Altilio, M.; Martínez Amezaga, M.; Nguyen, M.; Cribb, P.; Cybulski, L.E.; Ramírez, M.S.; Altabe, S.; Mussi, M.A. Light modulates metabolic pathways and other novel physiological traits in the human pathogen Acinetobacter baumannii. *J. Bacteriol.* **2017**, *199*, e00011-17. [CrossRef] [PubMed]

164. Yu, S.M.; Lee, Y.H. Effect of light quality on Bacillus amyloliquefaciens JBC36 and its biocontrol efficacy. *Biol. Control* **2013**, *64*, 203–210. [CrossRef]

165. Imada, K.; Tanaka, S.; Ibaraki, Y.; Yoshimura, K.; Ito, S. Antifungal effect of 405-nm light on Botrytis cinerea. *Lett. Appl. Microbiol.* **2014**, *59*, 670–676. [CrossRef] [PubMed]

166. Wilde, A.; Mullineaux, C.W. Light-controlled motility in prokaryotes and the problem of directional light perception. *FEMS Microbiol. Rev.* **2017**, *41*, 900–922. [CrossRef] [PubMed]

167. Rajalingam, N.; Lee, J.H. Effects of green light on the gene expression and virulence of the plant pathogen Pseudomonas cichorii JBC1. *Eur. J. Plant Pathol.* **2018**, *150*, 223–236. [CrossRef]

168. Someya, N.; Nakajima, M.; Hamamoto, H.; Yamaguchi, I.; Akutsu, K. Effects of light conditions on prodigiosin stability in the biocontrol bacterium *Serratia marcescens* strain B2. *J. Gen. Plant Pathol.* **2004**, *70*, 367–370. [CrossRef]

169. Guffy, J.S.; Wilborn, J. In vitro bactericidal effects of 405-nm and 470-nm blue light. *Photobiomodulation Photomed. Laser Surg.* **2006**, *24*, 684–688. [CrossRef]

170. Berrocal-Tito, G.; Sametz-Baron, L.; Eichenberg, K.; Horwitz, B.A.; Herrera-Estrella, A. Rapid blue light regulation of a Trichoderma harzianum photolyase gene. *J. Biol. Chem.* **1999**, *274*, 14288–14294. [CrossRef]

171. Papavizas, G.C. Trichoderma and Gliocladium: Biology, ecology and potential for biocontrol. *Ann. Rev. Phytopathol.* **1985**, *23*, 23–54. [CrossRef]

172. Van der Horst, M.A.; Hellingwerf, K.J. Photoreceptor proteins, "star actors of modern times": A review of the functional dynamics in the structure of representative members of six different photoreceptor families. *Acc. Chem. Res.* **2004**, *37*, 13–20. [CrossRef] [PubMed]

173. Van der Horst, M.A.; Key, J.; Hellingwerf, K.J. Photosensing in chemotrophic, non-phototrophic bacteria: Let there be light sensing too. *Trends Microbiol.* **2007**, *15*, 554–562. [CrossRef] [PubMed]

174. Jenal, U.; Malone, J. Mechanisms of Cyclic-di-GMP signaling in bacteria. *Ann. Rev. Genet.* **2006**, *40*, 385–407. [CrossRef] [PubMed]

175. Hengge, R.; Galperin, M.Y.; Ghigo, J.M.; Gomelsky, M.; Green, J.; Hughes, K.T.; Jenal, U.; Landini, P. Systematic Nomenclature for GGDEF and EAL Domain-Containing Cyclic Di-GMP Turnover Proteins of *Escherichia coli*. *J. Bacteriol.* **2016**, *198*, 7–11. [CrossRef]

176. Bumah, V.V.; Masson-Meyers, D.S.; Cashin, S.; Enwemeka, C.S. Optimization of the antimicrobial effect of blue light on methicillin-resistant Staphylococcus aureus (MRSA) in vitro. *Lasers Surg. Med.* **2015**, *47*, 266–272. [CrossRef] [PubMed]

177. Halstead, F.D.; Thwaite, J.E.; Burt, R.; Laws, T.R.; Raguse, M.; Moeller, R.; Webber, M.A.; Oppenheim, B.A. Antibacterial activity of blue light against nosocomial wound pathogens growing planktonically and as mature biofilms. *Appl. Environ. Microbiol.* **2016**, *82*, 4006–4016. [CrossRef] [PubMed]

178. Glantz, S.T.; Carpenter, E.J.; Melkonian, M.; Gardner, K.H.; Boyden, E.S.; Wong, G.K.S.; Chow, B.Y. Functional and topological diversity of LOV domain photoreceptors. *Proc. Natl. Acad. Sci. USA* **2016**, *113*, E1442–E1451. [CrossRef]

179. Zayner, J.P.; Antoniou, C.; French, A.R.; Hause, R.J., Jr.; Sosnick, T.R. Investigating models of protein function and allostery with a widespread mutational analysis of a light-activated protein. *Biophys. J.* **2013**, *105*, 1027–1036. [CrossRef]

180. Kraiselburd, I.; Moyano, L.; Carrau, A.; Tano, J.; Orellano, E.G. Bacterial photosensory proteins and their role in plant–pathogen interactions. *Photochem. Photobiol.* **2017**, *93*, 666–674. [CrossRef]

181. Masuda, S. Light detection and signal transduction in the BLUF photoreceptors. *Plant Cell Physiol.* **2013**, *54*, 171–179. [CrossRef]

182. Tschowri, N.; Busse, S.; Hengge, R. The BLUF-EAL protein YcgF acts as a direct anti-repressor in a blue-light response of Escherichia coli. *Genes Dev.* **2009**, *23*, 522–534. [CrossRef] [PubMed]

183. Meyer, T.E.; Yakali, E.; Cusanovich, M.A.; Tollin, F. Properties of a water soluble yellow protein isolated from a halophilic phototrophic bacterium that has photochemical activity analogous to sensory rhodopsin. *Biochemistry* **1987**, *26*, 418–423. [CrossRef] [PubMed]

184. Meyer, T.E.; Kyndt, J.A.; Memmi, S.; Moser, T.; Colon-Acevedo, B.; Devreese, B.; Van Beeumen, J. The growing family of photoactive yellow proteins and their presumed functional roles. *Photochem. Photobiol. Sci.* **2012**, *11*, 1495–1514. [CrossRef] [PubMed]

185. Imamoto, Y.; Kataoka, M. Structure and Photoreaction of Photoactive Yellow Protein, a Structural Prototype of the PAS Domain Superfamily†. *Photochem. Photobiol.* **2007**, *83*, 40–49. [CrossRef] [PubMed]

186. Nielsen, I.B.; Boyé-Péronne, S.; El Ghazaly, M.O.A.; Kristensen, M.B.; Brøndsted Nielsen, S.; Andersen, L.H. Absorption spectra of photoactive yellow protein chromophores in vacuum. *Biophys. J.* **2005**, *89*, 2597–2604. [CrossRef] [PubMed]

187. Hawkes, C.V.; Connor, E.W. Translating phytobiomes from theory to practice: Ecological and evolutionary considerations. *Phytobiomes* **2017**, *1*, 57–69. [CrossRef]

188. Werner, G.D.A.; Strassmann, J.E.; Ivens, A.B.F.; Engelmoor, D.J.P.; Verbryggen, E.; Queller, D.C.; Roë, R.; Collins Johnson, N.; Hammerstein, P.; Kiers, E.T. Evolution of microbial markets. *Proc. Natl. Acad. Sci. USA* **2014**, *111*, 1237–1244. [CrossRef]

189. Mercier, J.; Lindow, S.E. Role of leaf surface sugars in colonization of plants by bacterial epiphytes. *Appl. Environ. Microbiol.* **2000**, *66*, 369–374. [CrossRef]

190. Tukey, H.B.J.; Morgan, J.V. Injury to foliage and its effect upon the leaching of nutrients from above-ground plant parts. *Physiol. Plant.* **1963**, *16*, 557–564. [CrossRef]

80 kg ha^{-1} P$_2$O$_5$ and 120 kg ha^{-1} K$_2$O. During plant production, 40 kg ha^{-1} N were supplied three times at two-week intervals and, in the last N application, 7 kg ha^{-1} of P$_2$O$_5$ and of K$_2$O were also provided. Half of each fertilizer dose was applied just before transplanting and the remaining 50% by sidedressing at two week intervals. Drip irrigation was started at 80% soil available water. The organic management practices complied with EC Regulations 834/2007 and 889/2008. Plant protection was achieved by applying copper oxychloride against rust, and deltamethrin or azadirachtin in the conventional or organic systems, respectively, against aphids.

Harvests of mature plants were performed from 5 to 10 October, when the pseudo-stems had reached their maximum growth, and the leaf blades were trimmed to a 15 cm length for obtaining a marketable product. In each plot, determinations were made of the weight of the marketable product (pseudo-stems with 15 cm long leaf blades) and the mean pseudo-stem (with 15 cm long leaf blades) weight from twenty-plant samples. Further plant samples were collected, gently washed with water to remove surface contaminants and dried with filter paper. Pseudo-stems and leaves were separated, cut with a plastic knife, dried to a constant weight and homogenized. The resulting powders were subjected to laboratory analyses.

2.2. Dry Matter

The dry matter content in leaves and pseudo-stems of *A. porrum* was assessed after drying the fresh samples in an oven at 70 °C, until they reached constant weight.

2.3. Total Soluble Solids (TSS) and Sugars

Determination of total soluble solids was performed in filtered water extracts of leek leaves and pseudo-stems (1 g of dry sample per 100 mL) using a TDS-3 conductometer (HM Digital, Inc., Seoul, Korea).

Monosaccharides were determined using the ferricyanide colorimetric method, based on the reaction of monosaccharides with potassium ferrycianide [20]. Total sugars were determined after acidic hydrolysis of 50 mL of filtered water extracts with 5 mL of 20% hydrochloric acid. Fructose was used as an external standard.

2.4. Polyphenols

The concentrations of total phenolics in each sample of leaves and pseudo-stems were determined in filtered 70% ethanol extract (0.5 g of dry sample in 25 mL; 1 h at 80 °C) using the Folin-Ciocalteu colorimetric method, according to Golubkina et al. [21] using a Unico 2804 UV (Unico Inc., Wixom, MI, USA) spectrophotometer. The phenolic contents were calculated by using a calibration curve of gallic acid constructed with five concentrations of this compound (0–90 µg/mL). Phenolic contents were expressed as milligrams of gallic acid equivalents per 100 g of dry weight (mg GAE/100 g d.w.).

2.5. Ascorbic Acid

Ascorbic acid content of leek leaves and pseudo-stems was assessed by visual titration of fresh plant extracts in 6% trichloracetic acid with Tillmans reagent [22]. Five grams of fresh leek leaves were homogenized in porcelain mortar with 5 mL of 6% trichloracetic acid and quantitatively transferred to measuring cylinder. The volume was brought to 80 mL using trichloracetic acid, and the mixture was filtered through filter paper 15 min later. The ascorbic acid concentration was determined from the volume of Tillmans reagent which went into titration of the sample.

2.6. Antioxidant Activity

The antioxidant activity of leek leaves and pseudo-stems was assessed using a redox titration method [23], via titration of 0.01 N KMnO$_4$ solution with ethanolic extracts of leaves and pseudo-stems used for polyphenol determination (see Section 2.4). Reduction of KMnO$_4$ to colorless Mn^{+2} in this process reflects the concentration of antioxidants dissolvable in 70% ethanol. The values were expressed in mg GAE/100 g d.w. The use of $KMnO_4$ acidic solution is known to be successfully used for the determination of *Ocimum basilicum* antioxidant potential [24] and antioxidant capacity of serum [25].

2.7. Nitrates

Nitrate was assessed in fresh pseudo-stems using an ion selective electrode using a ionomer Expert-001 (Econix, Moscow, Russia). Five grams of fresh homogenized leek pseudo-stems were mixed with 50 mL of distilled water. Forty-five mL of filtered extract were mixed with 5 mL of 0.5 M potassium sulfate background solution (necessary for ionic strength regulation) and analyzed through the ionomer for nitrate determination.

2.8. Elemental Composition

Al, As, B, Ca, Cd, Co, Cr, Cu, Fe, Hg, I, K, Li, Mg, Mn, Na, Ni, P, Pb, Si, Sn, Sr, V and Zn contents of leek pseudo-stems were assessed using an ICP-MS on quadruple mass-spectrometer Nexion 300D (Perkin Elmer Inc., Shelton, CT 06484, USA) equipped with the 7-port FAST valve and ESI SC DX4 autosampler (Elemental Scientific Inc., Omaha, NE 68122, USA) in the Biotic Medicine Center (Moscow, Russia). Rhodium ^{103}Rh was used as an internal standard to eliminate instability during measurements. Quantitation was performed using external standard (Merck IV, multi-element standard solution), potassium iodide for iodine calibration and Perkin-Elmer standard solutions for P, Si and V. All the standard curves were obtained at 5 different concentrations.

For quality control purposes, internal controls and reference materials were tested together with the samples on a daily basis. Microwave digestion of samples was achieved according to the standard method [26] with sub-boiled HNO_3 diluted 1:150 with distilled deionized water (Fluka No. 02650 Sigma-Aldrich, Co., Saint Louis, MO, USA) in the Berghof SW-4 DAP-40 microwave system (Berghof Products + Instruments GmbH, 72800 Eningen, Germany). Trace levels of Hg in samples were not taken into account and, accordingly, they were excluded from the Tables below.

The instrument conditions and acquisition parameters were: plasma power and argon flow, 1500 and 18 L min^{-1} respectively; aux argon flow, 1.6 L min^{-1}; nebulizer argon flow, 0.98 L min^{-1}; sample introduction system, ESI ST PFA concentric nebulizer and ESI PFA cyclonic spray chamber (Elemental Scientific Inc., Omaha, NE 68122, USA); sampler and slimmer cone material, platinum; injector, ESI Quartz 2.0 mm I.D/; sample flow, 637 µL/min; internal standard flow, 84 µL/min; dwell time and acquisition mode, 10–100 ms and peak hopping for all analytes; sweeps per reading, 1; reading per replicate, 10; replicate number, 3; DRC mode, 0.55 L min^{-1} ammonia (294993-Aldrich Sigma-Aldrich, Co., St. Louis, MO 63103, USA) for Ca, K, Na, Fe, Cr, V, optimized individually for RPa and RPq; STD mode, for the rest of analytes at RPa = 0 and RPq = 0.25.

Se content of leek leaves and pseudo-stems was analyzed using the fluorimetric method previously described for tissues and biological fluids [27]. The method includes digestion of dried homogenized samples via heating with a mixture of nitric-chloral acids, subsequent reduction of Se^{+6} to Se^{+4} with a solution of 6 N HCl, and formation of a complex between Se^{+4} and 2,3-diaminonaphtalene. The Se concentration was assessed in triplicate by recording piazoselenol fluorescence values in hexane at 519 nm λ emission and 376 nm λ excitation. The results precision was checked using a reference standard-lyophilized cabbage at each determination with 150 µg/Kg Se concentration (Institute of Nutrition, Moscow, Russia).

2.9. Statistical Analysis

Data were processed by analysis of variance and mean separations were performed using the Duncan multiple range test, α = 0.05, using SPSS software version 21. The data expressed as a percentage were subjected to angular transformation before processing.

3. Results and Discussion

As the year of research had no significant effect on yield, quality and antioxidant variables examined, both as main factor or in interaction with the experimental factors management system or cultivar, the results are reported as average values of the two years of investigation.

3.1. Yield, Dry Matter, Sugars and Nitrates of Pseudo-Stems

The management system showed no significant effects on leek pseudo-stem yield and mean weight (Table 1). The experimental factor cultivar significantly affected the mean pseudo-stem weight and, accordingly, yield which ranged from 23.8×10^3 to 40.2×10^3 kg ha^{-1}, as an average management across systems; the variety Giraffe showed the highest yield, Premier and Kalambus the lowest. The coefficient of variation was rather low and reached 18.3%, which suggests a low genetic effect on this variable (Table 1).

Table 1. Yield, mean pseudo-stem weight, and content of dry matter, sugars and nitrates in leek.

Treatment	Yield 10^3 kg ha^{-1}	Mean Pseudo-Stem Weight g	Dry Matter %	Sugars g/100 g d.w.		Nitrates mg/kg f.w.
				Mono-	Total	
Crop management						
Organic	30.9	185.5	19.7	3.6	11.7	76
Conventional	31.2	187.0	17.8	3.2	10.4	102
	n.s.[z]	n.s.	*	*	*	*
Cultivar						
Goliath	31.1 [d,e,y]	188.0 [d,e]	12.4 ± 0.4 [e]	4.8 ± 0.3 [a]	7.3 ± 0.5 [e]	77 ± 3 [d]
Premier	23.8 [g]	142.4 [g]	15.2 ± 0.5 [d]	4.4 ± 0.3 [a,b]	10.5 ± 0.7 [c]	74 ± 3 [d]
Bandit	35.0 [bc]	209.8 [b,c]	15.3 ± 0.6 [d]	3.5 ± 0.2 [c]	8.6 ± 0.5 [d]	103 ± 4 [a]
Kalambus	24.0 [g]	143.6 [g]	17.6 ± 0.6 [c]	3.9 ± 0.2 [b,c]	10.3 ± 0.7 [c]	70 ± 3 [d]
Cazimir	26.9 [f]	160.9 [f]	18.9 ± 0.6 [c]	2.8 ± 0.2 [d]	10.7 ± 0.7 [b,c]	90 ± 4 [b,c]
Giraffe	40.2 [a]	242.5 [a]	20.5 ± 0.8 [b]	3.4 ± 0.2 [c]	11.0 ± 0.7 [bc]	86 ± 3 [c]
Camus	33.3 [c,d]	201.7 [c,d]	21.4 ± 0.7 [b]	2.5 ± 0.2 [d]	12.2 ± 0.8 [b]	105 ± 5 [a]
Vesta	29.0 [e,f]	175.0 [e,f]	23.4 ± 0.8 [a]	2.6 ± 0.2 [d]	14.3 ± 0.8 [a]	98 ± 4 [ab]
Summer Breeze	36.7 [b]	221.3 [b]	24.3 ± 0.9 [a]	2.8 ± 0.2 [d]	15.1 ± 0.9 [a]	98 ± 4 [ab]
M	31.1	187.2	18.8	3.86	11.11	89
SD	5.7	34.8	3.2	1.23	1.84	11
CV (%)	18.3	18.6	17.3	31.9	16.8	12.4
Concentration range	23.8–40.2	142.4–242.5	12.4–24.3	2.5–8.4	7.3–15.1	70–105

[z] n.s. not significant; * significant at $P \leq 0.05$. [y] Within each column, means followed by different letters are significantly different according to Duncan's Multiple Range Test at $P \leq 0.05$.

Organic management resulted in a higher dry matter and carbohydrate content in pseudo-stems, compared to the conventional managementsystem (Table 1). These results may be due to the enhancement of microbial biomass and activity leading to organic compound synthesis and, in this respect, similar trends were recorded in previous research carried out in greenhouses [28]. Moreover, dry matter content of the nine leek cultivars ranged from 12 to 24% (Table 1) which was much wider than leek grown in the Czech Republic (9–11%) [14]. These results identified varieties with high dry matter content (Summer Breeze and Vesta), which show a long shelf-life and areuseful for dry spice production, and genotypes with low dry matter content (Goliath, Premier and Bandit) more suitable for salad production, though dry matter is also dependent on water regime [29].

Cultivar differences in total sugar content were 2-fold lower than the coefficient of variation for monosaccharide content (Table 1). Therefore, the existence of a multidirectional nature of mono- and disaccharide accumulation in leek pseudo-stems may exist.

A significant positive correlation was observed between dry matter and disaccharides (r = 0.98 at $P < 0.01$), but there was a negative correlation (r = −0.86 at $P < 0.05$) between dry matter and monosaccharides (Figure 1). Indeed, the negative correlation between monosaccharides and dry matter content shown in Figure 1 is consistent with that previously found in onion [30]. Interestingly, total sugar content in onion did not differ among cultivars with low versus high dry matter content [31], in contrast to leek where the maximum value was twice the minimum. As observed in Figure 1, the highest concentrations of mono- and total saccharides in leek were associated with cultivars with high dry matter content. However, contrasting carbohydrate trends were found with the two varieties: Goliath had low dry matter content and a significantly higher concentration of monosaccharides compared to disaccharides; in Summer Breeze, the disaccharide content exceeded the monosaccharide content by 4.4-fold , with the ratio between the related total sugar contents at 2.1 (Figure 1).

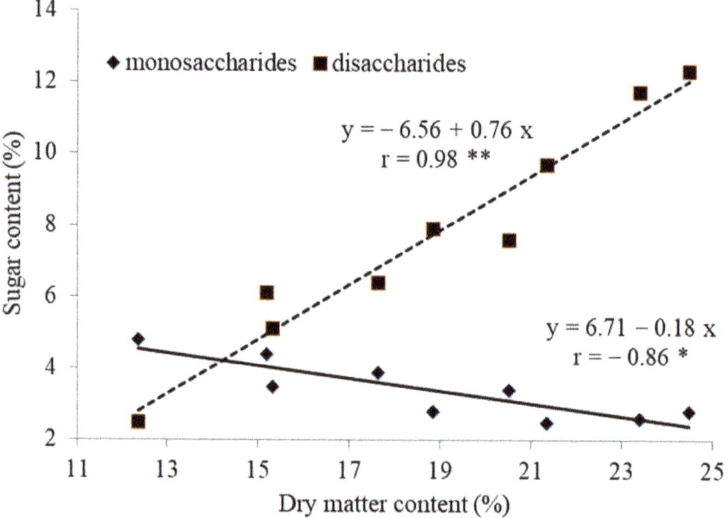

Figure 1. Correlations between dry matter and sugar content in leek pseudo-stems.

With regard to nitrate concentration in pseudo-stems (Table 1), lower values were recorded the organic management system compared to conventional management, consistent with previous reports [28]. Indeed, a slow nitrate release from organic fertilizers does not elicit abundant uptake of this anion in a short time period, which prevents its accumulation in plant parts [32].

Leek is characterized by a relatively high plant nitrate content [33], though it is much lower than in the top-accumulating species [32]. However, in our research the cultivars investigated had less than 105 mg nitrate per kg fresh weight (f.w.) and, moreover, the ascorbic acid content was sufficient to make the product safe and healthy, as ascorbic acid participates in producing essential nitrogen oxide for humans thus preventing nitrosamine formation [33]. Notably, low varietal differences in nitrate accumulation make this quality parameter the most stable among the cultivars grown in similar conditions.

3.2. Antioxidants

Polyphenols and vitamin C significantly affect plant antioxidant activity [34]. In our research, organic management resulted in higher vitamin C in leek pseudo-stems compared to conventional management, consistent with some previous investigations [35], but differently from others where no significant differences were observed between the two management systems [28]. The management system did not affect either polyphenol or selenium concentration in leek pseudo-stems and leaves.

There were significant varietal differences in ascorbic acid accumulation, in contrast to polyphenol content which was characterized by higher stability in both stems and especially in leaves (Table 2). Among the cultivars, Goliath showed the highest ascorbic acid content; notably, all of the cultivars from domestic selection had lower levels of polyphenols compared to prior reports, which may be related to different crop cycles and harvest times [36]. Similar findings have been reported from an investigation in Belgium on thirty leek cultivars [19], where the ascorbic acid concentration range was 90 to 350 mg/100 g d.w. and the polyphenol ranged from 7.3 to 11.3 mg GA/g d.w., whereas the domestic cultivars showed lower polyphenol values, i.e., 3.0 to 5.5 mg GA/g d.w. (Table 2).

Table 2. Concentrations of ascorbic acid, polyphenols and selenium in leek.

Treatment	Ascorbic Acid in Pseudo-Stems mg/100 g f.w.	Polyphenols mg GAE/100 g d.w.		Selenium µg/kg d.w.	
		Pseudo-Stems	Leaves	Pseudo-Stems	Leaves
Crop management					
Organic	47.2	375.4	696.7	76.3	64.4
Conventional	37.0	356.2	674.9	73.1	60.8
	* [z]	n.s.	n.s.	n.s.	n.s.
Cultivar					
Goliath	136.8 ± 8.1 [a,y]	555 ± 52 [a]	711 ± 19 [a]	107 ± 7 [a]	14 ± 1 [e]
Premier	58.6 ± 3.3 [b]	432 ± 34 [b]	650 ± 21 [b]	80 ± 5 [b]	65 ± 3 [c]
Bandit	40.5 ± 2.6 [c]	394 ± 26 [b,c]	647 ± 26 [b]	75 ± 5 [b,c]	48 ± 2 [d]
Kalambus	25.5 ± 2.3 [d,e]	319 ± 19 [d,e]	731 ± 46 [a]	72 ± 4 [b,c]	74 ± 3 [b]
Cazimir	30.2 ± 2.6 [d]	284 ± 20 [e]	728 ± 53 [a]	60 ± 3 [e]	76 ± 4 [a,b]
Giraffe	24.9 ± 2.0 [e,f]	331 ± 19 [d]	665 ± 41 [a,b]	73 ± 4 [b,c]	49 ± 2 [d]
Camus	21.1 ± 1.4 [f,g]	347 ± 21 [c,d]	684 ± 43 [a,b]	64 ± 3 [d,e]	81 ± 5 [a,b]
Vesta	19.2 ± 1.3 [g]	301 ± 19 [d,e]	740 ± 55 [a]	69 ± 3 [c,d]	84 ± 5 [a]
Summer Breeze	22.2 ± 1.9 [f,g]	329 ± 20 [d,e]	616 ± 42 [b]	72 ± 3 [b,c]	72 ± 4 [b,c]
M	42.1	317	686	74.7	62.5
SD	24.7	71	37	8.4	17
CV, %	58.7	22.4	5.4	11.2	27.2
Concentration range	19.2–136.8	284–555	616–740	60–107	14–84

[z] n.s. not significant; * significant at $P \leq 0.05$. [y] Within each column, means followed by different letters are significantly different according to Duncan's Multiple Range Test at $P \leq 0.05$.

A significant positive correlation between ascorbic acid and polyphenol concentration in leek stems is of particular interest (r = 0.94 at $P < 0.01$). The lack of a similar correlation in a study using thirty leek cultivars conducted in Belgium [19] was presumably related to the heterogeneity of the cultivars, which were selected on the basis of morphological types (light-green summer type, dark-green winter type and intermediate autumn type). Indeed, the autumn-grown Belgium varieties resulted in a similar correlation to ours between ascorbic acid and polyphenol content (r = 0.71 at $P < 0.05$).

The ratio between leaf and stem polyphenol concentration in leek plants was of interest. In the nine cultivars in our research polyphenol content was always higher in leaves compared to stems, with a negative correlation between the tissues (Figure 2). Otherwise, as found by Ben Arfa et al. [5], polyphenol levels in *A. porrum* leaves may be both higher or lower than those in stems. In this respect, the total polyphenol content per plant should be more stable that polyphenol concentrations recorded in stems or leaves.

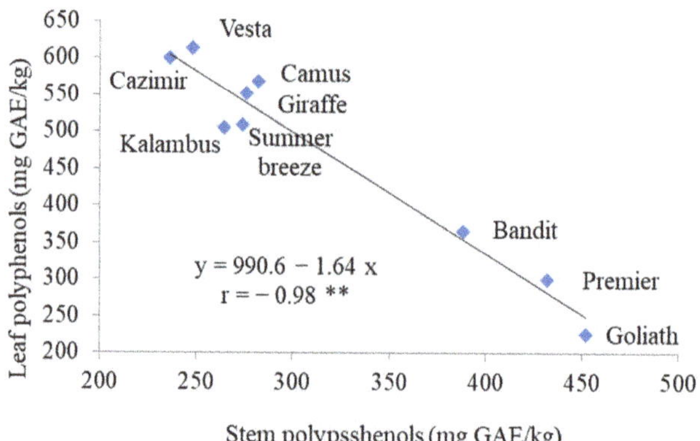

Figure 2. Correlation between leaf and stem polyphenol content.

Among the components of plant antioxidant systems, selenium plays a significant role. Indeed, though it is not an essential element for plants, selenium is able to provide a powerful antioxidant defense to plants against drought, salinity, frost, flooding, UV light and herbivores [37]. Notably, *Allium* species belong to the secondary selenium accumulators, which show a remarkable tolerance to high concentrations and consequent accumulation of this element due to Se ability to substitute for sulfur in natural compounds, as also reported by Turkish scientists in leek [17].

In our research, *A. porrum* grown in the Moscow region showed a Se accumulation range from 60 to 107 µg/kg d.w., which is much lower than the values recorded in Turkey [17]. This suggests the significant effect of selenium status in the environment on plant ability to concentrate the trace element. The negative correlation between selenium content in leaves and stems (r = −0.95 at $P < 0.01$; Figure 3), similar to that recorded for polyphenols, entails a rather stable level of selenium accumulation in the plant.

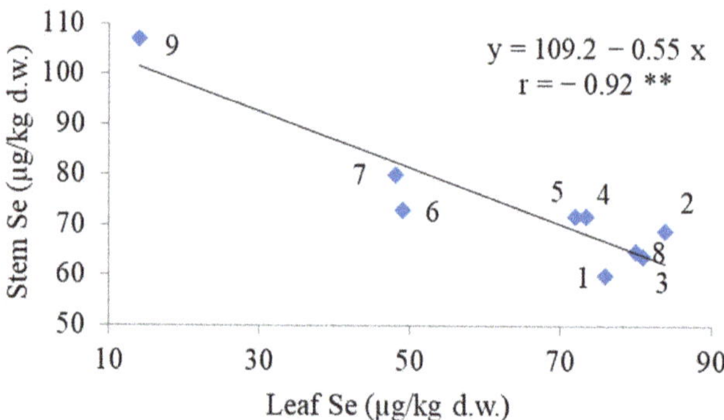

Figure 3. Correlation between leaf and stem selenium content in leek cultivars: (1) Cazimir, (2) Vesta, (3) Camus, (4) Kalambus, (5) Summer Breeze, (6) Giraffe, (7) Bandit, (8) Premier, (9) Goliath.

Reports relevant to selenium as plant secondary metabolites, as well as Se to polyphenols particularly in the absence of selenium loading are rather scarce and often controversial. In this respect, a positive correlation between selenium and polyphenol content was found in wheat [38] and an adverse correlation between quercetin and selenium was recorded in onion [30]. However, moderate doses of selenium are deemed to enhance the content of antioxidants such as polyphenols, flavonoids and carotenoids [39,40].

In our research, the leek genotypes investigated showed significant correlations between the components of the antioxidant system, i.e., selenium, ascorbic acid and polyphenols: Se and ascorbic acid (r = 0.93 at $P < 0.01$); Se and polyphenols (r = 0.92 at $P < 0.01$); ascorbic acid and polyphenols (r = 0.94 at $P < 0.01$). The latter correlations relevant to leek stems may be very useful for leek selection based on high antioxidant content.

3.3. Elemental Composition

The beneficial effect of many elements to human health has created unflagging interest in mineral composition of vegetable crops and in particular of leek [17,41]. Investigations of element content in leek have revealed that the plant is able to accumulate high concentrations of K and Fe. However, all investigations to date have restricted macro- and microelements (K, Ca, P, Na, Mg, Fe, Zn, Cu, Se) availability, giving no opportunity to evaluate either the leek total mineral profile or the varietal peculiarities of element accumulation.

Our research with nine leek cultivars has shown the existence of significant varietal differences in stem and leaf ash content (Figure 4). The ratio between leaf and stem ash content decreased as follows: Summer Breeze > Cazimir > Vesta > Giraffe > Bandit > Kalambus > Camus > Premier > Goliath.

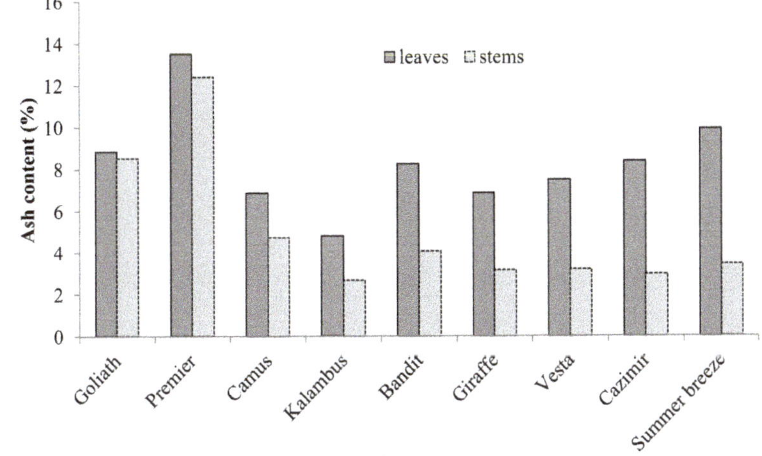

Figure 4. Varietal differences in leek leaf and stem ash content.

The ratio between leaf and stem element content was negatively correlated with stem polyphenol concentration (Figure 5), the latter being therefore related to both content and distribution of minerals in leek plants. As shown in Figure 5, the higher value of angular coefficient was associated with the water extract method, which provides higher mineral concentrations but lower polyphenol levels compared to an ethanol extract.

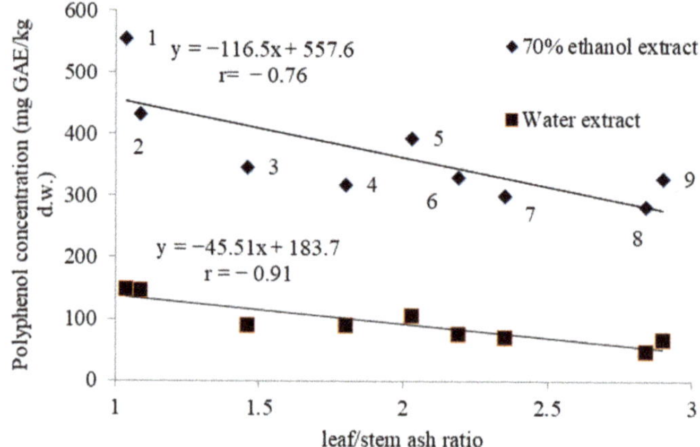

Figure 5. Correlations between leaf/stem ash content and stem polyphenol concentration: (1) Goliath; (2) Premier; (3) Camus; (4) Kalambus; (5) Bandit; (6) Giraffe; (7) Vesta; (8) Cazimir; 9) Summer Breeze.

The analysis of the content of twenty-five mineral elements in leek pseudo-stems has provided the opportunity to assess the varietal differences in elemental profile and has demonstrated that the ash content is directly connected with the concentration of K (Tables 3–5). The latter mineral showed a higher accumulation in pseudo-stems grown with organic management compared to those grown with conventional management (19.46 vs. 17.23 g/kg d.w.), whereas no significant differences were evident between the two management systems with regard to all the other elements analyzed.

Table 3. Macroelement concentration in *A. porrum* pseudo-stems (g/kg d.w.).

Element	Goliath	Cazimir	Premier	Vesta	Kalambus	Summer Breeze	Bandit	Giraffe	Camus
					Macro elements				
Ca	3.98 [b,c,z]	3.40 [c,d]	11.32 [a]	4.35 [b]	4.68 [b]	2.81 [d]	4.82 [b]	4.24 [b,c]	4.73 [b]
K	51.76 [a]	4.71 [e]	23.39 [b]	13.50 [c]	8.00 [d]	15.94 [c]	16.95 [c]	15.05 [c]	15.82 [c]
Mg	0.78 [c]	0.76 [c]	2.02 [a]	0.56 [d]	0.53 [d]	0.65 [c,d]	1.00 [b]	0.70 [c,d]	0.80 [c]
Na	0.34 [b,c]	0.39 [b]	0.81 [a]	0.16 [e]	0.16 [e]	0.12 [e]	0.17 [e]	0.19 [d,e]	0.28 [c,d]
P	2.95 [b]	2.61 [bc]	2.41 [c,d]	2.30 [c,d]	1.97 [d]	2.37 [cd]	3.85 [a]	2.64 [bc]	2.65 [b,c]

[z] Within each row, means followed by different letters are significantly different according to Duncan's Multiple Range Test at $P \leq 0.05$.

Table 4. Microelements concentration in *A. porrum* pseudo-stems (mg/kg d.w.).

Element	Goliath	Cazimir	Premier	Vesta	Kalambus	Summer Breeze	Bandit	Giraffe	Camus
B	21.25 [a,z]	15.21 [b,c]	16.75 [b]	9.68 [d,e]	8.61 [e]	9.55 [d,e]	9.73 [d,e]	12.79 [c,d]	11.14 [d]
Co	0.070 [c]	0.050 [d]	0.091 [b]	0.034 [d]	0.290 [a]	0.035 [d]	0.035 [d]	0.035 [d]	0.099 [b]
Cu	4.81 [d,f]	4.52 [e,f]	3.46 [g]	5.90 [b,c]	6.66 [a,b]	5.15 [c,e]	7.18 [a]	4.08 [f,g]	5.53 [c,d]
Fe	221 [a]	116 [c]	178 [b]	101 [b,d]	77 [e]	84 [d,e]	98 [c,e]	104 [c,d]	235 [a]
I	0.060 [a]	0.040 [d]	0.353 [a]	0.042 [c,d]	0.055 [b,d]	0.057 [b,d]	0.038 [d]	0.071 [b,c]	0.073 [b]
Li	0.110 [b]	0.040 [c]	0.160 [a]	0.025 [c]	0.014 [c]	0.032 [c]	0.023 [c]	0.029 [c]	0.109 [b]
Mn	12.57 [c]	12.18 [c]	23.15 [a]	9.87 [c]	6.39 [d]	9.69 [c]	10.93 [c]	19.97 [b]	22.51 [a,b]
Si	14.62 [c]	10.74 [b]	28.78 [a]	9.50 [e]	13.43 [c,d]	13.86 [c,d]	11.41 [d,e]	20.00 [b]	16.17 [c]
Sn	0.160 [e]	0.240 [c,d]	0.023 [f]	0.171 [e]	0.519 [b]	0.193 [d,e]	0.574 [a]	0.245 [c]	0.247 [c]
Zn	23.97 [a,b]	27.27 [a]	11.96 [f]	18.45 [de]	16.26 [e]	19.58 [ce]	21.96 [b,c]	22.83 [b,c]	21.72 [b,d]

[z] Within each row, means followed by different letters are significantly different according to Duncan's Multiple Range Test at $P \leq 0.05$.

Table 5. Heavy metal concentration in *A. porrum* pseudo-stems (mg/kg d.w.).

Element	Goliath	Cazimir	Premier	Vesta	Kalambus	Summer Breeze	Bandit	Giraffe	Camus
					Heavy metals				
Al	84.0 [c,z]	31.3 [d]	137.0 [a]	21.9 [d,f]	8.0 [g]	25.3 [d,e]	12.6 [f,g]	20.2 [e,f]	96.2 [b]
As	0.030 [b]	0.020 [c,d]	0.066 [a]	0.015 [c,e]	0.013 [d,e]	0.017 [c,e]	0.009 [e]	0.023 [b,c]	0.066 [a]
Cd	0.090 [c,d]	0.110 [b,c]	0.196 [a]	0.110 [b]	0.085 [d]	0.073 [d]	0.113 [b]	0.182 [a]	0.122 [b]
Cr	0.130 [c]	0.080 [g]	0.524 [a]	0.104 [d,f]	0.095 [e,g]	0.122 [c,d]	0.085 [f,g]	0.161 [b]	0.111 [c,e]
Ni	1.100 [a]	0.480 [c]	1.000 [a,b]	1.010 [a,b]	0.578 [c]	0.588 [c]	0.875 [b]	0.621 [c]	1.140 [a]
Pb	0.360 [b]	0.290 [b,c]	0.894 [a]	0.108 [e]	0.096 [e]	0.143 [d,e]	0.220 [c,d]	0.117 [e]	0.878 [a]
Sr	29.0 [a,b]	25.7 [c]	31.3 [a]	25.0 [c]	28.7 [a,b]	17.8 [d]	29.3 [a,b]	26.8 [b,c]	29.1 [a,b]
V	0.230 [b]	0.090 [c,d]	0.311 [a]	0.079 [c,d]	0.043 [e]	0.097 [c]	0.066 [d,e]	0.073 [c,d]	0.299 [a]

[z] Within each row, means followed by different letters are significantly different according to Duncan's Multiple Range Test at $P \leq 0.05$.

The comparison between the nine leek cultivars, in terms of elemental composition, has indicated three cultivars with contrasting features: Premier, Goliath and Cazimir. Indeed, Premier preferably accumulated Co, I, Al, As, Cd, Ni, Pb and Sr, but low levels of Cu and Zn. Goliath was characterized by the highest content of Fe, B, Zn, Se and K, and the lowest of Cd. Cazimir showed the highest concentration of Zn and Na, but the lowest of I, Se, K, Cr and Ni.

In our research, the nine leek cultivars have shown higher concentrations of most elements, compared to twenty varieties belonging to related species such as garlic grown in the same geochemical conditions [42]: Ca, Na, As, Cr, Ni, Co, Fe, I, Li, Pg, Sr, V, B, Co, Fe, I, Li, Sn; equal amounts of K, Mg, P, Cd, Cu, Mn, Se and Zn; and lower concentrations of Si. It is worth noting that, in conditions of marginal selenium deficiency in the Moscow region, the average levels of selenium accumulation by leek and garlic did not differ from each other (Figure 6).

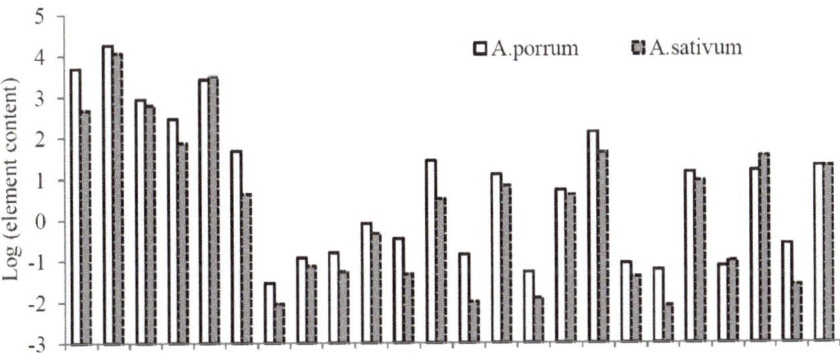

Figure 6. Comparative elemental profile of leek and winter garlic [42] grown in the same geochemical conditions of the Moscow region.

Interestingly, significant varietal differences in the content of most elements were evident (Figure 7). Though the values of the coefficient of variation (CV) relevant to macroelements were considerably lower than those of heavy metals and microelements, even among the former the CV reached 30 to 50%, with the exception of P having lower values. Among heavy metals, Al, Pb, V, As and Cr attained the highest coefficients of variation (40 to 80%), whereas among microelements Li, I, Sn, Co, Fe and Mn showed a CV > 40%.

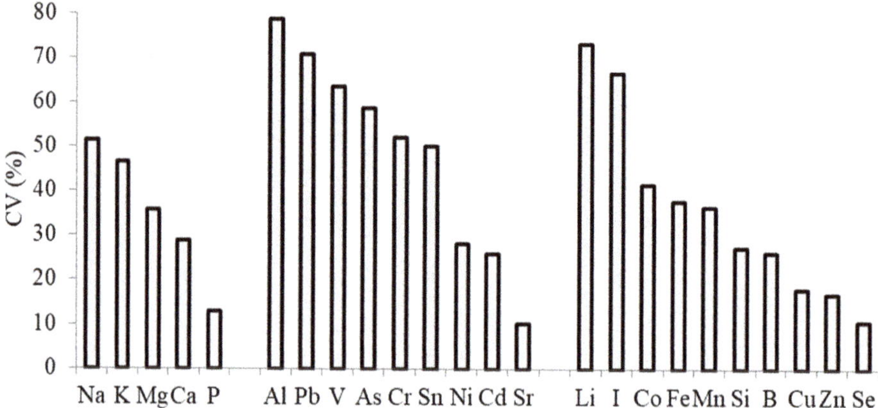

Figure 7. Coefficients of variation relevant to macro- and microelement content in *A.porrum*.

The several significant correlations recorded between the different elements revealed a higher complexity concerning the mineral dynamics in leek plants (Table 6), compared to related species such as garlic [42].

The highest number of significant correlations was recorded for Al, whose physiological role in plants has not been clearly determined so far, though this element is supposed to both activate some enzymes and control membrane permeability at low doses [43]. The significant intra- and interspecies variability in Al accumulation depends on plant tolerance thresholds to this element. The multiplicity of relationships between Al and other elements (Ca, Mg, Na, Co, Li, Fe, I, Cr, Mn, Cr, As, Pb, V) undoubtedly reflects the complex physiological functions of Al in leek. Among the minerals analyzed, the highest correlation coefficients were recorded between Al and As, Pb, V, Co and Li. As for heavy metals and As, highly significant correlations were found between V and Al, As, Co, Pb and Fe.

Following Al, Li showed wide varietal variation, consistent with previous reports [44]. According to the literature, leek is one of the least Li accumulating species. However, compared to Yalamanchali's [45] findings, our results suggest a wider concentration range of Li content in leek plants. The relationships between Li and other elements are in agreement with those reported, and, in particular, correlations between Li and Al, As, Ca, Co, Cr, Fe, I, Pb, Na, V, Co, Pb, V, Fe and Cr were found in five species grown both in ecological unpolluted and in oil polluted areas of Nigeria [46]. Investigations carried out in New Zealand showed correlations between Li, Fe and Ca only in ryegrass (*Lolium perenne*) [47], whereas such relationships were not recorded in lettuce (*Lactuca sativa*) and beet (*Beta vulgaris*) [48]. Positive correlations were also detected in a plant-soil system between Li and Fe, Al and Na.

Among the relationships involving Li, the interactions between this element, Al and Fe should be considered important, as the two latter minerals show similar atomic radius to lithium.

Although varietal differences in Se content in leek were rather low compared to other elements, the significant correlation between Se and K is a remarkable characteristic of this *Allium* species and it has been scarcely investigated so far.

Table 6. Correlation coefficients between mineral elements in leek.

	Al	As	B	Ca	Cd	Co	Cr	Cu	Fe	I	K	Li	Mg	Mn	Pb
As	0.93 *	1	0.38												
Ca	0.71 *	0.66	0.29	1											
Cd	0.47	0.54	0.26	0.72 *	1										
Co	0.95 ***	0.95 ***	0.53	0.55	0.37	1									
Cr	0.74 *	0.66	0.40	0.94 ***	0.73	0.53	1								
Fe	0.85 **	0.80 **	0.64	0.31	0.20	0.92 ***	0.30	−0.36	1						
I	0.77 **	0.70 *	0.37	0.96 ***	0.70 *	0.58	0.99 ***	−0.60	0.33	1					
K	0.50	0.23	0.75 e	0.13	−0.03	0.39	0.21	−0.23	0.63	0.17	1				
Li	0.99 ***	0.90 ***	0.68	0.70 *	0.46	0.93 ***	0.73 *	−0.59	0.86 **	0.76 *	0.56	1			
Mg	0.75 *	0.65	0.41	0.94 ***	0.69	0.59	0.93 ***	−0.50	0.36	0.94 ***	0.21	0.75 *	1		
Mn	0.74 *	0.86 **	0.36	0.58	0.81 **	0.77 *	0.62	−0.65	0.64	0.62	0.14	0.72 *	0.62	1	
Na	0.83 **	0.71 *	0.64	0.85 **	0.63	0.70	0.88 **	−0.59	0.50	0.90 ***	0.25	0.83 **	0.90 ***	0.61	
Ni	0.65	0.58	0.32	0.38	0.16	0.66	0.27	0.01	0.76 *	0.29	0.60	0.65	0.34	0.42	
Pb	0.92 ***	0.97 ***	0.38	0.67	0.47	0.97 ***	0.63	−0.42	0.80 **	0.68	0.22	0.90 ***	0.71 *	0.79 **	1
Se	0.35	0.05	0.69	0.14	−0.06	0.19	−0.72 *	−0.16	0.42	−0.71 *	0.95 ***	0.42	0.18	−0.03	0.04
Si	0.72 *	0.72 *	0.39	0.83 **	0.80 **	0.57	0.91 ***	−0.70	0.37	0.90 ***	0.23	0.72 *	0.81 **	0.77 *	0.64
Sn	−0.66	−0.57	−0.57	−0.37	−0.39	−0.53	−0.58	0.844 c	−0.47	−0.55	0.26	−0.64	−0.42	−0.52	−0.45
V	0.98 ***	0.94 ***	0.56	0.60	0.38	0.98 ***	0.61	−0.50	0.91 ***	0.65	0.51	0.97 ***	0.64	0.75 *	0.94 ***
Zn	−0.34	−0.33	0.21	−0.74 *	−0.31	−0.15	−0.69	0.05	0.07	−0.71 *	0.05	−0.31	−0.54	−0.14	−0.29

*** $P < 0.001$; ** $P < 0.01$; * $P < 0.05$.

In spinach (*Spinacia oleracea*) plants fertilization with sodium selenate increased K content in female but not male plants [21], whereas in other research [49] garlic biofortification led to selenium antagonistic activity towards K. Taking into account that K participates in plant protection against all forms of biotic and abiotic stresses along with Se and other components of antioxidant defense systems [50], the close relationship between the two minerals in leek suggests intensive interactions of all components of the defense system. The predominance of K in leek elemental composition and the significant correlation between polyphenol concentration and ash content (Figure 5) is in good agreement with the mentioned results. The correlation coefficient relevant to ash to K in leek was 0.78 at $P < 0.01$, whereas that related to K and polyphenols reached 0.96 at $P < 0.01$. The known ability of K to decrease the activity of polyphenol oxidase in plants and enhance polyphenol accumulation [51] may be a good explanation of the positive correlation between polyphenols and K in leek plants. The active participation of K in the antioxidant defense system of this *Allium* species was also characterized by a positive correlation of the element with the ascorbic acid content (r = 0.95 at $P < 0.01$). In this respect, the results of the present work revealed the close relationship between the main components of leek antioxidant systems including polyphenols, ascorbic acid as well as the macro- and trace elements Se and K.

In our research, the lowest negative correlation coefficients were Se with Cr and I. Se is known as an antagonist of Cr and its protective role towards Cr has been previously reported [24,52,53]. The interaction between Se and I is more complex; neither element is essential for plants, but at low concentrations, they may improve plant growth, development and protection from biotic and abiotic stresses [54]. However, the rather scarce and contradictory data about Se and I interactions in plants do not allow clear conclusions. Separate plant fortification with Se and I showed the possibility of mutual stimulation by the two elements in some but not all cases [40]. The selective accumulation of selenium in male spinach plants and of iodine in female spinach plants suggests the participation of phytohormones in the interactions between Se and I [21]. A negative correlation between Se and I in leek plants has not yet been reported.

4. Conclusions

From research carried out in the Moscow region with the aim of evaluating the performance of nine leek (*A. porrum*) cultivars grown in greenhouses under either organic or conventional management systems, interesting clues have been drawn. The varieties showed a uniform behavior under both management systems: no yield differences were recorded between organic and conventional systems. When cultivated with organic procedures, all cultivars attained higher dry matter, sugar, ascorbic acid and potassium content but lower nitrates in the pseudo-stems than for conventional management, but with the same ranking as for conventional management. Moreover, in contrast to related species, highly significant correlations between the antioxidants and mineral elements in leek plants provide opportunities for obtaining genotypes with improved quality features.

Author Contributions: N.A.G. and G.C. designed the experimental protocol; T.M.S. and G.C.T. were concerned with the crop organic management; N.A.G., M.S.A. and O.V.K. performed analytical measurements; G.C. and N.A.G. equally performed the data statistical processing and manuscript writing.

Funding: This research received no external funding.

Conflicts of Interest: The authors declare no conflict of interest.

References

1. Ozgur, M.; Akpinar-Bayaziy, A.; Ozcan, T.; Afolayan, A.J. Effect of dehydration on several physic-chemical properties and the antioxidant activity of leeks (*Allium porrum* L.). *Not. Bot. Hort. Agrobot. Cluj-Napoca* **2011**, *39*, 144–151. [CrossRef]
2. Steinmetz, A.K.; Potter, D.J. Vegetables, fruit, and cancer prevention: A review. *J. Am. Diet. Assoc.* **1996**, *96*, 1027–1039. [CrossRef]

3. Fattorusso, E.; Lanzotti, V.; Taglialatela-Scafati, O.; Cicala, C. The flavonoids of leek, *Allium porrum*. *Phytochemistry* **2001**, *57*, 565–569. [CrossRef]

4. Galeone, C.; Pelucchi, C.; Levi, F.; Negri, E.; Fraceschi, S.; Talamini, R.; Giacosa, A.; La Vecchia, C. Onion and garlic use and human cancer. *Am. J. Clin. Nutr.* **2006**, *84*, 1027–1032. [CrossRef] [PubMed]

5. Ben Arfa, A.; Najjaa, H.; Yahia, B.; Tlig, A.; Neffati, M. Antioxidant capacity and phenolic composition as a function of genetic diversity of wild Tunisian leek (*Allium ampeloprasum* L.). *Acad. J. Biotechnol.* **2015**, *3*, 15–26. [CrossRef]

6. Griffiths, G.; Trueman, L.; Crowther, T.; Thomas, B.; Smith, B. Onions as global benefit to health. *Phytother. Res.* **2002**, *16*, 603–615. [CrossRef] [PubMed]

7. Radovanović, B.; Mladenović, J.; Radovanović, A.; Pavlović, R.; Nikolić, V. Phenolic composition, antioxidant, antimicrobial and cytotoxic activites of *Allium porrum* L. (Serbia) extracts. *J. Food Nutr. Res.* **2015**, *3*, 564–569. [CrossRef]

8. Bianchini, F.; Vainio, H. *Allium* vegetables and organosulfur compounds: Do they help prevent cancer? *Environ. Health Perspect.* **2001**, *109*, 893–902. [CrossRef] [PubMed]

9. Hsing, A.W.; Chokkalingam, A.P.; Gao, Y.T.; Madigan, M.P.; Deng, J.; Gridley, G.; Fraumeni, J.F. *Allium* vegetables and risk of prostate cancer: A population based study. *J. Natl. Cancer Inst.* **2002**, *94*, 1648–1651. [CrossRef] [PubMed]

10. Kyoung-Hee, K.; Hye-Joung, K.; Myung-Woo, B.; Hong-Sun, Y. Antioxidant and antimicrobial activities of ethanol extract from 6 vegetables containing different sulfur compounds. *J. Korean Soc. Food Sci. Nutr.* **2012**, *41*, 577–583.

11. Vergawen, R.; Van Leuven, F.; Van Laere, A. Purification and characterization of strongly chitin-binding chitinases from salicylic acid-treated leek (Allium porrum). *Physiol. Plant.* **1998**, *104*, 175–182. [CrossRef]

12. Yin, M.C.; Tsao, S.M. Inhibitory effect of seven Allium plants upon three Aspergillus species. *Int. J. Food Microbiol.* **1999**, *49*, 49–56. [CrossRef]

13. Mnayer, D.; Fabiano-Tixier, A.-S.; Petitcolas, E.; Hamieh, T.; Nehme, N.; Ferrant, C.; Fernandez, X.; Chemat, F. Chemical composition, antibacterial and antioxidant activities of six essential oils from the *Alliaceae* family. *Molecules* **2014**, *19*, 20034–20053. [CrossRef] [PubMed]

14. Lundegardh, B.; Botek, P.; Schulzo, V.; Hajšlo, V.J.; Stromberg, V.A.; Andersson, H.C. Impact of different green manures on the content of S-Alk(en)yl-L-cysteine sulfoxides and L-Ascorbic acid in leek (*Allium porrum*). *J. Agric. Food Chem.* **2008**, *56*, 2102–2111. [CrossRef] [PubMed]

15. Naem-Rana, K.; Had-Noora, A. The antimicrobial activity of *Allium porrum* water extract against pathogenic bacteria. *J. Kerbala Univ.* **2012**, *10*, 45–49.

16. Sekara, A.; Pokluda, R.; Del Vacchio, L.; Somma, S.; Caruso, G. Interactions among genotype, environment and agronomic practices on production and quality of storage onion (*Allium cepa* L.). *Rev. Hortic. Sci.* **2017**, *44*, 21–42. [CrossRef]

17. Koca, I.; Tasci, B. Mineral composition of leek. *Acta Hortic.* **2016**, *1143*, 147–151. [CrossRef]

18. Conti, S.; Villari, G.M.; Amico, E.; Caruso, G. Effects of production system and transplanting time on yield, quality and antioxidant content of organic winter squash (*Cucurbita moschata* Duch.). *Sci. Hortic.* **2015**, *183*, 136–143. [CrossRef]

19. Bernaert, N.; De Paepe, D.; Bouten, C.; De Clercq, H.; Stewart, D.; Van Bockstaele, E.; De Loose, M.; Van Droogenbroeck, B. Antioxidant capacity, total phenolic and ascorbate content as a function of the genetic diversity of leek (*Allium ampeloprasum* var. porrum). *Food Chem.* **2012**, *134*, 669–677. [CrossRef] [PubMed]

20. Swamy, P.M. *Laboratory Manual on Biotechnology*; Rastogi Publications: Meerut, India, 2008; 617p.

21. Golubkina, N.A.; Kosheleva, O.V.; Krivenkov, L.V.; Nadezhkin, S.M.; Dobrutskaya, H.G.; Caruso, G. Intersexual differences in plant growth, yield, mineral composition and antioxidants of spinach (*Spinacia oleracea* L.) as affected by selenium form. *Sci. Hortic.* **2017**, *225*, 350–358. [CrossRef]

22. AOAC. *The Official Methods of Analysis of AOAC (Association Official Analytical Chemists) International*; AOAC: Arlington, VA, USA, 2012; Volume 22.

23. Maximova, T.V.; Nikulina, I.N.; Pakhomov, V.P.; Shkarina, H.I.; Chumakova, Z.V.; Arzamastsev, A.P. Method of Antioxidant Activity Determination. RF Patent No. 2.170,930, 20 July 2001.

24. Srivastava, S.; Adholeya, A.; Conlan, X.A.; Cahill, D.M. Acidic potassium permanganate chemiluminescence for the determination of antioxidant potential in three cultivars of *Ocimum basilicum*. *Plant Foods Hum. Nutr.* **2015**, *70*. [CrossRef] [PubMed]

25. Zhan, M.G.; Liu, N.; Liu, H. Determination of the total mass of antioxidant substances and antioxidant capacity per unit mass in serum using redox titration. *Bioinorg. Chem. Appl.* **2014**, 928595. [CrossRef]

26. Skalny, A.V.; Lakarova, H.V.; Kuznetsov, V.V.; Skalnaya, M.G. *Analytical Methods in Bioelementology*; Saint Petersburg-Science: St. Petersburg, Russia, 2009.

27. Alfthan, G.V. A micromethod for the determination of selenium in tissues and biological fluids by single-test-tube fluorimetry. *Anal. Chim. Acta* **1984**, *65*, 187–194. [CrossRef]

28. Caruso, G.; Villari, G.; Borrelli, C.; Russo, G. Effects of crop method and harvest seasons on yield and quality of green asparagus under tunnel in southern Italy. *Adv. Hort. Sci.* **2012**, *26*, 51–58.

29. Biswas, S.K.; Khair, A.; Sarker, P.K.; Alom, M.S. Yield and storability of onion (*Allium cepa* L.) as affected by various levels of irrigation. *Bangladesh J. Agric. Res.* **2010**, *35*, 247–255. [CrossRef]

30. Golubkina, N.A.; Kekina, H.G.; Antoshkina, M.S.; Agafonov, A.F.; Nadezhkin, S.M. Inter varietal differences in accumulation of biologically active compounds by *Allium cepa* L. *Messenger Russ. Agric. Sci.* **2016**, *2*, 51–55.

31. Sinclair, P.J.; Blakeney, A.B.; Barlow, E.W. Relationships between bulb dry matter content, soluble solids concentration and non-structural carbohydrate composition in the onion (*Allium cepa*). *J. Sci. Food Agric.* **1995**, *69*, 203–209. [CrossRef]

32. Caruso, G.; Conti, S.; La Rocca, G. Influence of crop cycle and nitrogen fertilizer form on yield and nitrate content in different species of vegetables. *Adv. Hort. Sci.* **2011**, *25*, 81–89.

33. Santamaria, P. Nitrates in vegetables: Toxicity content, intake and EC regulation. *J. Food Agric.* **2006**, *86*, 10–17. [CrossRef]

34. Proteggente, A.R.; Pannala, A.S.; Pagana, G.; Van Buren, L.; Wagner, E.; Wiseman, S. The antioxidant activity of regularly consumed fruit and vegetables reflects their phenolic and vitamin C composition. *Free Radic. Res.* **2002**, *36*, 217–233. [CrossRef] [PubMed]

35. Conti, S.; Villari, G.; Faugno, S.; Melchionna, G.; Somma, S.; Caruso, G. Effects of organic vs. conventional management system on yield and quality of strawberry grown as an annual or biennial crop in southern Italy. *Sci. Hortic.* **2014**, *180*, 63–71. [CrossRef]

36. Biesiada, A.; Kolota, E.; Adamczewska-Sowinska, K. The effect of maturity stage on nutritional value of leek, zucchini and kohlrabi. *Res. Bull.* **2007**, *66*, 39–45. [CrossRef]

37. Malagoli, M.; Schiavon, M.; dall'Acqua, S.; Pilon-Smits, E.A.H. Effects of selenium biofortification on crop nutritional quality. *Front. Plant Sci.* **2015**, *6*, 280. [CrossRef] [PubMed]

38. Lachman, J.; Miholová, D.; Pivec, V.; Jírů, K.; Janovská, D. Content of phenolic antioxidants and selenium in grain of einkorn (*Triticum monococcum*), emmer (*Triticum dicoccum*) and spring wheat (*Triticum aestivum*) varieties. *Plant Soil Environ.* **2011**, *57*, 235–243. [CrossRef]

39. Kavalcová, P.; Bystrická, J.; Trebichalský, P.; Volnová, B.; Kopernická, M. The influence of selenium on content of total polyphenols and antioxidant activity of onion (*Allium cepa* L.). *J. Microbiol. Biotechnol. Food Sci.* **2014**, *3*, 238–240.

40. Golubkina, N.; Kekina, H.; Caruso, G. Foliar biofortification of Indian mustard (*Brassica juncea* L.) with selenium and iodine. *Plants* **2018**, *7*, 80. [CrossRef] [PubMed]

41. Sajid, M.; Butt, M.S.; Shehzad, A.; Tanwer, S. Chemical and mineral analysis of garlic, a golden herb. *Pak. J. Food Sci.* **2014**, *24*, 108–110.

42. Seredin, T.M.; Agafonov, A.F.; Gerasimova, L.I.; Krivenkov, L.V. Element composition of winter garlic. *Veg. Crop.* **2015**, *3–4*, 81–85.

43. Ahn, S.-J.; Matsumoto, H. The role of the plasma membrane in the response of plant roots to aluminum toxicity. *Plant Signal Behav.* **2006**, *1*, 37–45. [CrossRef] [PubMed]

44. Kabata-Pendias, A.; Pendias, H. *Trace Elements in Soils and Plants*; CRC Press: Boca Raton, FL, USA, 2010.

45. Yalamanchali, R.C. Lithium, an Emerging Environmental Contaminant, Is Mobile in the Soil-Plant System. Ph.D. Thesis, Lincoln University, Lincoln, UK, 2012.

46. Essiett, U.A.; Effiong, G.S.; Ogbemudia, F.O.; Bruno, E.J. Heavy metal concentrations in plants growing in crude oil contaminated soil in Akwaıbom State, South-Eastern Nigeria. *Afr. J. Pharm. Pharmacol.* **2010**, *4*, 465–470.

47. Crush, J.R.; Evans, J.P.M.; Cosgrove, G.P. Chemical composition of ryegrass (*Lolium perenne* L.) and prairie grass (*Bromus willdenowii* Kunth) pastures. *N. Z. J. Agric. Res* **1989**, *32*, 461–468. [CrossRef]

48. Bozokalfa, K.; Yağmur, B.; Aşçıoğul, T.; Eşiyok, D. Diversity in nutritional composition of Swiss chard (*Beta vulgaris* subsp. L. var. cicla) accessions revealed by multivariate analysis. *Plant Genet. Resour.* **2011**, *9*, 557–566. [CrossRef]

49. Põldma, P.; Tõnutare, T.; Viitak, A.; Luik, A.; Moor, U. Effect of Selenium treatment on mineral nutrition, bulb size, and antioxidant properties of garlic (*Allium sativum* L.). *J. Agric. Food Chem.* **2011**, *59*, 5498–5503.

50. Wang, M.; Zheng, Q.; Shen, Q.; Guo, S. The critical role of potassium in plant stress response. *Int. J. Mol. Sci.* **2013**, *14*, 7370–7390. [CrossRef] [PubMed]

51. Mudau, F.N.; Soundy, P.; du Toit, E.S. Effects of nitrogen, phosphorus, and potassium nutrition on total polyphenol content of bush tea (Athrixia phylicoides L.) leaves in shaded nursery environment. *Hort Sci.* **2007**, *42*, 334–338.

52. Qing, X.; Zhao, X.; Hu, C.; Wang, P.; Zhang, Y.; Zhang, X.; Wang, P.; Shi, H.; Jia, F.; Qu, C. Selenium alleviates chromium toxicity by preventing oxidative stress in cab-bage (*Brassica campestris* L. ssp. Pekinensis) leaves. *Cotoxicol. Environ. Saf.* **2015**, *114*, 179–189. [CrossRef] [PubMed]

53. Belokobylsky, A.I.; Ginturi, E.I.; Kuchava, N.E.; Kirkesali, E.I.; Mosulishvili, L.M.; Frontasyeva, M.V.; Pavlov, S.S.; Aksenova, N.G. Accumulation of selenium and chromium in the growth dynamics of *Spirulina platensis*. *J. Radioanal. Nucl. Chem.* **2004**, *259*, 65–68. [CrossRef]

54. Pilon-Smits, E. Selenium in plants. In *Progress in Botany*; Luttge, U., Beyschlag, W., Eds.; Springer International Publishing: Basel, Switzerland, 2015; pp. 93–107.

 *horticulturae*

Article

Growth Responses and Root Characteristics of Lettuce Grown in Aeroponics, Hydroponics, and Substrate Culture

Qiansheng Li [1], Xiaoqiang Li [2], Bin Tang [3] and Mengmeng Gu [1,*]

[1] Department of Horticultural Sciences, Texas A & M AgriLife Extension Service, Texas A & M University, College Station, TX 77843, USA; qianshengli@tamu.edu
[2] Shanghai Jieyou Agriculture Sci & Tech Co., Ltd., Shanghai 201210, China; xql992@126.com
[3] Spraying Systems (Shanghai) Co., Shanghai 201611, China; billtb@spray.com.cn
* Correspondence: mgu@tamu.edu; Tel.: +1-979-845-8567

Received: 4 September 2018; Accepted: 16 October 2018; Published: 24 October 2018

Abstract: Aeroponics is a relatively new soilless culture technology which may produce food in space-limited cities or on non-arable land with high water-use efficiency. The shoot and root growth, root characteristics, and mineral content of two lettuce cultivars were measured in aeroponics, and compared with hydroponics and substrate culture. The results showed that aeroponics remarkably improved root growth with a significantly greater root biomass, root/shoot ratio, and greater total root length, root area, and root volume. However, the greater root growth did not lead to greater shoot growth compared with hydroponics, due to the limited availability of nutrients and water. It was concluded that aeroponics systems may be better for high value true root crop production. Further research is necessary to determine the suitable pressure, droplet size, and misting interval in order to improve the continuous availability of nutrients and water in aeroponics, if it is to be used to grow crops such as lettuce for harvesting above-ground parts.

Keywords: soilless culture; root growth; root/shoot ratio

1. Introduction

Soilless culture, including aeroponics, aquaponics, and hydroponics, is considered one of the more innovative agricultural strategies to produce more from less, in order to feed the estimated 11 billion people in the world by 2100 [1]. Aeroponics is a promising technology that grows plants with their root systems exposed to a nutrient mist in a closed chamber [2]. Plants grow well in aeroponics, primarily because of the highly aerobic environment it creates. It is even possible to control the root-zone atmosphere when it is combined with a gas delivery system [3]. Integrated vertical aeroponic farming systems with manipulation of temperature and CO_2 in the root-zone environment can achieve more efficient use of land area to secure a vegetable supply in space-limited cities [4]. Aeroponics is also an excellent option for space mission life support systems that require optimum control of growth parameters [5].

Aeroponics has been widely used in plant physiology research, but is not as commonly used as hydroponic methods on a commercial scale [6]. However, aeroponics has been increasingly used for growing numerous vegetable crops such as lettuce, cucumber, melon, tomato, herbs, potato, and floral crops, and especially for those crops where roots are harvested as the end product. Seed potato production may be the most successful application of aeroponics on a commercial scale, done mostly in China, Korea, South America, and African countries in recent years [7–10]. Aeroponics is able to produce large numbers of minitubers in one generation that can be harvested sequentially, eliminating the need for field production, thereby reducing costs and saving time [7]. This technique was applied to effectively produce minitubers of yam (*Dioscorea rotundata* and *D. alata*) in Nigeria and Ghana [11,12]. Aeroponics could be an alternative production system for other high-value root and rhizome crops, such as great burdock (*Arctium lappa*) [13], ginger (*Zingiber officinale*) [14], medicinal crops, such as *Urtica dioica* and *Anemopsis californica* [15] and saffron (*Crocus sativus*) [16]. Essential oil production of herbs like valerian (*Valeriana officinalis*) grown using aeroponics has also been reported [17]. Aeroponics has also been reported as an economic method for rapid root induction and clonal propagation of three endangered and medicinally important plants [18]. Aeroponics was used to produce tree saplings (*Acacia mangium*) with arbuscular mycorrhiza (AM) fungi inoculation [19]. The well-aerated root environment of aeroponics was beneficial for root initiation and subsequent root growth in woody (*Ficus*) and herbaceous (*Chrysanthemum*) cuttings [20].

Many studies have clearly shown that aeroponics promotes plant growth rates through optimization of root aeration because the plant is totally suspended in air, giving the plant stem and root systems access to 100% of the available oxygen in the air [7]. Droplet size and frequency of exposure of the roots to the nutrient solution are the critical factors which may affect oxygen availability [2]. Large droplets lead to less oxygen being available to the root system, while fine droplets produce excessive root hair without developing a lateral root system for sustained growth [10]. Three broad categories are generally used to classify droplet forming systems and droplet size; regular spray nozzles with droplet size >100 μm (spray), compressed gas atomizers with droplet size between 1 to 100 μm (fog), and ultrasonic systems with droplet size 1 to 35 μm (mist) [21]. The most common type is when the nutrient solution is compressed through nozzles by a high pressure pump, forming a fine mist in the growth chambers [7]. An ultrasonic misting system was adopted in a sterile aeroponics culture system for in vitro propagation [22].

In this study, air atomizing nozzles (1/4J Series) were employed for the aeroponics system. The air atomizing nozzles require a single air source for atomizing the air and to provide independent control of liquid, atomizing air, and fan air pressure for fine tuning of the flow rate, droplet size, spray distribution, and coverage. These air atomizing nozzles were equipped with clean-out needles to eliminate clogging and ensure optimum performance. The objectives of the present study were to compare shoot and root growth, root characteristics, and mineral contents of two lettuce cultivars grown in aeroponics, hydroponics (nutrient film technique, NFT) and substrate culture.

2. Materials and Methods

2.1. Cultivation Systems

The experiment was carried out in a 12.8 × 24 m experimental Venlo type glasshouse, which was equipped with outside and inside shade nets, fans and pad, and misting system. The aeroponics units were built in an A-frame shape, 1.4 m wide, 1.4 m high, and 6 m long; both sides were covered with multiple Styrofoam panels at 60° angles. The planting density was 25 plants per m^2 at a spacing of 20 × 20 cm. Six nozzles (AEROJSUMAX-6SS, Spraying Systems (Shanghai) Co., Shanghai, China) were placed horizontally at the end and middle of the A-frame. The nozzles can be operated under an air pressure from 0.7×10^5 to 4×10^5 Pa, with liquid capacity from 7.6 to 63 L/h. AutoJet® Spray System (Spraying Systems (Shanghai) Co.) and was installed to monitor and automatically adjust the spray pattern, flow rate, droplet size, liquid pressure, and atomizing air pressure. Misting lasted 20 s with a 30 s interval before misting again. The droplet size was adjusted to 50 μm and the nutrient solution was recycled.

The NFT hydroponics system consisted of a PVC trough on a slope of 1 percent. The trough was 10 cm wide, 5 cm deep, and 6 m long. The cascade troughs were suspended one above the other, up to 7 levels. The nutrient solution entered the high end of the slightly sloping top trough, exited at the low end of that trough into the high end of the next one, and so forth, and back to the reservoir from where it was pumped. The flow rate was set at 5 L/min. Plants were set 20 cm apart in each trough.

The substrate culture system was conducted using square PVC plastic pots, which were 46 cm long, 40 cm wide, and 18 cm high. Six plants were planted in each pot at a 20 × 20 cm spacing. A mix of 50% peat and 50% perlite was used as the substrate. The substrate depth was approximately 18 cm. The nutrient solution of 1.2 L/day was supplied through 3 drip lines in each pot twice a day.

2.2. Planting and Experimental Arrangement

Lettuce seeds of cultivars 'Nenglv naiyou' and 'Dasusheng' (Institute of vegetable and flower, CAAS, Beijing, China) were sown in 72-cell polystyrene trays. Each cell was filled with a hydroponic planting basket with a sponge for support. At the two true leaves stage, all plants were watered with a half strength Hoagland's nutrient solution [23] until the seedlings were ready for transplanting. Four weeks after sowing, lettuce seedlings were transplanted to the aeroponics, hydroponics, and substrate culture systems. Plants were supplied with full strength Hoagland's nutrient solution (containing N 210 mg/L, K 235 mg/L, Ca 200 mg/L, P 31 mg/L, S 64 mg/L, Mg 48 mg/L, B 0.5 mg/L, Fe 1 to 5 mg/L, Mn 0.5 mg/L, Zn 0.05 mg/L, Cu 0.02 mg/L, Mo 0.01 mg/L). Three A-shape aeroponics units, 12 hydroponic troughs, and 24 substrate culture pots were planted for the comparison experiment.

2.3. Harvesting and Measurement

2.3.1. Biomass and Root/Shoot Ratio

Nine plants from each cultivation system were harvested and washed with tap water. Substrates in the roots of the plants from the substrate cultivation treatment were gently washed off. The fresh weight of shoots and roots was recorded immediately after removing the free surface moisture with soft paper towels. Shoot and root samples were then oven dried at 85 °C for 48 h, and weighed for dry weight on a scale accurate to 0.0001 g. The root/shoot ratio was calculated as the root dry weight/shoot dry weight.

2.3.2. Root Characteristics

Five plants from each cultivation treatment were randomly sampled for measurement of root characteristics. Washed roots were immersed and spread out in a 40 × 25 × 10 cm square blue plastic container which was filled with tap water to a depth of 3 cm. The entire root system was photographed from above with a digital camera (Nikon D90, Nikon Corporation, Tokyo, Japan) and saved using the jpeg format (Figure 1A). The photographs were re-cropped, scaled (Figure 1B), and processed with GiA Roots software (Georgia Tech Research Corporation and Duke University, USA) to obtain a threshold image (Figure 1C) for measuring the characteristics of all the roots. The measured root characteristics included average root diameter (width), root length (network length), root area (network surface area), root volume (network volume), maximum number of roots, median number of roots, and network perimeter [24].

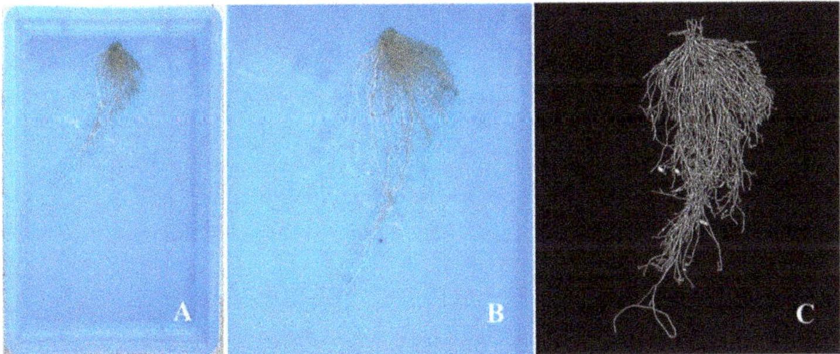

Figure 1. The entire root was immersed and spread out in a 40 × 25 × 10 cm square plastic container filled with tap water to a depth of 3 cm (**A**), the image was re-cropped after being scaled (**B**), and changed to a threshold image (**C**) with GiA Roots software, for root characteristics measurements.

2.3.3. Plant Leaf Nitrogen, Phosphorus, and Potassium Content

Three plants from each cultivation treatment were randomly sampled for mineral nutrient analysis and oven dried as above. The dried shoots from the different treatments were milled to passed through a 1 mm screen. The ground dry material (~0.2000 g) was wet digested using a H_2SO_4-H_2O_2 solution. Nitrogen content was determined using the Kjeldahl method [25]. Phosphorus was determined by the ascorbic acid molybdenum blue method [26]. Potassium was determined by flame emission spectrophotometry [27].

2.4. Statistical Analysis

All results were subjected to a two-way analysis of variance (ANOVA) using SPSS Statistics 19.0; the effects of the cultivation system, genotypes (cultivars), and their interaction were analyzed. Within each cultivar, means were separated using Duncan's multiple range test at $P = 0.05$. The results were expressed as means $\pm$ SE.

3. Results

3.1. Plant Growth and Biomass

The cultivation systems significantly influenced the growth of both lettuce cultivars. Lettuce grown in hydroponics had larger above-ground parts, while the aeroponic lettuce had greater root dry weight and root/shoot ratio, and plants from substrate cultivation had the smallest size (Figure 2, Table 1).

Figure 2. Whole plants of lettuce cultivars 'Dasushen' (**A**) and 'Nenglv naiyou' (**B**) 45 days after transplanting in aeroponic, hydroponic, and substrate cultivation systems.

The two-way ANOVA showed significant effects of cultivation system on shoot and root fresh weight, shoot and root dry weight, and root/shoot ratio (Table 1). However, genotype only showed significant effects on shoot and root dry weight. An interaction between cultivation system and genotype was significant on fresh weight and dry weight of shoot and root, but not on the root/shoot ratio.

In both cultivars, the shoot fresh and dry weights of hydroponic lettuce were approximately twice that of aeroponic and substrate cultivated lettuce (Table 1). Root fresh weights of aeroponics and hydroponics lettuce were significantly higher than that of substrate cultivated lettuce. The root dry weights of both cultivars in aeroponics were significantly higher than that of hydroponics and substrate cultivation. The most remarkable difference between the three growing methods was the root/shoot ratio. In both cultivars, the root/shoot ratio of aeroponics lettuce was almost three times that of the hydroponics lettuce, and was also significantly higher than that of the substrate culture (Table 1).

Table 1. Shoot and root fresh weight (FW), dry weight (DW), and root to shoot ratio of two lettuce cultivars grown in three cultivation systems.

	Shoot FW (g)	Root FW (g)	Shoot DW (g)	Root DW (g)	Root/Shoot Ratio
	Lactuca sativa 'Dasusheng'				
aeroponics	37.8 ± 2.67 [bz]	8.67 ± 1.20 [a]	2.40 ± 0.17 [b]	0.80 ± 0.15 [a]	0.32 ± 0.04 [a]
hydroponics	88.8 ± 9.47 [a]	8.78 ± 1.24 [a]	4.86 ± 0.54 [a]	0.59 ± 0.06 [b]	0.12 ± 0.01 [c]
substrate	49.2 ± 2.34 [b]	6.92 ± 0.43 [b]	3.23 ± 0.13 [b]	0.69 ± 0.03 [b]	0.22 ± 0.02 [b]
	Lactuca sativa 'Nenglv Naiyou'				
aeroponics	50.9 ± 2.60 [b]	10.3 ± 0.46 [a]	2.58 ± 0.11 [b]	0.77 ± 0.07 [a]	0.30 ± 0.03 [a]
hydroponics	96.1 ± 4.23 [a]	11.5 ± 1.07 [a]	4.80 ± 0.16 [a]	0.54 ± 0.03 [b]	0.11 ± 0.01 [b]
substrate	39.4 ± 1.72 [c]	3.9 ± 0.35 [b]	2.03 ± 0.06 [c]	0.26 ± 0.03 [c]	0.13 ± 0.01 [b]
	Significance				
Cultivation system (CS)	*** [y]	***	***	**	***
Genotype (G)	Ns	Ns	*	*	ns
CS × G	*	**	*	*	ns

[z] Values are mean ± SE (n = 9). In the same cultivar, values followed by the same superscript letter are not significantly different ($P \leq 0.05$). [y] *** = $P < 0.001$; ** = $P < 0.01$; * = $P < 0.05$; ns, not significant at $P \geq 0.05$.

3.2. Root Characteristics

Analysis of the root characteristics by GiA Roots software revealed details of the influence of growing methods on root growth. The root length, area, volume, and network perimeter of aeroponic lettuce (both cultivars) were significantly greater than that of hydroponic and substrate cultivated lettuce (Table 2). In particular, for the cultivar 'Dasusheng', the root length, root area, root volume, and perimeter in aeroponic cultivation were four to five times that of the hydroponic and substrate cultivation. However, the average root diameter did not significantly differ among treatments in the cultivar 'Dasusheng'. Average root diameter of hydroponically-grown 'Nenglv naiyou' was significantly greater than that from substrate cultivation. The maximum and median numbers of roots of aeroponic 'Dasusheng' lettuce were two to three times higher than that from hydroponic and substrate cultivation; such differences in maximum root number were not found for the cultivar 'Nenglv naiyou' where only median root number for aeroponic cultivation was greater than hydroponic but not substrate cultivation.

Table 2. Root characteristics of lettuce grown in aeroponics, hydroponics, and substrate culture systems.

	Average Root Diameter (mm)	Root Length (cm)	Root Area (cm²)	Root Volume (cm³)	Maximum No. of Roots	Median No. of Roots	Network Perimeter (cm)
	Lactuca sativa 'Dasusheng'						
aeroponics	0.501 ± 0.017 [az]	3043 ± 231 [a]	479 ± 42 [a]	7.24 ± 0.77 [a]	75.0 ± 5.8 [a]	39.0 ± 4.1 [a]	6019 ± 473 [a]
hydroponics	0.551 ± 0.025 [a]	581 ± 113 [b]	100 ± 20 [b]	1.64 ± 0.35 [b]	24.6 ± 2.9 [b]	13.8 ± 1.6 [b]	1164 ± 221 [b]
substrate	0.501 ± 0.007 [a]	724 ± 126 [b]	114 ± 21 [b]	1.67 ± 0.35 [b]	32.6 ± 5.3 [b]	17.4 ± 2.2 [b]	1437 ± 245 [b]
	Lactuca sativa 'Nenglv Naiyou'						
aeroponics	0.511 ± 0.0023 [ab]	2634 ± 260 [a]	424 ± 46 [a]	6.63 ± 0.84 [a]	63.0 ± 7.4 [a]	34.4 ± 3.1 [a]	5197 ± 557 [a]
hydroponics	0.554 ± 0.0009 [a]	1688 ± 239 [b]	296 ± 46 [b]	4.91 ± 0.85 [ab]	54.4 ± 3.0 [a]	19.0 ± 4.8 [b]	3379 ± 473 [b]
substrate	0.487 ± 0.0006 [b]	1378 ± 58 [b]	211 ± 10 [b]	2.99 ± 0.15 [b]	51.2 ± 1.6 [a]	37.0 ± 2.6 [a]	2755 ± 113 [b]
	Significance						
Cultivation system (CS)	ns [y]	***	**	*	***	***	*
Genotype (G)	Ns	***	***	***	***	**	***
CS × G	*	*	ns	Ns	ns	*	*

[z] Values are mean ± SE (n = 9). In the same cultivar, values followed by the same superscript letter are not significantly different ($P \leq 0.05$). [y] *** = $P < 0.001$; ** = $P < 0.01$; * = $P < 0.05$; ns, not significant at $P \geq 0.05$.

The two-way ANOVA indicated that both the cultivation system and genotype significantly affected all the root characteristics except the average root diameter, i.e., average root diameter, root length, median number of roots, and root perimeter (Table 2).

3.3. Leaf N, P, and K Contents

In both cultivars, the leaf N content of hydroponic lettuce was significantly higher than that of aeroponic and substrate cultivated lettuce, but there was no difference between aeroponic and substrate cultivation (Figure 3A). The leaf P content of hydroponic lettuce was significantly higher than that of the aeroponic and substrate cultivated lettuce (Figure 3B). Leaf K content of both aeroponic and hydroponic lettuce was significantly higher than that of substrate cultivated lettuce, but there was no difference between the aeroponic and hydroponic lettuce (Figure 3C).

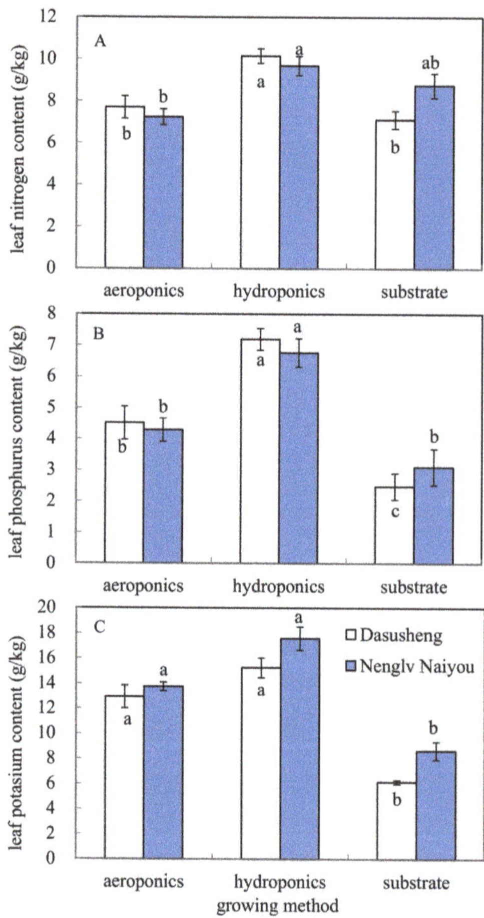

Figure 3. Leaf nitrogen (**A**), phosphorus (**B**), and potassium (**C**) content of two lettuce cultivars grown on aeroponics, hydroponics, and substrate cultivation systems. Values are mean ± SE ($n = 3$). In the same cultivar, values followed by the same letter are not significantly different ($P \leq 0.05$).

The two-way ANOVA indicated that the cultivation system significantly affected leaf N, P, and K contents (Table 3), while genotype only had a significant effect on the leaf K content, and there was no significant interaction between cultivation system and genotype.

Table 3. *F*-values from a two-way analysis of variance (ANOVA) of leaf nitrogen, phosphorus, and potassium content as affected by three cultivation systems and two genotypes.

	N	P	K
Cultivation system (CS)	11.6 *** [z]	50.6 ***	84.6 **
Genotype (G)	0.28 ns	0.01 ns	10.7 **
CS × G	2.51 ns	0.88 ns	0.84 ns

[z] *** = $P < 0.001$; ** = $P < 0.01$; ns, not significant, $P \geq 0.05$.

4. Discussion

The most impressive result of this study was the significant improvement of root growth of lettuce in the aeroponic system. The cultivation systems also affected the fresh weight, dry weight, and root/shoot ratio, while the genotypes only had significant effects on dry weight, and the interaction between them also significantly affected the biomass but not the root/shoot ratio. In both cultivars, the root dry weight of aeroponic lettuce was significantly higher than that of hydroponic and substrate cultivation, and the root/shoot ratio of aeroponic lettuce was two to three times that of the other two systems. The two-way ANOVA results indicated that the root characteristics were more dependent on genotype; however, the cultivation systems also had significant effects on the root characteristics except on root diameter. The greater total root length, root area, and root volume further proved that aeroponics was beneficial to root growth. However, the greater root system of aeroponics did not lead to more shoot biomass (yield) than hydroponics. Instead, shoot biomass of aeroponic lettuce was significantly less than that of hydroponics. This may be due to sufficient nutrient and water supply when the root system was submerged continuously in nutrient solution. The cultivation systems had significant influences on leaf N, P, and K content, while genotype only showed significant effects on K content, and there was no cultivation system by genotype interaction on the mineral contents.

In aeroponics, the nutrient solution was only sprayed as fine droplets at intervals, which may limit shoot growth and improve root growth, as the plant's response may be to adapt to the relative deficit of water and nutrients during the intervals. In valerian (*Valeriana officinalis*) cultivation trials, it was also found that both leaf area and biomass production in an aeroponic system were lower than in floating raft hydroponic and substrate cultivation systems; it was concluded that this may be caused by the higher proliferation of roots inside the frame reducing the performance of nozzles [17]. The root number of saffron (*Crocus sativus*) plants was also significantly greater in aeroponics than that in hydroponics and soil culture, but no significant difference in shoot growth was found [16]. The larger distance between misting sprayers and roots restricted root access to the water micro droplets, resulted in decreasing nutrient availability and absorbance. In this case, plants were forced to compensate by increasing root surface area and weight [28]. Thus, the droplet size and the misting interval will have a great effect on plant growth in aeroponic culture.

Good aeration of the root environment is the most important advantage of aeroponics. Aeroponics significantly improved adventitious root formation in rapid root induction and clonal propagation of three endangered and medicinally important plants over soil grown stem cuttings [18]. Aeroponics showed higher yield and better size distribution in potato minituber production, but growth was influenced by such factors as the genotype, the availability of nutrients, the stretching of the cycle, and the culture density [8]. Higher root vitality of plants was observed in aeroponics and aerohydroponics than that of deep water culture [29]. If the roots of plants in aeroponics can absorb nutrients and water readily, better growth of the above-ground part will result. However, the major disadvantage of aeroponics is the possibility of irreversible damage or complete loss, since there is no substrate at all (neither solid nor water) that could enable the plants to survive in the event of a technical or power failure [7,9]. During our experiment, there was a one-day mechanical breakdown of the aeroponic system, which could also have affected growth.

5. Conclusions

From this study, it can be concluded that aeroponics is beneficial to root growth, with significantly greater root/shoot ratio, root length, root area, and root volume. Thus, an aeroponic system may be superior for producing high value, true root crops, particularly for medicinal plants as suggested by Hayden [13,15]. When growing root crops in aeroponics, clean products may be harvested sequentially. To grow crops like lettuce in aeroponics for harvesting above-ground parts, further research is necessary to determine suitable pressure, droplet size, and misting interval in order to improve the continuous availability of nutrients and water so that growth of above-ground parts of plants can be optimized.

Author Contributions: Conceptualization, Q.L.; data curation, Q.L.; formal analysis, Q.L.; investigation, X.L.; methodology, Q.L.; resources, B.T.; supervision, M.G.; writing—original draft, Q.L.; writing—review and editing, M.G.

Funding: This project was funded by Spraying Systems (Shanghai) Co. research project J2017-30.

Acknowledgments: We thank Xue Zhang and Minmin Wan for their help with measurements.

Conflicts of Interest: The authors declare no conflict of interest.

References

1. Lal, R. Feeding 11 billion on 0.5 billion hectare of area under cereal crops. *Food Energy Secur.* **2016**, *5*, 239–251. [CrossRef]
2. Jones, J.B., Jr. *Complete Guide for Growing Plants Hydroponically*; CRC Press: Boca Raton, FL, USA, 2014.
3. Kratsch, H.A.; Graves, W.R.; Gladon, R.J. Aeroponic system for control of root-zone atmosphere. *Environ. Exp. Bot.* **2006**, *55*, 70–76. [CrossRef]
4. He, J. Farming of Vegetables in Space-Limited Environments. *Cosmos* **2015**, *11*, 21–36. [CrossRef]
5. Clawson, J.; Hoehn, A.; Stodieck, L.; Todd, P.; Stoner, R. *Re-Examining Aeroponics for Spaceflight Plant Growth*; SAE Technical Paper 0148-7191; SAE International: Warrendale, PA, USA, 2000.
6. Resh, H.M. *Hydroponic Food Production: A Definitive Guidebook for the Advanced Home Gardener and the Commercial Hydroponic Grower*; CRC Press: Boca Raton, FL, USA, 2012.
7. Buckseth, T.; Sharma, A.K.; Pandey, K.K.; Singh, B.P.; Muthuraj, R. Methods of pre-basic seed potato production with special reference to aeroponics—A review. *Sci. Hortic.* **2016**, *204*, 79–87. [CrossRef]
8. Tierno, R.; Carrasco, A.; Ritter, E.; de Galarreta, J.I.R. Differential Growth Response and Minituber Production of Three Potato Cultivars Under Aeroponics and Greenhouse Bed Culture. *Am. J. Potato Res.* **2013**, *91*, 346–353. [CrossRef]
9. Mateus-Rodriguez, J.R.; de Haan, S.; Andrade-Piedra, J.L.; Maldonado, L.; Hareau, G.; Barker, I.; Chuquillanqui, C.; Otazú, V.; Frisancho, R.; Bastos, C.; et al. Technical and Economic Analysis of Aeroponics and other Systems for Potato Mini-Tuber Production in Latin America. *Am. J. Potato Res.* **2013**, *90*, 357–368. [CrossRef]
10. Margaret, C. Potential of aeroponics system in the production of quality potato (*Solanum tuberosum* L.) seed in developing countries. *Afr. J. Biotechnol.* **2012**, *11*, 3993–3999.
11. Oteng-Darko, P.; Kyei-Baffour, N.; Otoo, E.; Agyare, W.A. Growing Seed Yams in the Air: The Agronomic Performance of Two Aeroponics Systems Developed in Ghana. *Sustain. Agric. Res.* **2017**, *6*, 106–116. [CrossRef]
12. Maroya, N.; Balogun, M.; Asiedu, R.; Aighewi, B.; Kumar, P.L.; Augusto, J. Yam propagation using aeroponics technology. *Annu. Res. Rev. Biol.* **2014**, *4*, 3894–3903. [CrossRef]
13. Hayden, A.; Yokelsen, T.; Giacomelli, G.; Hoffmann, J. Aeroponics: An alternative production system for high-value root crops. *Acta Hortic.* **2004**, *629*, 207–213. [CrossRef]
14. Hayden, A.L.; Brigham, L.A.; Giacomelli, G.A. Aeroponic cultivation of ginger (*Zingiber officinale*) rhizomes. *Acta Hortic.* **2004**, *659*, 397–402. [CrossRef]
15. Hayden, A.L. Aeroponic and hydroponic systems for medicinal herb, rhizome, and root crops. *Hortscience* **2006**, *41*, 536–538.
16. Souret, F.F.; Weathers, P.J. The Growth of Saffron (*Crocus sativus* L.) in Aeroponics and Hydroponics. *J. Herbs Spices Med. Plants* **2000**, *7*, 25–35. [CrossRef]

17. Tabatabaei, S.J. Effects of Cultivation Systems on the Growth, and Essential Oil Content and Composition of Valerian. *J. Herbs Spices Med. Plants* **2008**, *14*, 54–67. [CrossRef]

18. Mehandru, P.; Shekhawat, N.S.; Rai, M.K.; Kataria, V.; Gehlot, H.S. Evaluation of aeroponics for clonal propagation of Caralluma edulis, Leptadenia reticulata and Tylophora indica—Three threatened medicinal Asclepiads. *Physiol. Mol. Biol. Plants* **2014**, *20*, 365–373. [CrossRef] [PubMed]

19. Martin-Laurent, F.; Lee, S.-K.; Tham, F.-Y.; Jie, H.; Diem, H.G. Aeroponic production of Acacia mangium saplings inoculated with AM fungi for reforestation in the tropics. *For. Ecol. Manag.* **1999**, *122*, 199–207. [CrossRef]

20. Soffer, H.; Burger, D.W. Effects of dissolved oxygen concentrations in aero-hydroponics on the formation and growth of adventitious roots. *J. Am. Soc. Hortic. Sci.* **1988**, *113*, 218–221.

21. Weathers, P.; Liu, C.; Towler, M.; Wyslouzil, B. Mist reactors: Principles, comparison of various systems, and case studies. *Electron. J. Integr. Biosci.* **2008**, *3*, 29–37.

22. Tisserat, B.; Jones, D.; Galletta, P.D. Construction and use of an inexpensive in vitro ultrasonic misting system. *HortTechnology* **1993**, *3*, 75–78.

23. Hoagland, D.R.; Arnon, D.I. *The Water-Culture Method for Growing Plants without Soil*, 2nd ed.; Circular 347; California Agricultural Experiment Station: Berkeley, CA, USA, 1950.

24. Galkovskyi, T.; Mileyko, Y.; Bucksch, A.; Moore, B.; Symonova, O.; Price, C.A.; Topp, C.N.; Iyer-Pascuzzi, A.S.; Zurek, P.R.; Fang, S. GiA Roots: Software for the high throughput analysis of plant root system architecture. *BMC Plant Biol.* **2012**, *12*, 116. [CrossRef] [PubMed]

25. Horneck, D.A.; Miller, R.O. Determination of total nitrogen in plant tissue. In *Handbook of Reference Methods for Plant Analysis*; Kalra, Y.P., Ed.; CRC Press LLC: Boca Raton, FL, USA, 1998; pp. 75–83.

26. Jones, J.B., Jr. *Laboratory Guide for Conducting Soil Tests and Plant Analysis*; CRC Press: Boca Raton, FL, USA, 2001.

27. Horneck, D.; Hanson, D. Determination of potassium and sodium by flame emission spectrophotometry. In *Handbook of Reference Methods for Plant Analysis*; Kalra, Y.P., Ed.; CRC Press LLC: Boca Raton, FL, USA, 1998; pp. 153–155.

28. Salachas, G.; Savvas, D.; Argyropoulou, K.; Tarantillis, P.; Kapotis, G. Yield and nutritional quality of aeroponically cultivated basil as affected by the available root-zone volume. *Emir. J. Food Agric.* **2015**, *27*, 911–918. [CrossRef]

29. Chang, D.C.; Park, C.S.; Kim, S.Y.; Lee, Y.B. Growth and Tuberization of Hydroponically Grown Potatoes. *Potato Res.* **2012**, *55*, 69–81. [CrossRef]

 *horticulturae*

Article

Assessing Quantitative Criteria for Characterization of Quality Categories for Grafted Watermelon Seedlings

Filippos Bantis [1], Athanasios Koukounaras [1,*], Anastasios Siomos [1], Georgios Menexes [1], Christodoulos Dangitsis [2] and Damianos Kintzonidis [2]

1. School of Agriculture, Aristotle University of Thessaloniki, 54124 Thessaloniki, Greece; fbanths@gmail.com (F.B.); siomos@agro.auth.gr (A.S.); gmenexes@agro.auth.gr (G.M.)
2. Agris S.A., Kleidi, 59300 Imathia, Greece; cdaggitsis@agris.gr (C.D.); damianos@agris.gr (D.K.)
* Correspondence: thankou@agro.auth.gr; Tel.: +30-2310-994123

Received: 26 December 2018; Accepted: 28 January 2019; Published: 2 February 2019

Abstract: Vegetable grafting is a practice employed worldwide since it helps prevent biotic and abiotic disorders, and watermelon is one of the most important species grafted. The objective of this study was to set critical limits for the characterization of quality categories for grafted watermelon seedlings. Specifically, watermelon (scion) seedlings were grafted onto squash (rootstock) seedlings, moved into a healing chamber for 7 days, and then transferred into a greenhouse for seven more days. At 7 and 14 days after grafting, experienced personnel assessed grafted seedling quality by categorizing them. The categories derived were Optimum and Acceptable for both time intervals, plus Not acceptable at 14 days after grafting. Optimum seedlings showed greater leaf area, and shoot and root fresh and dry weights at both time intervals. Moreover, they had greater stem diameter, root-to-shoot ratio, shoot dry weight-to-length ratio and Dickson's quality index compared to the other category at 14 days after grafting. Therefore, Optimum seedlings would likely develop into marketable plants of high quality, with better establishment in the field. Not acceptable seedlings showed considerably inferior development, while Acceptable seedlings were between the other categories, but were still marketable.

Keywords: *Citrullus lanatus*; vegetable grafting; optimal production; marketable seedlings; quality indicators

---

1. Introduction

The use of grafting for vegetable seedlings is a well-established practice worldwide [1–3]. The important advantage of using grafted seedlings to prevent significant crop loss due to biotic [3,4] and abiotic [5,6] factors (soil-borne diseases, salinity, low temperatures, etc.), as well as the reduction of the use of agrochemical products, provide grafting as an environmentally-friendly practice [3]. Applications of grafting mainly focus on species of the Cucurbitaceae and Solanaceae families, particularly watermelon and tomato, respectively [2,7]. In Greece the use of grafted watermelon seedlings is almost 99% of growers using low tunnel protected cultivation in order to achieve early yield.

To enjoy the advantages of grafting, use of high quality grafted seedlings is a prerequisite. As a result, a rapid development and expansion of a vegetable nursery industry is in progress. Successful grafting requires good connection between the rootstock and the scion for healthy uniform growth and development of the grafted plants [8]. The most important stage for seedling evaluation for the grower is at the time of purchase and transplanting to avoid possibly negative results during subsequent cultivation. However, the definition of what constitutes a high-quality seedling is a very complicated issue [8].

Generally, high-quality seedlings could be defined as plants uniform in size and traits, proper size or height with a thick healthy stem with large thick leaves, a well-developed root system, good root-to-shoot ratio, and a good ratio of shoot dry weight divided by shoot length [3,8]. However, most of the above characteristics could be suitable for some species such as tomatoes, peppers, and eggplants, but could be adapted for the cucurbits [3]. Moreover, most seedling quality parameters have been qualitatively determined by experienced personnel. Therefore, it is critical to define quantitative parameters that could be easily and objectively applied by anyone in this chain (industry, grower, etc.).

Therefore, the aim of this study was to set critical limits for objective measurements of grafted watermelon seedling quality categories as well as to suggest the most accurate and convenient among them for application by the industry and growers. Evaluations were conducted at 7 and 14 days after grafting. Grafted watermelon seedlings exit the healing chamber at 7 days after grafting and therefore quality assessment at that time is essential for possible later research. By 14 days, seedlings are considered "final product" and therefore quality evaluation at that time is valuable to assess the product marketability.

2. Materials and Methods

2.1. Plant Material

The experiments were conducted in the facilities of Agris S.A. in Kleidi, Imathia, Greece. All measurements were executed at Aristotle University of Thessaloniki, Greece. During the experiment, standard commercial practices were applied.

Watermelon (scion—*Citrullus lanatus*) "Celine" (HM. Clause SA, Portes-Les-Valence, France) and squash (rootstock—*Cucurbita moschata*) "TZ-148" (HM. Clause SA, Portes-Les-Valence, France) were used for the production of grafted seedlings. Watermelon seeds were sown in plastic 171-cell plug trays, while squash seeds were sown in plastic 128-cell plug trays (both types: 67×33 cm, G.K. Rizakos S.A., Lamia, Greece). Both plug tray types were filled with a 5:1:2 mixture of peat, perlite, and vermiculite.

2.2. Germination, Grafting, Healing, and Acclimatization

Following planting, the plug trays of scions and rootstocks were moved into a growth chamber ($25\,^\circ$C, 95–98% relative humidity (RH)) until germination. Watermelon and squash germinated after 72 and 48 h, respectively, and afterwards they were moved to a glass greenhouse for 9 (scion) and 10 (rootstock) days at a 21.5 $^\circ$C minimum night temperature for both species, and 100 ± 10 µmol $m^{-2}\,s^{-1}$ photosynthetic photon flux density (PPFD) emitted by high-pressure sodium (HPS) lamps (MASTER GreenPower 600 W, 400 V E40, Philips Lighting, Eindhoven, The Netherlands) with an 18 h photoperiod only for watermelon. The natural light photoperiod during the experiment was 12.5 h from sunrise to sunset. Supplemental lighting is commonly practiced for the production of watermelon seedlings in order to achieve high quality product. On the other hand squash seedlings have adequate development under natural light conditions and no supplemental lighting is employed.

Grafting was performed with the "splice grafting" technique, 12 days after sowing. Using a razor blade, the scion was diagonally cut just below the cotyledons while the rootstock was diagonally cut on the cotyledon level leaving only one cotyledon. The rootstock was also cut just above ground level, which is a commonly practiced technique for cucurbit rootstocks in order to achieve increased grafting efficiency [9]. Afterwards, the grafted seedlings were placed in polystyrene 72-cell plug trays (50×30 cm, G.K. Rizakos S.A., Lamia, Greece) filled with a 3:1:1 mixture of peat, perlite, and vermiculite. Grafting was performed by experienced personnel to minimize critical errors.

Following grafting, healing and acclimatization of grafted seedlings was achieved during 7 days in a growth chamber at 25 $^\circ$C, recirculating air, 45 µmol $m^{-2}\,s^{-1}$ PPFD emitted by fluorescent tubes (Fluora 58W, Osram, GmbH, Munich, Germany) for an 18 h photoperiod, and RH of 98% for days 1–4, 93% for day 5, and 89% for days 6 and 7. RH was high at the beginning of healing in order to prevent leaf dehydration and it was gradually decreased in order for the seedlings to get acclimated to lower RH conditions. The growth conditions were monitored using a climate control system (Priva SA, De Lier, The Netherlands).

After 7 days in the healing chamber the grafted seedlings were placed in a glass greenhouse (21.5 °C minimum night temperature, 60 ± 10 µmol m^{-2} s^{-1} PPFD emitted by HPS lamps for an 18 h photoperiod). The high RH was applied to prevent leaf dehydration due to water loss.

2.3. Quality Categorizing and Measurements

Seedling quality categorizing, sampling and measurements were conducted at two times, 7 and 14 days after grafting. Experienced personnel categorized quality. The quality categories are listed in Table 1. Critical parameters for categorizing seedling quality were true leaf and cotyledon area, cotyledon color, and root system development (personal communication with Agris S.A., Kleidi, Imathia, Greece). In total, 50 seedlings per category were sampled and the number of samples was equally distributed throughout the three periods of production

Table 1. Seedling quality categories derived seven and 14 days after grafting of watermelon seedlings.

Days after Grafting	Quality Categories	Marketable
7	Optimum	Yes
	Acceptable	Yes
14	Optimum	Yes
	Acceptable	Yes
	Not Acceptable	No

A digital caliper (Powerfix, Milomex, Pulloxhill, UK) was used to measure shoot height, stem diameter (about 1 cm above the substrate surface), and thickness of true leaves, scion cotyledons or rootstock cotyledons. Leaf area of true leaves, scion cotyledons or rootstock cotyledons were measured using a leaf area meter (LI-3000C, LI-COR Biosciences, Lincoln, NE, USA). Fresh and dry weights of shoots (stem and leaves) and roots were determined. Dry weights were obtained after three days of drying in an oven. Moreover, root-to-shoot (R/S) dry weight ratio, shoot dry weight-to-length (DW/L) ratio, and Dickson's quality index (DQI) were estimated. DQI was calculated as follows [10]:

$$\text{Quality index} = \frac{\text{Seedling total dry weight (g)}}{\dfrac{\text{Height (mm)}}{\text{Stem diameter (mm)}} + \dfrac{\text{Shoot dry weight(g)}}{\text{Root dry weight (g)}}} \tag{1}$$

Relative chlorophyll content was measured using a portable chlorophyll meter (CCM-200 plus, Opti-Sciences, USA). Maximum quantum yield of primary photochemistry, variable to maximal fluorescence of dark-adapted leaves (F_v/F_m), was measured with a fluorometer (Pocket-PEA, Hansatech Instruments, Norflock, UK). Finally, the color of true leaves, scion cotyledons, and rootstock cotyledons was characterized using the colorimetric coordinates lightness (L*), Hue (h°), Chroma (C*), a*/b* (a*: red/green coordinate; b*: yellow/blue coordinate), and Hue (h°) obtained from a digital colorimeter (CR-400 Chroma Meter, Konica Minolta Inc., Tokyo, Japan) according to McGuire. [11].

2.4. Statistical Analysis

Statistical analysis was performed using IBM SPSS software (SPSS 23.0, IBM Corp., Armonk, NY, USA). Data measured at 7 days after grafting were analyzed using a t-test ($P \leq 0.05$), since at that time point the seedlings were grouped into only two quality categories (Optimum and Acceptable). Data measured at 14 days after grafting were analyzed within the methodological frame of one-way analysis of variance (ANOVA), since at that time point the seedlings were grouped at three quality categories (Optimum, Acceptable and not Acceptable). In this case, mean comparisons were conducted using the Scott-Knott procedure [12], at a significance level of $\alpha = 0.05$, using the StatsDirect v.2.8.0. statistical software (StatsDirect, Ltd., Grantchester, Cambridge, UK). The choice of the Scott-Knott method was based on its important and unique characteristic that does not present any overlapping in its grouping results. The above is critical in order to obtain quality indices that segregate the different quality categories and, therefore, overlapping results between groups must not occur.

3. Results

Seven days after grafting, shoots of Optimum seedlings were significantly longer than Acceptable ones (Table 2). However, stem diameter, shoot DW/L ratio, DQI and thickness of leaves and cotyledons (both scion and rootstock) did not show any differences between Optimum and Acceptable seedlings. Relative chlorophyll content of true leaves and cotyledons, as well as F_v/F_m were also similar in the two quality categories tested (Table 2). Colorimetric parameters of true leaves also did not show differences between the two categories, while scion and rootstock cotyledons showed significant differences in parameters such as lightness (L*), Hue (h°), a*/b* (a*: red/green coordinate; b*: yellow/blue coordinate), and h° (Table 3). Moreover, Optimum seedlings developed significantly greater true leaves and scion cotyledons compared to the Acceptable ones, while no differences were observed in rootstock cotyledons (Figure 1A). Similarly, fresh and dry weight production of shoots and roots were significantly greater for Optimum compared to Acceptable characterized seedlings (Figure 2A,B). However, R/S ratio did not exhibit any significant differences between the different categories (Figure 2C).

Table 2. Morphological and developmental parameters of grafted watermelon seedlings from two quality categories 7 days after grafting.

Parameters		Quality Categories	
		Optimum	**Acceptable**
Height (mm)		57.82 ± 1.14 a [y]	47.38 ± 1.03 b
Stem diameter (mm)		4.34 ± 0.04 a	4.23 ± 0.05 a
DW/L [z]		$0.004 \pm {<}0.001$ a	$0.004 \pm {<}0.001$ a
DQI		$0.012 \pm {<}0.001$ a	0.012 ± 0.001 a
F_v/F_m		$0.82 \pm {<}0.01$ a	$0.82 \pm {<}0.01$ a
Thickness (mm)	True leaf	0.62 ± 0.02 a	0.59 ± 0.02 a
	Scion cot.	0.69 ± 0.01 a	0.69 ± 0.01 a
	Roots. cot.	1.20 ± 0.03 a	1.15 ± 0.03 a
Relative chl. content	True leaf	27.17 ± 1.26 a	29.14 ± 1.04 a
	Scion cot.	44.25 ± 1.38 a	43.37 ± 1.22 a
	Roots. cot.	56.50 ± 1.88 a	57.16 ± 2.31 a

[z] shoot dry weight-to-length ratio; DQI: Dickson's quality index; F_v/F_m: maximum quantum yield of primary photochemistry of a dark-adapted leaf; [y] Mean values ($\pm$SE) (n = 50), within a row, followed by different letters are significantly different by t-test ($P \leq 0.05$).

Table 3. Colorimetric parameters of grafted watermelon seedlings from two quality categories 7 days after grafting.

Plant Tissue	Colorimetric Parameters	Quality Categories	
		Optimum	**Acceptable**
True leaves	L* [z]	41.75 ± 1.39 a [y]	42.30 ± 1.96 a
	C*	24.76 ± 4.08 a	24.18 ± 3.69 a
	h°	129.05 ± 1.64 a	128.32 ± 1.62 a
	a*/b*	-0.81 ± 0.05 a	-0.79 ± 0.05 a
Scion cotyledons	L*	42.85 ± 1.53 b	44.12 ± 2.11 a
	C*	20.42 ± 2.31 a	20.61 ± 3.15 a
	h°	128.05 ± 1.21 a	126.54 ± 1.98 b
	a*/b*	-0.78 ± 0.03 b	-0.74 ± 0.05 a
Rootstock cotyledons	L*	39.32 ± 1.87 a	39.87 ± 2.00 a
	C*	19.66 ± 1.79 a	19.69 ± 2.54 a
	h°	130.07 ± 1.11 a	129.26 ± 1.68 b
	a*/b*	-0.84 ± 0.03 a	-0.82 ± 0.05 a

[z] lightness; C*: chroma; h°: hue angle; a*: red/green coordinate; b*: yellow/blue coordinate; [y] Mean values ($\pm$SE) (n = 50), within a row, followed by different letters are significantly different by t-test ($P \leq 0.05$).

Quite similar results were obtained between the quality categories at 14 days after grafting. Specifically, shoot height was significantly greater for Optimum seedlings compared to Acceptable ones, while shoot DW/L ratio was significantly greater for the two marketable categories compared to Not Acceptable seedlings. Stem diameter and DQI were greater for Optimum seedlings compared to the other categories. Nevertheless, thickness of leaves and cotyledons (both scion and rootstock) were similar for all categories. Relative chlorophyll content and F_v/F_m were not different between the three quality categories, similar to the 7 day measurements (Table 4). However, differences between the three categories were observed in colour, particularly parameters such as h° and a*/b* for the true leaf, and C*, L*, h°, and a*/b* for rootstock cotyledons (Table 5). At 7 days after grafting, Optimum scion cotyledons had a darker (lower L*) and greener (lower a*/b*) color compared to Acceptable seedlings. At 14 days after grafting, scion cotyledons did not show any color differences but rootstock cotyledons categorized as Optimum had darker (lower L*) and greener (lower a*/b*) color compared to the rest of the quality categories. True leaves of Optimum seedlings had significantly greater area compared to the other categories, but scion and rootstock leaf area were not different (Figure 1B). Moreover, fresh and dry biomass of shoots and roots, as well as R/S ratio were significantly greater for Optimum seedlings compared to the other categories (Figure 2D–F).

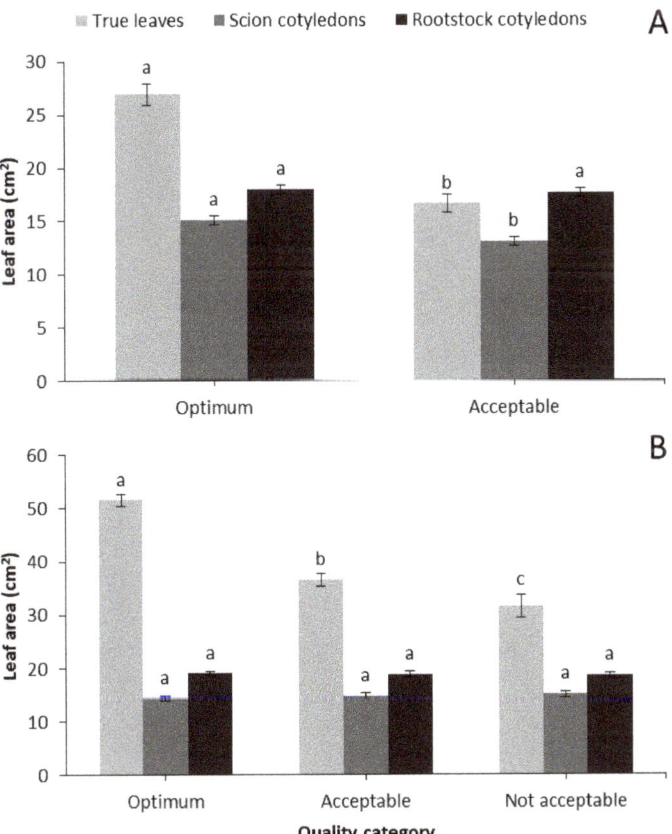

Figure 1. Leaf area of true leaves, scion cotyledons and rootstock cotyledons of grafted watermelon seedlings from the quality categories derived 7 days (**A**) or 14 days (**B**) after grafting. Each data point is a mean value ± standard error (SE) of the mean (n = 50). Bars of the same color (same tissue type) across categories with different letters are significantly different ($P \leq 0.05$) according to the results of the t-test (**A**) or the Scott-Knott method (**B**).

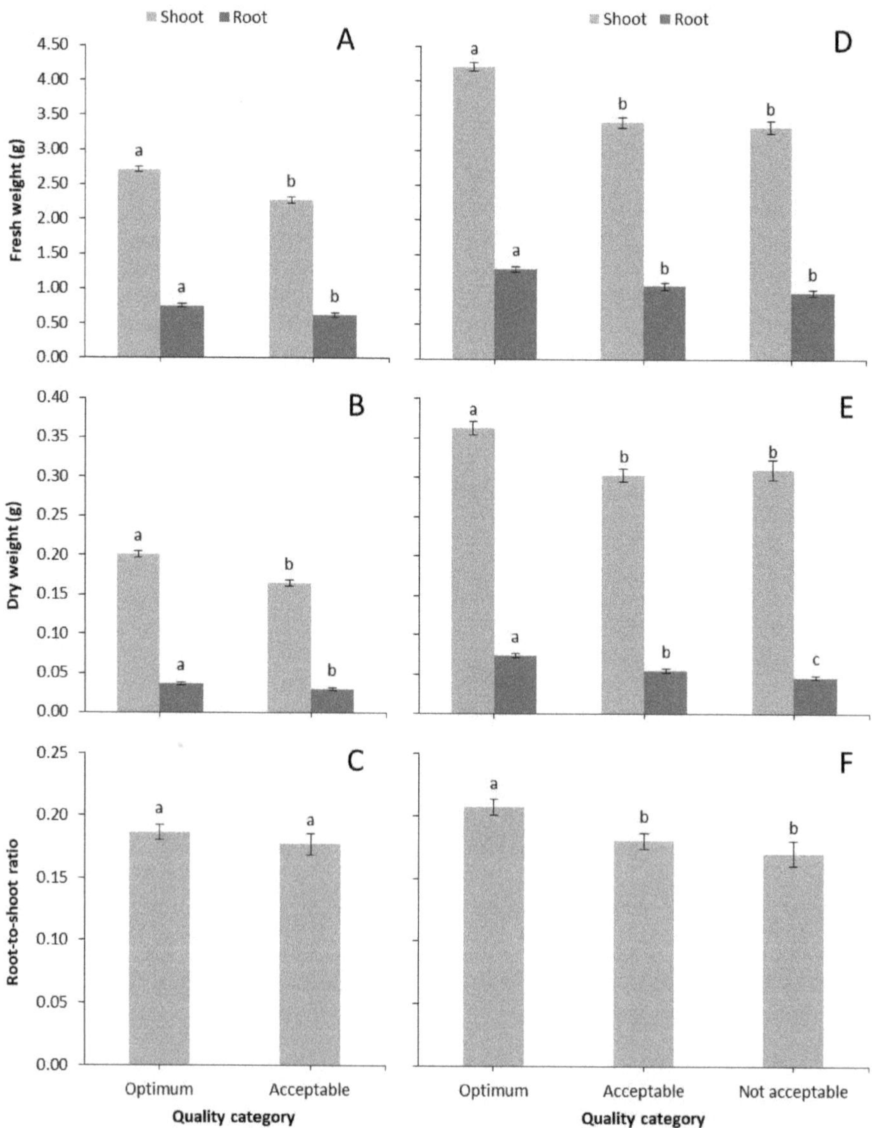

Figure 2. Fresh weight (**A**) and dry weight (**B**) of shoots and roots, and root-to-shoot ratio (**C**) of grafted watermelon seedlings from the quality categories derived 7 days after grafting. Fresh weight (**D**) and dry weight (**E**) of shoots and roots, and root-to-shoot ratio (**F**) of grafted watermelon seedlings from the quality categories derived 14 days after grafting. Each data point is a mean value of 50 observations. Error bars correspond to the standard error (SE) of the mean. Bars of the same colour (same tissue type) across categories followed by different letters are significantly different ($P \leq 0.05$) according to the results of the t-test (**A–C**) or the Scott-Knott method (**D–F**).

Table 4. Morphological and developmental parameters of grafted watermelon seedlings from two quality categories 14 days after grafting.

Parameters		Quality Categories		
		Optimum	Acceptable	Not Acceptable
Height (mm)		59.10 ± 1.19 a [y]	52.00 ± 1.20 b	56.80 ± 1.59 a
Stem diameter (mm)		4.65 ± 0.05 a	4.49 ± 0.07 b	4.35 ± 0.05 c
DW/L [z]		$0.006 \pm {<}0.001$ a	$0.006 \pm {<}0.001$ a	$0.005 \pm {<}0.001$ b
DQI		0.025 ± 0.001 a	0.021 ± 0.001 b	0.018 ± 0.001 c
F_v/F_m		$0.84 \pm {<}0.01$ a	$0.84 \pm {<}0.01$ a	$0.84 \pm {<}0.01$ a
Thickness (mm)	True leaf	0.69 ± 0.03 a	0.62 ± 0.02 a	0.64 ± 0.02 a
	Scion cot.	0.72 ± 0.01 a	0.73 ± 0.01 a	0.74 ± 0.01 a
	Roots. cot.	1.17 ± 0.03 a	1.12 ± 0.02 a	1.18 ± 0.03 a
Relative chl. content	True leaf	32.60 ± 1.13 a	31.91 ± 1.40 a	31.76 ± 1.17 a
	Scion cot.	27.56 ± 1.10 a	30.32 ± 1.12 a	29.21 ± 1.48 a
	Roots. cot.	51.11 ± 1.98 a	46.51 ± 2.45 a	50.16 ± 2.98 a

[z] shoot dry weight-to-length ratio; DQI: Dickson's quality index; F_v/F_m: maximum quantum yield of primary photochemistry of a dark-adapted leaf; [y] Mean values ($\pm$SE) (n = 50), within a row, followed by different letters are significantly different ($P \leq 0.05$) according to the results of the Scott-Knott method.

Table 5. Colorimetric parameters of grafted watermelon seedlings from two quality categories derived 14 days after grafting.

Plant Tissue	Colorimetric Parameters	Quality Categories		
		Optimum	Acceptable	Not Acceptable
True leaves	L* [z]	40.00 ± 0.46 a [y]	41.26 ± 0.49 a	41.30 ± 0.53 a
	C*	19.39 ± 0.51 a	20.73 ± 0.53 a	21.23 ± 0.67 a
	$h°$	129.40 ± 0.42 b	129.40 ± 0.42 b	128.56 ± 0.48 b
	a*/b*	-0.86 ± 0.01 b	-0.83 ± 0.01 a	-0.80 ± 0.01 a
Scion cotyledons	L*	44.70 ± 0.47 a	43.87 ± 0.42 a	44.89 ± 0.39 a
	C*	23.29 ± 0.69 a	23.34 ± 0.55 a	23.74 ± 0.51 a
	$h°$	126.54 ± 0.41 a	126.35 ± 0.37 a	126.20 ± 0.30 a
	a*/b*	-0.74 ± 0.01 as	-0.74 ± 0.01 a	-0.73 ± 0.01 a
Rootstock cotyledons	L*	40.52 ± 0.43 b	41.92 ± 0.43 a	42.53 ± 0.59 a
	C*	19.47 ± 0.43 b	20.92 ± 0.46 a	22.10 ± 0.71 a
	$h°$	129.46 ± 0.29 a	128.07 ± 0.33 b	127.82 ± 0.40 b
	a*/b*	-0.82 ± 0.01 b	-0.79 ± 0.01 a	-0.78 ± 0.01 a

[z] lighting; C*: chroma; $h°$: hue angle; a*: red/green coordinate; b*: yellow/blue coordinate; [y] Mean values ($\pm$SE) (n = 50), within a row, followed by different letters are significantly different ($P \leq 0.05$) according to the results of the Scott-Knott method.

4. Discussion

Seedling quality is one of the major concerns among farmers, and grafted watermelon seedlings are mainly produced by professional nurseries instead of individual farmers. Many factors influence quality evaluation and, therefore, it is difficult to define and categorize seedlings of different qualities. Seedlings of high-quality should have uniformity in terms of size and traits [3].

During healing, grafted seedlings remained in an environmentally controlled growth chamber where microclimate was almost identical for all seedlings. Nevertheless, two quality categories derived 7 days after grafting (i.e., after healing). The seedlings from both categories showed promising potential to develop into marketable plants of high quality. However, the days between the exit from the healing chamber and planting by the grower (i.e., between day 6 and day 14 after grafting) are crucial for maintaining high seedling quality. Many seedlings suffer during the period of acclimatization which might lead to quality deterioration [3]. In our study, three quality categories were developed at 14 days after grafting: two marketable categories (Optimum and Acceptable) and one non-marketable category (Not acceptable).

After 7 days, shoot height was highly distinguishable between the two quality categories, however, it was not confirmed after 14 days. Therefore, shoot height alone cannot be used as an efficient index of seedling quality before going to market. On the other hand, stem diameter was greater for Optimum seedlings at 14 days after grafting. Even though the differences were very slight and could not easily be detected by eye, this parameter is reliable for distinguishing the quality of grafted seedlings before going to market [13–15]. Color differences were not visually detectable, but colorimetry revealed slight distinctions at both times. However, relative chlorophyll content was similar at all measurement dates and quality categories. Additionally, no differences were detected between the quality categories in parameters such as leaf or cotyledon thickness and F_v/F_m, both seven and 14 days after grafting.

Seven days after grafting, one of the greatest morphological parameters that distinguished the Optimum and Acceptable seedlings was the area of true leaves, which leads to greater absorption of incident light in the first very important days of seedling development. Therefore, leaf area is a valuable indicator not only between marketable and not marketable seedlings, but also between the different quality categories of grafted watermelon.

As discussed above, Optimum seedlings were defined by faster leaf development, i.e., larger photosynthetic area in a shorter amount of time. Subsequently, this quality category contained seedlings with greater fresh and dry biomass production compared to the other categories, both 7 and 14 days after grafting, proving that these parameters can be used as index of marketable seedlings. Seven days is considered a short amount of time for grafted seedlings to develop a vigorous root system, especially when the original roots were completely removed, as in our case. However, the quality categories were also distinguished by root biomass.

R/S ratio is a parameter related to the possibility of successful seedling establishment in the field which depends on the proper allocation of biomass between the above and below ground parts. Fourteen days after grafting, R/S ratio was greater for Optimum seedlings which developed a vigorous root system. Shoot DW/L ratio, which is a good indicator of seedling quality [3] revealed that Optimum plants were of higher value. Since this parameter was comprised of shoot dry weight and length (which is similar for the quality categories), biomass accumulation is decisive for the production of high quality seedlings at the nursery. DQI is commonly used for the evaluation of forest or fruit tree seedlings, but recently it has also been employed for assessing horticultural species such as cucumber, muskmelon and tomato [16–18]. Even though the parameter incorporates a number of destructive measurements, it is a useful indicator of seedling quality and plantation performance [19] since its values was correlated with the quality categories in our study.

Optimum seedlings have a better chance of developing into high quality marketable plants, with better establishment in the field, since they excelled in almost all tested parameters, including the essential leaf area and root dry weight. This superiority was also highlighted by shoot DW/L ratio and DQI. Acceptable seedlings were on the border between the other two categories. Parameters, such as stem diameter, shoot DW/L ratio, and DQI, were valuable for identifying them as marketable or not. Not acceptable seedlings after 14 days had inferior development with smaller leaves and considerably weaker root systems compared to the marketable seedlings. Their lower chance of successful establishment and slower development do not favor these seedlings as marketable.

It is concluded that leaf and cotyledon area of scion, stem diameter, shoot and root dry weights as well as shoot DW/L and DQI are good indicators for categorizing grafted watermelon seedlings. Specifically, seedlings of the highest quality must have a leaf area of about 25 cm^2 and 50 cm^2 at 7 and 14 days after grafting, respectively. Scion (watermelon cv. Celine) cotyledons must be fully expanded (about 15 cm^2) 7 days after grafting, while rootstock cotyledons were not good indicators of seedling quality. Shoot height was a weak quality indicator 14 days after grafting, since Not acceptable values were similar to Optimum seedlings. However, stem diameter was a good quality index even though values between the categories were very close. Moreover, shoot and root dry weights, as well as shoot

DW/L ratio and DQI, proved valuable indicators of grafted watermelon seedling quality. The benefits of grafting is associated with the use of high quality watermelon seedlings, and their categorization could help both the nursery industry and growers.

Author Contributions: Conceptualization, methodology, and data analysis: F.B., A.K., A.S., and G.M.; experimental measurements: F.B., C.D. and D.K.; writing—original draft preparation: F.B. and A.K.; writing—review and editing: F.B., A.K, A.S., and G.M., supervision and project administration: A.K.

Funding: This research has been co-financed by the European Union and Greek national funds through the Operational Program Competitiveness, Entrepreneurship and Innovation, under the call RESEARCH–CREATE–INNOVATE (project code: T1EDK-00960, LEDWAR.gr).

Conflicts of Interest: The authors declare no conflict of interest. The founding sponsors had no role in the design of the study; in the collection, analyses, or interpretation of data; in the writing of the manuscript; or in the decision to publish the results.

References

1. Lee, J.-M.; Bang, H.-J.; Ham, H.-S. Grafting of vegetables. *J. Jpn. Soc. Hort. Sci.* **1998**, *67*, 1098–1104. [CrossRef]
2. Davis, A.R.; Perkins-Veazie, P.; Hassell, R.; Levi, A.; King, S.R.; Zhang, X. Grafting Effects on Vegetable Quality. *HortScience* **2008**, *43*, 1670–1672. [CrossRef]
3. Lee, J.-M.; Kubota, C.; Tsao, S.J.; Bie, Z.; Hoyos Echevarria, P.; Morra, L.; Odag, M. Current status of vegetable grafting: Diffusion, grafting techniques, automation. *Sci. Hortic.* **2010**, *127*, 93–105. [CrossRef]
4. Louws, F.J.; Rivard, C.L.; Kubota, C. Grafting fruiting vegetables to manage soilborne pathogens, foliar pathogens, arthropods and weeds. *Sci. Hortic.* **2010**, *127*, 127–146. [CrossRef]
5. Savvas, D.; Colla, G.; Rouphael, Y.; Schwarz, D. Amelioration of heavy metal and nutrient stress in fruit vegetables by grafting. *Sci. Hortic.* **2010**, *127*, 156–161. [CrossRef]
6. Schwarz, D.; Rouphael, Y.; Colla, G.; Venema, J.H. Grafting as a tool to improve tolerance of vegetables to abiotic stresses: Thermal stress, water stress and organic pollutants. *Sci. Hortic.* **2010**, *127*, 162–171. [CrossRef]
7. Kyriacou, M.C.; Rouphael, Y.; Colla, G.; Zrenner, R.; Schwarz, D. Vegetable Grafting: The Implications of a Growing Agronomic Imperative for Vegetable Fruit Quality and Nutritive Value. *Front. Plant Sci.* **2017**, *8*, 741. [CrossRef] [PubMed]
8. Lee, S.G. Production of high quality vegetable seedling grafts. *Acta Hortic.* **2007**, *759*, 169–174. [CrossRef]
9. Lee, J.M.; Oda, M. Grafting of herbaceous vegetable and ornamental crops. In *Horticultural Review*; Janick, J., Ed.; John Wiley & Sons: New York, NY, USA, 2003; pp. 61–124.
10. Dickson, A.; Leaf, A.L.; Hosner, J.F. Quality appraisal of white spruce and white pine seedling stock in nurseries. *For. Chron.* **1960**, *36*, 10–13. [CrossRef]
11. McGuire, R.G. Reporting of objective color measurements. *HortScience* **1992**, *27*, 1254–1255. [CrossRef]
12. Scott, A.J.; Knott, M. A Cluster analysis method for grouping means in the analysis of variance. *Biometrics* **1974**, *30*, 507–512. [CrossRef]
13. Mattsson, A. Predicting field performance using seedling quality assessment. *New For.* **1996**, *13*, 223–248.
14. Rawat, J.S.; Singh, T.P. Seedling indices of four tree species in nursery and their correlations with field growth in Tamil Nadu, India. *Agrofor. Syst.* **2000**, *49*, 289–300. [CrossRef]
15. Davis, A.S.; Jacobs, D.F. Quantifying root system quality of nursery seedlings and relationship to outplanting performance. *New For.* **2005**, *30*, 295–311. [CrossRef]
16. Guisolfi, L.P.; Lo Monaco, P.A.V.; Haddade, I.R.; Krause, M.R.; Meneghelli, L.A.M.; Almeida, K.M. Production of cucumber seedlings in alternative substrates with different compositions of agricultural residues. *Rev. Caatinga* **2018**, *31*, 791–797. [CrossRef]
17. Vendruscolo, E.P.; Campos, L.F.C.; Nascimento, L.M.; Seleguini, A. Quality of muskmelon seedlings treated with thiamine in pre-sowing and nutritional supplementation. *Sci. Agrar.* **2018**, *19*, 164–171.

18. Costa, E.; Leal, P.A.M.; Benett, C.G.S.; Benett, K.S.S.; Salamene, L.C.P. Production of tomato seedlings using different substrates and trays in three protected environments. *Eng. Agríc.* **2012**, *32*, 822–830. [CrossRef]
19. Bayala, J.; Dianda, M.; Wilson, J.; Ouedraogo, S.J.; Sanon, K. Predicting field performance of five irrigated tree species using seedling quality assessment in Burkina Faso, West Africa. *New For.* **2009**, *38*, 309–322. [CrossRef]

 *horticulturae*

Article

Evaluation of Two Wild Populations of Hedge Mustard (*Sisymbrium officinale* (L.) Scop.) as a Potential Leafy Vegetable

Marta Guarise [1,*], Gigliola Borgonovo [2], Angela Bassoli [2] and Antonio Ferrante [1]

[1] Department of Agriculture and Environmental Science—Production, Landscape, Agroenergy, Università degli Studi di Milano, 20133 Milano, Italy; antonio.ferrante@unimi.it

[2] Department of Food, Environmental and Nutritional Science, Università degli Studi di Milano, 20133 Milano, Italy; gigliola.borgonovo@unimi.it (G.B.); angela.bassoli@unimi.it (A.B.)

* Correspondence: marta.guarise@unimi.it; Tel.: +39-0250-316593

Received: 5 October 2018; Accepted: 11 January 2019; Published: 1 February 2019

Abstract: The minimally processed industry is always looking for produce innovation that can satisfy consumer needs. Wild leafy vegetables can be a good source of bioactive compounds and can be attractive for the consumer in term of visual appearance and taste. In this work, *Sisymbrium officinale* (L.) Scop., commonly called hedge mustard, was grown in a greenhouse and evaluated as a potential leafy vegetable. Two wild populations, Milano (MI) and Bergamo (BG), were grown in peat substrate and harvested at the commercial stage for the minimally processing industry. Leaf pigments such as chlorophyll and carotenoids were determined as well as chlorophyll *a* fluorescence parameters. Total sugars, antioxidant compounds such as ascorbic acid, phenolic index, total phenols, anthocyanins, and nitrate were determined at harvest. Significant differences between wild populations were found in April with higher nitrate content in BG, 2865 mg/kg FW than in MI, 1770 mg/kg FW. The nitrate levels of *S. officinale* measured in the present study are significantly lower than the maximum NO_3 level allowed in other fresh leafy vegetables. Ascorbic acid measured in November was higher in MI compared BG with values of 54.4 versus 34.6 mg/100 g FW, respectively. The chlorophyll *a* fluorescence data showed that BG reached optimal leaf functionality faster than MI. Overall results indicated that *Sisymbrium officinale* (L.) Scop. can be suggested as a potential leafy vegetable for the minimally processed industry.

Keywords: *Sisymbrium officinale*; *Brassicaceae*; hedge mustard; leafy vegetables

1. Introduction

Minimally processed leafy vegetable production has been evolving in recent years by providing new produce with beneficial effect on human health. There are several wild species that can be considered as potential leafy vegetables. The introduction of new species can be useful for diet enrichment and diversification. Moreover, wild plants can be highly adaptative to different environments.

Sisymbrium officinale (L.) Scop., synonym *Erysimum officinale*, commonly known as hedge mustard in English, *erísimo* in Spanish, *erisimo* or *erba cornacchia* in Italian, and *velar* in French, is a medicinal plant that belongs to the Brassicaceae family. This species could have potential for introduction into the leafy vegetable production for the minimally processed or fresh-cut industry.

S. officinale is a terophyte scapose plant with a reddish-violet erect trunk, that present a lot of trichomes and many branches. Basal leaves are different from the upper ones with a dentate shape. Hedge mustard has a linear racemose inflorescence; each flower has four small (1–2 mm) yellow petals; the fruit is a tiny siliqua, close-fitting to the trunk. Flowering occurs in Spring–Summer, from May to

July–August, depending on the climate. Siliqua pods usually are pubescent, once they reach maturity they release seeds. Seeds are very small, each siliqua can contain from 10 to 20 seeds. *S. officinale* is endemic in the Eurasian continent and widespread in all Italian regions from 0 to 1000 m. above sea level (a.s.l.), and rarely up to 2400 m a.s.l. [1]. This annual or biennial herbaceous plant is described as ruderal, growing on disturbed sites such as field margins and roadsides [2].

Flowers and leaves of hedge mustard are commonly used as a traditional medicinal herb for the treatment of sore throats, coughs, and hoarseness [3–5] under specific indication based upon long-standing use [6] and recent clinical studies [7]. For that reason, *S. officinale* is largely known as "singer's plant" and is used among singers, actors, and professionals who use the voice for working. The therapeutic activity of this plant is attributed to its sulfurated components. Dried flowering aerial parts contain: total glucosinolates (0.63–0.94%), mucilage (13.5–10.9%), total thiols (8.9–10.2%), and total flavonoids (0.50–0.56%). The main glucosinolate in *S. officinale* is glucoputranjivine [8]. It represents 58.3% of total glucosinolates on a fresh weight basis. This percentage declined to 32.5% after autolysis [6].

Brassicaceae is one of the most important botanical families in horticultural production in Mediterranean countries, due to their great diversity expressed both in spontaneous and cultivated species. In Italy, horticultural Brassicaceae are widespread on about 40,000 ha, in particular in the center-southern region [9].

In spite of its long, traditional therapeutic use for treating voice discomfort as dried plants (including leaves, stem, and flowers) for preparing decoctions, tinctures, or propolis, *S. officinale* has barely been investigated for its beneficial proprieties, and there are no data about its possible use and consumption as a fresh leafy vegetable. Its low agronomic requirements allow the cultivation in different Mediterranean environments.

In order to evaluate the possibility of recommending this species as a potential leafy vegetable, the aim of present study was to investigate production of two different wild populations, one collected in Milan and the other one in Bergamo, Italy. Cultivation was performed in pots containing fertilized substrate in a greenhouse. Total chlorophyll content and chlorophyll *a* fluorescence were measured for evaluating photosynthetic activity. The most common quality parameters that are usually considered for leafy vegetable evaluation were determined such as ascorbic acid, carotenoids, phenols, anthocyanins, nitrates, and total sugar. Furthermore, to evaluate the production of leaves, fresh and dry biomass were measured at the baby leaf stage which is usually the developmental stage for leafy vegetables destined to the fresh-cut industry.

2. Materials and Methods

2.1. Plant Material

Seeds of two wild populations of *Sisymbrium officinale* (L.) Scop., or hedge mustard, respectively named MI (Milan) and BG (Bergamo), obtained from controlled seed reproduction at Fondazione Minoprio (Como, Italy) during summer in 2017, were sown separately in polystyrene panels using common horticultural fertilized substrate under controlled conditions in a greenhouse at the Faculty of Agricultural and Food Science of Milan, 16 January 2018, for the first evaluation, and 8 October, for the second evaluation (Supplementary Materials Table S1). Cultivation was performed in the greenhouse of the Agricultural Faculty, which was a single gable covered with glass and provided with a cooling system and supplemental light only for the second growing cycle. The supplemental lighting was provided for 16 h from the 7:00 a.m. to 11:00 p.m. with 400 W/m^2 High Pressure Sodium lamps. The environmental parameters are reported in the Supplementary Materials Figure S1.

Plantlets were transplanted and grown in complete substrate (Vigorplant, Italy) containing the following components: 21% Baltic peat, 22% dark peat, 26% Irish peat, 13% volcanic peat, 18% calibrated peat at a pH of 6.5 in 10 cm diameter plastic pots. The plant density was 16–18 plants/pot or 80–90 plant/m^2.

Harvest was performed at the end of each cultivation cycle, on 3 May, and on 7 November, when the plant reached the commercial baby leaf stage, which corresponded to plants at a 15 cm height with 4–6 fully expanded true leaves. Plants were randomly chosen from each pot and sampled for the analyses. Plants were not supplied with extra-fertilizers in either experimental period and were watered every day to maintain optimal water availability.

2.2. Non-Destructive Analyses

Chlorophyll a Fluorescence

For the characterization of the two *S. officinale* wild populations, non-destructive analyses were conducted on fresh leaf tissue. Each week, starting from 18 of April to the first week of May 2018 and 7 November, chlorophyll *a* fluorescence was measured using a hand-portable fluorometer (Handy PEA, Hansatech, Kings Lynn, United Kingdom). Leaves were dark-adapted for 30 minutes using leaf clips. After this time, a rapid pulse of high-intensity light of 3000 μmol m^{-2} s^{-1} (600 W m^{-2}) was administered to the leaf inducing fluorescence. Fluorescence parameters were calculated automatically by the device: variable fluorescence to maximum fluorescence (Fv/Fm). From the fluorescence parameters, JIP analyses were performed to determine the following indices: Performance Index (PI), dissipation energy per active reaction center (DIo/RC), and density of reaction centers (RC/CSm).

2.3. Destructive Analyses

To evaluate qualitative characteristics of the two wild populations of *S. officinale*, small samples of fresh leaves, about 1 g for each sample, were sampled one month after transplanting and at the beginning of May for the first cultivation cycle, and on 7 November for the second cycle. Some leaf samples were immediately stored at $-20\,^{\circ}$C to prevent tissue degradation.

To evaluate the yield at the baby leaf stage, plant fresh weight and dry weight were recorded at the end of biological cycle, in May for the first cultivation cycle and in November for the second cycle.

2.3.1. Chlorophyll and Carotenoids

Chlorophyll and carotenoids were extracted from fresh leaves. Leaf disks of 5 mm diameter (or 20–30 mg) in 5 mL 99.9% methanol as solvent were kept in a dark cold room at 4 $^{\circ}$C for 24 h. Quantitative chlorophyll determinations were carried out immediately after extraction. Absorbance readings were measured at 665.2 and 652.4 nm for chlorophyll pigments and 470 nm for total carotenoids. Chlorophylls and carotenoid concentrations were calculated by Lichtenthaler's formula [10].

2.3.2. Phenolic Index, Total Phenols, and Anthocyanins

For the following analyses, fresh leaf tissue (disks of 5 mm diameter, or 20–30 mg) was extracted in 3 mL 1% methanolic HCl. The Phenolic Index of leaf tissue was determined spectrophotometrically by direct measurement of leaf extract absorbance at 320 nm. After overnight incubation the supernatant was read at 320 nm. The values were expressed as ABS$_{320nm}$/g FW.

Total phenols were determined spectrophotometrically following the Folin-Ciocalteu reagent method [11] using 200 μL of each sample extract 7.8 mL of distilled water, 0.5 mL of Folin-Ciocalteu reagent and 1.5 mL of 20% Na$_2$CO$_3$. Samples were extracted for 2 h in the dark and then read at 760 nm. Total phenols were calculated using a standard curve performed with gallic acid.

Anthocyanin content was determined spectrophotometrically. Sample extracts were incubated overnight at 4 $^{\circ}$C in darkness. The concentration of cyanidin-3-glucoside equivalents was determined spectrophotometrically at 535 nm using an extinction coefficient (ε) of 29,600 [12].

2.3.3. Ascorbic Acid Determination

Only the November samples were analyzed for ascorbic acid content. For analysis, about 1 g of frozen leaves (frozen at -80 °C) were homogenized in a mortar with 1.3 mL of cold 6% (*w/v*) metaphosphoric acid and centrifuged at 10.000 $\times$ *g* at 4 °C. The pellet obtained by centrifugation was washed with 1.06 mL of cold metaphosphoric acid solution and centrifuged again. The supernatants were combined and 6% metaphosphoric acid was added to make a final volume of 3.3 mL.

After filtration through nylon filter, a 10 µL sample aliquot was injected onto an Inertsil ODS-3 GL Science column at 20 °C attached to a Series 200 LC pump. Peaks were converted to concentrations by using the dilution of stock ascorbic acid to construct a standard curve. Chromatographic data were stored and processed with a PerkinElmer TotalChrom 6.3 data Processor (PerkinElmer, Norwalk, CT, USA) [13].

2.3.4. Nitrate Determination

Nitrate concentration was measured by the salicyl sulfuric acid method [14]. About 1 g of fresh leaves was ground in 5 mL of distilled water. The extracts were centrifuged at 4000 rpm for 15 min. After centrifugation, the supernatant was collected for colorimetric determinations. Twenty µL of sample were collected and 80 µL of 5% (*w/v*) salicylic acid in concentrated sulfuric acid were added. After the reaction, 3 mL of NaOH 1.5 N were added. Each sample was cooled, and absorbance was measured at 410 nm. Nitrate concentration was calculated referring to a KNO_3 standard calibration curve.

2.3.5. Total Sugar Determination

To determine total sugar levels, extracts were prepared as above for the determination of nitrate levels. Total sugars were determined using the anthrone assay [15] with slight modification. The anthrone reagent was prepared using 0.1 g of anthrone dissolved in 50 mL of 95% H_2SO_4. The reagent was left 40 min before use; then, 200 µL of extract was added to 1 mL of anthrone, put in ice for 5 min and vortexed. The reaction was heated at 95 °C for 5 min. Samples were cooled and absorbance was read at 620 nm. Total sugar concentration was calculated referring to a glucose standard calibration curve.

2.4. Statistical Analyses

Data from the first cultivation cycle were subjected to two-way ANOVA and differences among means were determined using Tukey's post-test ($P < 0.05$). Data from the second cultivation cycle were analyzed using a t-test ($P < 0.05$). The number of replicate samples used in each analysis or measurement is reported in the legend of the figures or tables.

3. Results

3.1. Total Chlorophylls, Carotenoids, Phenols, and Anthocyanins

The yield at the baby leaf stage, a 15 cm height and 4–6 leaves, for both wild populations ranged from 2.41 g to 4.00 g for the winter-spring season and from 2.26 g to 2.45 g for the autumn season. The dry weight percentage ranged from 10.64 to 10.76 in May and from 8.07 to 7.87 in November. No significant differences were found in FW or DW between values of MI and BG harvested in May and in November (Table 1).

The plants showed different leaf pigment contents at different sampling times. In May a higher chlorophyll content was observed in MI than in April, while no significant differences between MI and BG were found in either cycle (Table 2).

Table 1. Fresh weight and % of dry matter in MI and BG hedge mustard wild populations at the baby leaf stage after two production cycles. Data are expressed as means of five plants (*n* = 5).

Cycle		Wild Population	Fresh Weight (g/plant)		Dry Matter (%)	
I	May	MI	2.41	ns	10.64	ns
		BG	4.00		10.76	
II	November	MI	2.26	ns	8.07	ns
		BG	2.45		7.87	

Data were analyzed using a *t*-test (*P* < 0.05). ns means no statistical differences.

Table 2. Chlorophyll (a, b) and total carotenoid content in MI and BG hedge mustard wild populations after two production cycles. Data are reported as mg/g FW (*n* = 4).

Cycle		Wild Population	Chl a (mg/g FW)	Chl b (mg/g FW)	Total Carotenoids (mg/g FW)	
I	April	MI	0.63 b	0.19 b	0.120	
		BG	0.93 ab	0.37 ab	0.143	ns
	May	MI	1.22 a	0.42 a	0.215	
		BG	0.79 ab	0.27 ab	0.157	ns
II	November	MI	1.56	0.51	0.382	
		BG	1.44 (ns)	0.47 (ns)	0.379	ns

Data of cycle I were subjected to two-way ANOVA and differences among wild populations and dates within a cycle were determined using Tukey's test (*P* < 0.05). Data of cycle II were analysed using a *t*-test (*P* < 0.05). ns indicates no statistical differences.

In cycle I in April the total phenols and anthocyanins did not differ between populations. In May higher anthocyanins values were observed in BG compared to MI, while no significant differences were observed for total phenol among the two wild populations. Significant differences were found in phenolic index between the two wild populations (MI and BG) only in November with a higher concentration for BG. In November ascorbic acid determination revealed significant differences between MI and BG with a higher concentration for MI (Table 3). Among the antioxidant compounds, ASA and phenolics showed opposite differences between the two *S. officinale* populations cultivated in November, ASA higher and lower phenolic index in MI.

Table 3. Ascorbic acid (ASA), phenolic index, total phenols, and anthocyanin content in MI and BG hedge mustard wild populations (*n* = 4).

Cycle		Wild Population	ASA (mg/100 g FW)	Phenolic Index (ABS$_{320nm}$/g FW)		Total Phenols		Anthocyanins (mg/100 g FW)	
I	April	MI	-	27.45		1.13		32.13 ab	
		BG	-	26.72	ns	1.09	ns	26.38 b	
	May	MI	-	19.05		0.75		14.88 c	
		BG	-	32.60	ns	1.35	ns	38.79 a	
II	November	MI	54.45	42.51		1.62		29.53	
		BG	34.57 (*)	55.44 (*)		2.37	ns	31.51	ns

Data of cycle I were subjected to two-way ANOVA and differences among wild populations and dates within cycles were determined using Tukey's test (*P* < 0.05). Data of cycle II were analysed using a *t*-test (*P* < 0.05). An asterisk (*) indicates a significant difference and ns indicates no statistical differences.

3.2. Chlorophyll a Fluorescence Measurements

From the chlorophyll *a* fluorescence data, four parameters were considered: Fv/Fm (maximum quantum yield of PSII), PI (Performance Index), DI0/RC (rate of energy dissipated by PSII per reaction center), and RC/CSm (active RCs per excited cross-section). No significant differences were found between the two wild populations for Fv/Fm and DIo/RC parameters. The Fv/Fm ratio after 26 April in both wild populations showed values higher than 0.83.

PI did not show significant differences between wild populations during cycle I. Significant differences were found between BG measured on 18 April and 2 May. PI index increased from 18 April to 2 May. Then, it decreased by the following measurement on 5 May but remained higher than the initial measurement.

Like the Fv/Fm ratio, the DIo/RC values did not show significant differences between wild populations and measurement times. However, this index declined in both wild populations during cultivation.

Significant differences were found for RC/CSm in MI between 18 April and 2 May and between 26 April and 2 May. In addition, significant differences were found in the BG wild population between 18 April and 26 April (Figure 1).

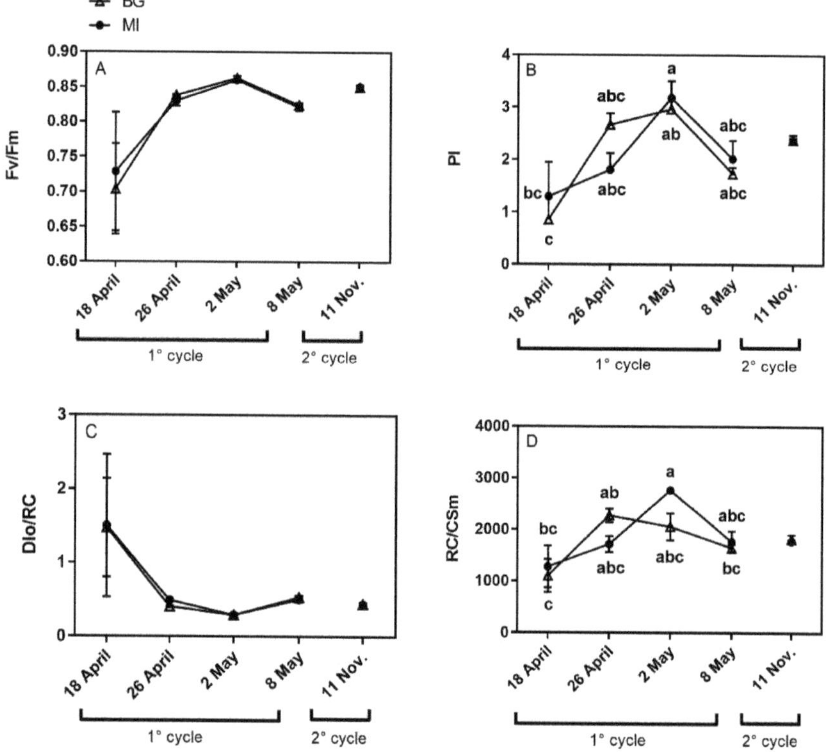

Figure 1. Chlorophyll *a* fluorescence parameters (Fv/Fm (**A**), PI (**B**), DIo/RC (**C**), and RC/CSm (**D**)) in leaves of two hedge mustard wild populations, MI and BG. Data are means with standard errors (n = 4 for April and May, n = 5 for November). Data of cycle I were subjected to two-way ANOVA and differences among wild populations and dates within cycles were determined using Tukey's test (P < 0.05). Data of cycle II were analysed using a *t*-test (P < 0.05). Different letters indicate statistical differences, and no letters indicate no significant differences.

3.3. Nitrate and Total Sugars

The quality of hedge mustard grown as baby leaf vegetables was also evaluated in terms of nitrate accumulation and total sugars. The nitrate content was statistically different between wild populations in April, with higher nitrate content in BG, 2865 mg/kg FW than in MI (Figure 2). In May, the two wild populations showed lower and significantly different values compared with those measured in April, nitrate content ranged from 199 to 256 mg/kg, but there was no difference between them. In November, both wild populations showed a nitrate content that ranged from 1756 mg/kg FW to 1683 mg/kg FW, with no significant differences between them (Figure 2).

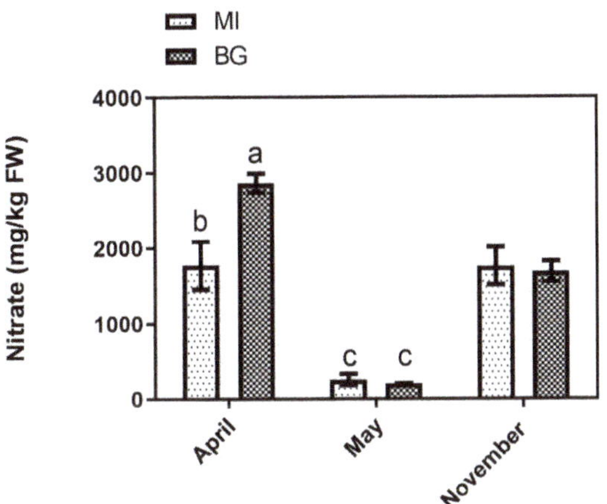

Figure 2. Nitrate content in leaves of two hedge mustard wild populations, BG and MI. Data are means with standard errors ($n = 3$ for April and May, $n = 4$ for November). Data of cycle I were subjected to two-way ANOVA and differences among wild populations and dates within cycles were determined using Tukey's test. Data of cycle II were analysed using a *t*-test ($P < 0.05$). Different letters indicate statistical differences for $P < 0.05$. Population means did not differ in November.

Total sugars were not statistically different between wild populations (MI and BG) or by dates within cycle I (April, May). The total sugar content ranged from to 5.84 mg Glu eq./g FW observed in MI in November, to 3.1 mg Glu eq./g FW in the April (Figure 3).

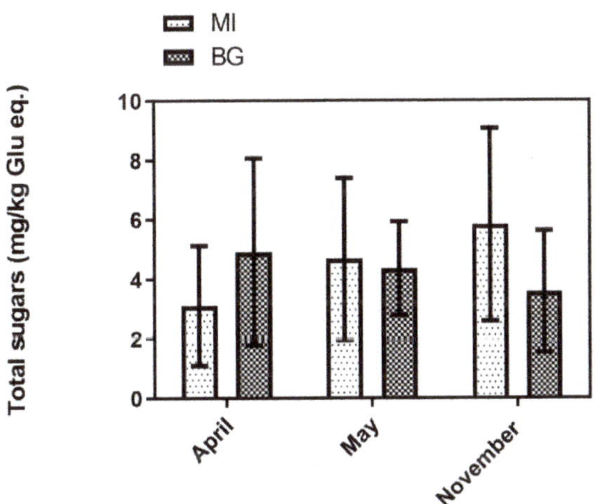

Figure 3. Total sugar content in leaves of two wild hedge mustard populations, BG and MI. Data are means with standard errors ($n = 3$ for April and May, $n = 4$ for November). Data of cycle I were subjected to two-way ANOVA. Data of cycle II were analysed using a *t*-test ($P < 0.05$). There were no significant differences between wild populations or times.

4. Discussion

The Brassicaeae family includes a wide number of species that can be used for vegetable production. Some of them have been described as potential vegetables and sources of antioxidant compounds for the Mediterranean area [16,17]. Hedge mustard is a wild Brassicaceae species widely dispersed and, therefore, it has been evaluated as a potential leafy vegetable for the minimally processed industry. The nutritional components were similar to other leafy vegetables and could provide a good quantity of ascorbic acid that is higher than lettuce, which shows values ranging from 10 to 30 mg/100 g [18]. The leaf pigments observed in hedge mustard were similar to other leafy vegetables such as rocket (*Eruca vesicaria* subsp. *sativa*) [19], lamb's lettuce (*Valerianella locusta*) [20], and lettuce (*Lactuca sativa*) [21]. It is well known that leaf pigments are important parameters because they contribute to leaf color and visual appearance [22]. The leaf color is very important in minimally processed leaf vegetables because it is the first quality parameter that consumers evaluate at purchase.

During the growing period lower chlorophyll concentrations were observed in spring than in autumn. Usually the leaf pigments are higher at lower light intensity; in our experimental conditions the higher values could be due to the low light conditions in the greenhouse during autumn, even though supplementary lighting was provided. This relationship between lower light availability and higher pigments have been found in different leafy vegetables [23].

Chlorophyll *a* fluorescence-derived parameters were used for evaluating the PSII activity of the two *S. officinale* wild populations (MI and BG). The Fv/Fm ratio indicated the maximal efficiency of PSII photochemistry, and it did not significantly change during the experiments. Values of Fv/Fm below 0.83 are usually considered as indicative of stressful conditions in plants [24]. The Fv/Fm ratio increased at the end of April, while the values were slightly lower at the earlier measurement. This result may be due to Fv/Fm increasing with leaf development until reaching the fully expanded stage, when the leaves are fully photosynthetically active. Values above 0.83 during May demonstrated that plants were under optimal growing conditions as was also observed in the second cycle performed in autumn. The higher values of PI in May also indicated higher light use efficiency and better performance of the plants [23]. The BG wild population seemed to have had faster adaptation and reached optimal leaf functionality earlier. In fact, BG also had higher RC/CSm values in April and these results were repeated in the second cycle performed in November

Nitrate plays a crucial role in the nutrition and function of plants and naturally occurs as a compound in the nitrogen cycle. In plants, nitrate levels are higher in leaves, whereas lower levels occur in tubers and seeds. In fact, leaf crops such as spinach (*Spinacea oleracea* L.), lettuce and rocket have high nitrate concentration [25]. Nitrate and nitrite are also commonly used as preservatives in food. Nitrate is non-toxic, but its metabolites and reaction products (nitrite, nitric oxide, and N-nitroso compounds) could be dangerous for human health inducing methaemoglobinaemia or carcinogenesis [26]. In addition to nitrate, leafy vegetables also provide several bioactive compounds with beneficial effects on health, and are widely recommended in the diet.

The European Union, in order to limit the nitrate supply in human nutrition, has defined the maximum nitrate levels permitted in some vegetables considered to have the highest levels of this compound. Nitrate concentrations are directly correlated with light availability. This may explain the lower values in May and higher values in April and November. However, the nitrate levels of *S. officinale* measured in the present study are significantly lower than the maximum NO_3 level allowed in rocket salad, another species from *Brassicacea* family, fixed at 6000 or 7000 mg NO_3/kg FW by the UE Commission [27], depending on the harvest time. Rocket plants grown in different cultivation systems in greenhouses such as soil, substrate or floating, have higher nitrate levels [28,29]. However, the higher ascorbic acid concentration compared to other leafy salad contributes a reduced risk of nitrosamine formation and carcinogenic effects of the nitrate in the diet. In fact, it has been reported that ascorbic acid is a nitrosation inhibitor and could inhibit nitrate reduction [30]. The low nitrate content of *S. officinale* is a good quality trait for possible use of this species as a leafy vegetable. Moreover, under greenhouse cultivation, the level of ascorbic acid is similar or slightly higher than those found in plants grown in the wild [31].

The leaf sugar content is related to photosynthetic activity and biomass production. It is also an important parameter for the storage of the product. Higher sugar content can be potentially associated with higher shelf life, because sugars are used for the basal metabolism and maintaining quality of the product. The total sugar content measured in the present study (maximum level of 5.84 mg/g) is lower if compared to the average sugar content observed in rocket leaves, where the total sugar reached 6.30 mg/g in the first harvest and 7.61 mg/g in second harvest [32]. The total sugars were similar among the two wild populations which also corresponds to biomass production and dry matter percentage.

5. Conclusions

Sysimbrium officinale is a wild species from the *Brassicaceae* family, quite common in all temperate Euroasiatic areas. Our results indicated that this species can be successfully grown in a greenhouse with nutritional components as well as quality parameters such as nitrate, chlorophyll, and sugar content similar to the most common commercial leafy vegetables. It has a good concentration of ascorbic acid, higher than common leafy vegetables. Although these results suggest that *Sysimbrium officinale* can be grown as a leafy vegetable, further investigation will be required for evaluating quality during postharvest storage and handling.

Supplementary Materials: The following are available online at http://www.mdpi.com/2311-7524/5/1/13/s1, Figure S1: Temperature minimum and maximum and solar radiation (RG) during the cultivation period, Table S1: Cultivation period including sowing, transplanting, sampling, and harvest time.

Author Contributions: The authors contributed to the work as follows: conceptualization: A.F. and A.B.; methodology: M.G.; formal analysis: M.G.; investigation: M.G.; resources: G.B.; data curation: M.G.; writing—original draft preparation: M.G., A.F.; writing—review and editing: A.F., A.B.; supervision: A.F.; project administration: A.B.; funding acquisition: A.B.

Funding: This research was funded by Fondazione Cariplo, ERISIMO A MILANO project, no. 2017-1653.

Conflicts of Interest: The authors declare no conflict of interest.

References

1. Pignatti, S. *Flora d'Italia*; Edizioni Edagricole: Bologna, Italy, 1982; Volume 2, p. 377.
2. Bouwmeester, H.J.; Karseen, C.M. Annual changes in dormancy and germination in seeds of Sisymbriumofficinale (L.) Scop. *New Phytol.* **1993**, *124*, 179–191. [CrossRef]
3. Politi, M.; Braca, A.; Altinier, G.; Sosa, S.; Ndjoko, K.; Wolfender, J.L.; Hostettmann, K.; Jimenez-Barbero, J. Different approaches to study the traditional remedy of "hierba del canto" Sisymbriumofficinale (L.) Scop. *Bol. Latinoam. Caribe Plantasmedicinales Aromàt.* **2008**, *7*, 30–37.
4. Blažević, I.; Radonić, A.; Mastelić, J.; Zekić, M.; Skočibusić, M.; Maravić, A. Hedge Mustard (Sisymbriumofficinale): Chemical diversity of volatiles and their antimicrobial activity. *Chem. Biodivers.* **2010**, *7*, 2023–2034. [CrossRef] [PubMed]
5. Di Sotto, A.; Vitalone, A.; Nicoletti, M.; Piccin, A.; Mazzanti, G. Pharmacological and phytochemicalstudy on Sisymbriumofficinale Scop. Extract. *J. Ethnopharmacol.* **2010**, *127*, 731–736. [CrossRef] [PubMed]
6. EMA—European Medicines Agency. *Assessment report on Sisymbriumofficinale (L.) Scop., Herba*; EMA: London, UK, 15 January 2014.
7. Calcinoni, O. Sisymbrium "Singers' Plant" Efficacy in Reducing Perceived Vocal Tract Disability. *J. Otolaryngol. Ent Res.* **2017**, *8*, 00243. [CrossRef]
8. Carnat, A.; Fraisse, D.; Carnat, A.P.; Groubert, A.; Lamaison, J.L. Normalization of hedge mustard, Sisymbriumofficinale L. *Ann. Pharm. Fr.* **1998**, *56*, 36–39.
9. Pardossi, A. *Orticoltura. Principi e Pratica*; Edizioni Edagricole: Bologna, Italy, 2018; pp. 261–276.
10. Lichtenthaler, H.K. Chlorophylls and carotenoids: Pigmentsof photosynthetic membranes. *Methods Enzymol.* **1987**, *148*, 350–382.

11. Singleton, V.L.; Orthofer, R.; Lamuela-Raventos, R.M. Analysis of total phenols and other oxidation substrates and antioxidants by means of folin-ciocalteu reagent. *Methods Enzymol.* **1999**, *299*, 152–178.

12. Giusti, M.M.; Rodríguez-Saona, L.E.; Wrolstad, R.E. Molar absorptivity and color characteristics of acylated and non-acylated pelargonidin-based anthocyanins. *J. Agric. Food Chem.* **1999**, *47*, 4631–4637. [CrossRef] [PubMed]

13. Rizzolo, A.; Brambilla, A.; Valsecchi, S.; Eccher-Zerbini, P. Evaluation of sampling and extraction procedures for the analysis of ascorbic acid from pear fruit tissue. *Food Chem.* **2002**, *77*, 257–262. [CrossRef]

14. Cataldo, C.A.; Maroon, M.; Schrader, L.E.; Youngs, V.L. Rapid colorimetric determination of nitrate in plant tissue by titration of salicylic acid. *Commun. Soil Sci. Plant Anal.* **1975**, *6*, 71–80. [CrossRef]

15. Yemm, E.W.; Willis, A.J. The estimation of carbohydrates in plant extracts by anthrone. *Biochem. J.* **1954**, *57*, 508–514. [CrossRef] [PubMed]

16. Branca, F. Studies on some wild Brassicaceae species utilizable as vegetables in the Mediterranean areas. *Plant Genet. Resour. Newsl.* **1995**, *104*, 6–9.

17. Branca, F.; Li, G.; Goyal, S.; Quiros, C.F. Survey of aliphatic glucosinolates in Sicilian wild and cultivated Brassicaceae. *Phytochemistry* **2002**, *59*, 717–724. [CrossRef]

18. Albrecht, J. Ascorbic Acid Content And Retention In Lettuce1. *J. Food Qual.* **1993**, *16*, 311–316. [CrossRef]

19. Žnidarčič, D.; Ban, D.; Šircelj, H. Carotenoid and chlorophyll composition of commonly consumed leafy vegetables in Mediterranean countries. *Food Chem.* **2011**, *129*, 1164–1168. [CrossRef]

20. Ferrante, A.; Maggiore, T. Chlorophyll a fluorescence measurements to evaluate storage time and temperature of Valeriana leafy vegetables. *Postharvest Biol. Technol.* **2007**, *45*, 73–80. [CrossRef]

21. Fallovo, C.; Rouphael, Y.; Rea, E.; Battistelli, A.; Colla, G. Nutrient solution concentration and growing season affect yield and quality of Lactuca sativa L. var. acephala in floating raft culture. *J. Sci. Food Agric.* **2009**, *89*, 1682–1689. [CrossRef]

22. Ferrante, A.; Incrocci, L.; Maggini, R.; Serra, G.; Tognoni, F. Colour changes of fresh-cut leafy vegetables during storage. *J. Food Agric. Environ.* **2004**, *2*, 40–44.

23. Colonna, E.; Rouphael, Y.; Barbieri, G.; De Pascale, S. Nutritional quality of ten leafy vegetables harvested at two light intensities. *Food Chem.* **2016**, *199*, 702–710. [CrossRef]

24. Bulgari, R.; Cola, G.; Ferrante, A.; Franzoni, G.; Mariani, L.; Martinetti, L. Micrometeorological environment in traditional and photovoltaic greenhouses and effects on growth and quality of tomato (*Solanum lycopersicum* L.). *Ital. J. Agrometeorol.* **2015**, *2*, 27–38.

25. Maxwell, K.; Johnson, G.N. Chlorophyll fluorescence—A practical guide. *EXBOTJ* **2000**, *51*, 659–668. [CrossRef]

26. Alberici, A.; Quattrini, E.; Penati, M.; Martinetti, L.; Gallina, P.M.; Ferrante, A.; Schiavi, M. Effect of the Reduction of Nutrient Solution Concentration on Leafy Vegetables Quality Grown in Floating System. *Acta Hortic.* **2008**, *801*, 1167–1176. [CrossRef]

27. EFSA—European Food Safety Authority. *Nitrate in Vegetables. Scientific Opinion of the Panel on Contaminants in the Food Chain*; Question N. EFSA-Q-2006-071; EFSA: Parma, Italy, 2008; Volume 689, pp. 1–79.

28. European Commission. *Maximum Levels for Nitrates in Foodstuffs*; Regulation (EU) N. 1258/2011 Amending Regulation (EC) N. 1881/2006; Official Journal European Union: Brussel, Belgium, 2011; pp. 15–17.

29. Hanafy Ahmed, A.H.; Khalil, M.K.; Farrag, A.M. Nitrate accumulation, growth, yield and chemical composition of Rocket (*Eruca vesicaria* subsp. sativa) plant as affected by NPK fertilization, kinetin and salicylic acid. *Ann. Agric. Sci. Cairo* **2002**, *47*, 1–26.

30. Ferrante, A.; Incrocci, L.; Maggini, R.; Serra, G.; Tognoni, F. Preharvest and postharvest strategies for reducing nitrate content in rocket (*Eruca sativa* L.). *Acta Hortic.* **2003**, *628*, 153–159. [CrossRef]

31. Tannenbaum, S.R.; Wishnok, J.S.; Leaf, C.D. Inhibition of nitrosamine formation by ascorbic acid. *Am. J. Clin. Nutr.* **1991**, *53*, 247S–250S. [CrossRef]

32. Shad, A.A.; Shah, H.U.; Bakht, J. Ethnobotanical assessment and nutritive potential of wild food plants. *J. Anim. Plant Sci.* **2013**, *23*, 92–97.

MDPI

St. Alban-Anlage 66

4052 Basel

Switzerland

Tel. +41 61 683 77 34

Fax +41 61 302 89 18

www.mdpi.com

Horticulturae Editorial Office

E-mail: horticulturae@mdpi.com

www.mdpi.com/journal/horticulturae